Chaos in dynamical systems

Edward Ott

University of Maryland
College Park, Maryland, USA

CAMBRIDGE
UNIVERSITY PRESS

Published by the Press Syndicate of the University of Cambridge
The Pitt Building, Trumpington Street, Cambridge CB2 1RP
40 West 20th Street, New York, NY 10011-4211, USA
10 Stamford Road, Oakleigh, Melbourne 3166, Australia

First published 1993

Printed in Canada

A catalogue record for this book is available from the British Library

Library of Congress cataloguing in publication data available

ISBN 0 521 43215 4   hardback
ISBN 0 521 43799 7   paperback

VN

Contents

CHAPTER ONE

Introduction and overview

1.1 Some history

Chaotic dynamics may be said to have started with the work of the French mathematician Henri Poincaré at about the turn of the century. Poincaré's motivation was partly provided by the problem of the orbits of three celestial bodies experiencing mutual gravational attraction (e.g., a star and two planets). By considering the behavior of orbits arising from *sets* of initial points (rather than focusing on *individual* orbits). Poincaré was able to show that very complicated (now called chaotic) orbits were possible. Subsequent noteworthy early mathematical work on chaotic dynamics includes that of G. Birkhoff in the 1920s, M. L. Cartwright and J. E. Littlewood in the 1940s, S. Smale in the 1960s, and Soviet mathematicians, notably A. N. Kolmogorov and his coworkers. In spite of this work, however, the possibility of chaos in real physical systems was not widely appreciated until relatively recently. The reasons for this were first that the mathematical papers are difficult to read for workers in other fields, and second that the theorems proven were often not strong enough to convince researchers in these other fields that this type of behavior would be important in their systems. The situation has now changed drastically, and much of the credit for this can be ascribed to the extensive numerical solution of dynamical systems on digital computers. Using such solutions, the chaotic character of the time evolutions in situations of practical importance has become dramatically clear. Furthermore, the complexity of the dynamics cannot be blamed on unknown extraneous experimental effects, as might be the case when dealing with an actual physical system.

In this chapter, we shall provide some of the phenomenology of chaos and will introduce some of the more basic concepts. The aim is to provide a motivating overview[1] in preparation for the more detailed treatments to be pursued in the rest of this book.

1.2   **Examples of chaotic behavior**

Most students of science or engineering have seen examples of dynamical behavior which can be fully analyzed mathematically and in which the system eventually (after some transient period) settles either into periodic motion (a limit cycle) or into a steady state (i.e., a situation in which the system ceases its motion). When one relies on being able to specify an orbit analytically, these two cases will typically (and falsely) appear to be the only important motions. The point is that chaotic orbits are also very common but cannot be represented using standard analytical functions. Chaotic motions are neither steady nor periodic. Indeed, they appear to be very complex, and, when viewing such motions, adjectives like wild, turbulent, and random come to mind. In spite of the complexity of these motions, they commonly occur in systems which themselves are not complex and are even surprisingly simple. (In addition to steady state, periodic and chaotic motion, there is a fourth common type of motion, namely quasiperiodic motion. We defer our discussion of quasiperiodicity to Chapter 6.)

Before giving a definition of chaos we first present some examples and background material. As a first example of chaotic motion, we consider an

Figure 1.1 The apparatus of Moon and Holmes (1979).

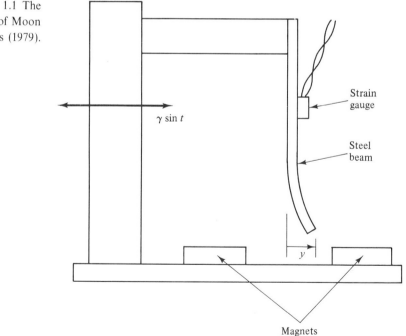

experiment of Moon and Holmes (1979). The apparatus is shown in Figure 1.1. When the apparatus is at rest, the steel beam has two stable steady-state equilibria: either the tip of the beam is deflected toward the left magnet or toward the right magnet. In the experiment, the horizontal position of the apparatus was oscillated sinusoidally with time. Under certain conditions, when this was done, the tip of the steel beam was observed to oscillate in a very irregular manner. As an indication of this very irregular behavior, Figure 1.2(a) shows the output signal of a strain gauge attached to the beam (Figure 1.1). Although the apparatus appears to be very simple, one might attribute the observed complicated motion to complexities in the physical situation, such as the excitation of higher order vibrational modes in the beam, possible noise in the sinusoidal shaking device, etc. To show that it is not necessary to invoke such effects, Moon and Holmes considered a simple model for their experiment, namely, the forced Duffing equation in the following form,

$$\frac{d^2 y}{dt^2} + v \frac{dy}{dt} + (y^3 - y) = g \sin t. \tag{1.1}$$

In Eq. (1.1), the first two terms represent the inertia of the beam and dissipative effects, while the third term represents the effects of the magnets and the elastic force. The sinusoidal term on the right-hand side represents the shaking of the apparatus. In the absence of shaking ($g = 0$), Eq. (1.1) possesses two stable steady states, $y = 1$ and $y = -1$, corresponding to the two previously mentioned stable steady states of the beam. (There is also an unstable steady state $y = 0$.) Figure 1.2(b) shows the results of a digital computer numerical solution of Eq. (1.1) for a particular choice of v and g. We observe that the results of the physical experiment are qualitatively similar to those of the numerical solution. Thus, it is unnecessary to invoke complicated physical processes to explain the observed complicated motion.

Figure 1.2(a) Signal from the strain gauge. (b) Numerical solution of Eq. (1.1) (Moon and Holmes 1979).

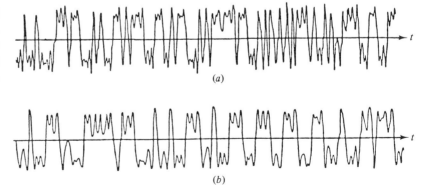

(a)

(b)

As a second example, we consider the experiment of Shaw (1984) illustrated schematically in Figure 1.3. In this experiment, a slow steady inflow of water to a 'faucet' was maintained. Water drops fall from the faucet, and the times at which successive drops pass a sensing device are recorded. Thus, the data consists of the discrete set of times $t_1, t_2, t_n, \ldots$ at which drops were observed by the sensor. From these data, the time intervals between successive drops can be formed, $\Delta t_n \equiv t_{n+1} - t_n$. When the inflow rate to the faucet is sufficiently small, the time intervals Δt_n are all equal. As the inflow rate is increased, the time interval sequence becomes periodic with a short interval Δt_a followed by a longer interval Δt_b, so that the sequence of time intervals is of the form $\ldots, \Delta t_a, \Delta t_b, \Delta t_a, \Delta t_b, \Delta t_a, \ldots$. We call this a period two sequence since $\Delta t_n = \Delta t_{n+2}$. As the inflow rate is further increased, periodic sequences of longer and longer periods were observed, until, at sufficiently large inflow rate, the sequence $\Delta t_1, \Delta t_2, \Delta t_3, \ldots$ apparently has no regularity. This irregular sequence is argued to be due to chaotic dynamics (see Section 2.4.3).

As a third example, we consider the problem of chaotic Rayleigh–Benard convection, originally studied theoretically and computationally in the seminal paper of Lorenz (1963) and experimentally by, for example, Ahlers and Behringer (1978), Gollub and Benson (1980), Bergé *et al.* (1980) and Libchaber and Maurer (1980). In Rayleigh–Benard convection, one considers a fluid contained between two rigid plates and subjected to gravity, as shown in Figure 1.4. The bottom plate is maintained at a higher temperature $T_0 + \Delta T$ than the temperature T_0 of the top plate. As a result, the fluid near the warmer lower plate expands, and buoyancy creates a tendency for this fluid to rise. Similarly, the cooler more dense fluid near the top plate has a tendency to fall. While Lorenz's equations are too idealized a model to describe the experiments accurately, in the case where the experiments were done with vertical bounding

Figure 1.3 Schematic illustration of the experiment of Shaw (1984).

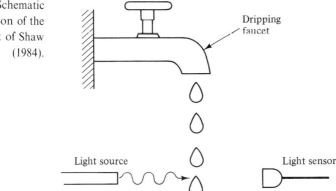

side-walls situated at a spacing of two to three times the distance between the horizontal walls, there was a degree of qualitative correspondence between the model and the experiments. In particular, in this case, for some range of values of the temperature difference ΔT, the experiments show that the fluid will execute a *steady* convective cellular flow, as shown in the figure. At a somewhat larger value of the temperature difference, the flow becomes time-dependent, and this time dependence is chaotic. This general behavior is also predicted by Lorenz's paper.

From these simple examples, it is clear that chaos should be expected to be a very common basic dynamical state in a wide variety of systems. Indeed, chaotic dynamics has by now been shown to be of potential importance in many different fields including fluids,[2] plasmas,[3] solid state devices,[4] circuits,[5] lasers,[6] mechanical devices,[7] biology,[8] chemistry,[9] acoustics,[10] celestial mechanics,[11] etc.

In both the dripping faucet example and the Rayleigh–Benard convection example, our discussions indicated a situation as shown schematically in Figure 1.5. Namely, there was a system parameter, labeled p in Figure 1.5, such that, at a value $p = p_1$, the motion is observed to be nonchaotic, and at another value $p = p_2$, the motion is chaotic. (For the faucet example, p is the inflow rate, while for the example of Rayleigh–Benard convection, p is the temperature difference ΔT.) The natural question raised by Figure 1.5 is *how does chaos come about as the parameter p is varied continuously from p_1 to p_2?* That is, how do the dynamical motions of the system evolve with continuous variation of p from p_1 to p_2? This question of the *routes to chaos*[12] will be considered in detail in Chapter 8.

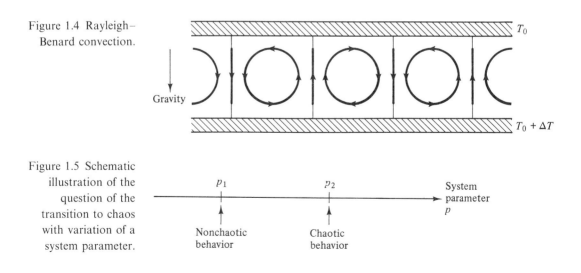

Figure 1.4 Rayleigh–Benard convection.

Figure 1.5 Schematic illustration of the question of the transition to chaos with variation of a system parameter.

1.3   Dynamical systems

A *dynamical system* may be defined as a deterministic mathematical prescription for evolving the state of a system forward in time. Time here either may be a continuous variable, or else it may be a discrete integer-valued variable. An example of a dynamical system in which time (denoted t) is a continuous variable is a system of N first-order, autonomous, ordinary differential equations,

$$\left.\begin{array}{l} dx^{(1)}/dt = F_1(x^{(1)}, x^{(2)}, \ldots, x^{(N)}), \\ dx^{(2)}/dt = F_2(x^{(1)}, x^{(2)}, \ldots, x^{(N)}), \\ \quad\vdots \\ dx^{(N)}/dt = F_N(x^{(1)}, x^{(2)}, \ldots, x^{(N)}), \end{array}\right\} \tag{1.2}$$

which we shall often write in vector form as

$$d\mathbf{x}(t)/dt = \mathbf{F}[\mathbf{x}(t)], \tag{1.3}$$

where $\mathbf{x}$ is an N-dimensional vector. This is a dynamical system because, for any initial state of the system $x(0)$, we can in principle solve the equations to obtain the future system state $x(t)$ for $t > 0$. Figure 1.6 shows the path followed by the system state as it evolves with time in a case where $N = 3$. The space $(x^{(1)}, x^{(2)}, x^{(3)})$ in the figure is referred to as *phase space*, and the path in phase space followed by the system as it evolves with time is referred to as an *orbit* or *trajectory*. Also, it is common to refer to a continuous time dynamical system as a *flow*. (This latter terminology is apparently motivated by considering the trajectories generated by *all* the initial conditions in the phase space as roughly analogous to the paths followed by the particles of a flowing fluid.)

Figure 1.6 An orbit in a three-dimensional ($N = 3$) phase space.

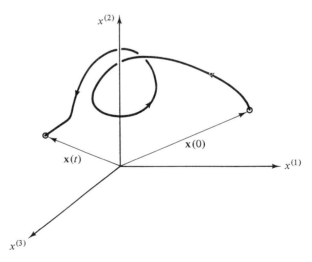

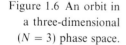

In the case of discrete, integer-valued time (with n denoting the time variable, $n = 0, 1, 2, \ldots$), an example of a dynamical system is a map, which we write in vector form as

$$\mathbf{x}_{n+1} = \mathbf{M}(\mathbf{x}_n), \tag{1.4}$$

where $\mathbf{x}_n$ has N components, $\mathbf{x}_n = (x_n^{(1)}, x_n^{(2)}, \ldots, x_n^{(N)})$. Given an initial state $\mathbf{x}_0$, we obtain the state at time $n = 1$ by $\mathbf{x}_1 = \mathbf{M}(\mathbf{x}_0)$. Having determined $\mathbf{x}_1$, we can then determine the state at $n = 2$ by $\mathbf{x}_2 = \mathbf{M}(\mathbf{x}_1)$, and so on. Thus, given an initial condition $\mathbf{x}_0$, we generate an orbit (or trajectory) of the discrete time system: $\mathbf{x}_0, \mathbf{x}_1, \mathbf{x}_2, \ldots$. As we shall see, a continuous time system of dimensionality N can often profitably be reduced to a discrete time map of dimensionality $N - 1$ via the Poincaré surface of section technique.

It is reasonable to conjecture that the complexity of the possible structure of orbits can be greater for larger system dimensionality. Thus, a natural question is *how large does N have to be in order for chaos to be possible*? For the case of N first-order autonomous ordinary differential equations, the answer is that

$$N \geq 3 \tag{1.5}$$

is sufficient.[13] Thus, if one is given an autonomous first-order system with $N = 2$, chaos can be ruled out immediately.

Example: Consider the forced damped pendulum equation (cf. Figure 1.7)

$$\frac{\mathrm{d}^2\theta}{\mathrm{d}t^2} + v\frac{\mathrm{d}\theta}{\mathrm{d}t} + \sin\theta = T\sin(2\pi f t), \tag{1.6a}$$

where the first term represents inertia, the second, friction at the pivot, the third, gravity, and the term on the right-hand side represents a sinusoidal torque applied at the pivot. (This equation also describes the behavior of a

Figure 1.7 Forced, damped pendulum.

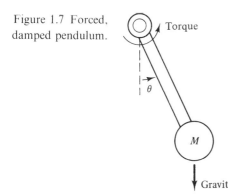

simple Josephson junction circuit.) We ask: is chaos ruled out for the driven damped pendulum equation? To answer this question, we put the equation (which is second-order and nonautonomous) into first-order autonomous form by the substitution

$$x^1 = d\theta/dt,$$
$$x^{(2)} = \theta,$$
$$x^{(3)} = 2\pi ft.$$

(Note that, since both $x^{(2)}$ and $x^{(3)}$ appear in Eq. (1.6a) as the argument of a sine function, they can be regarded as angles and may, if desired, be defined to lie between 0 and 2π.) The driven damped pendulum equation then yields the following first-order autonomous system,

$$\left.\begin{array}{l} dx^{(1)}/dt = T \sin x^{(3)} - \sin x^{(2)} - vx^{(1)}, \\ dx^{(2)}/dt = x^{(1)}, \\ dx^{(3)}/dt = 2\pi f. \end{array}\right\} \tag{1.6b}$$

Since $N = 3$, chaos is not ruled out. Indeed, numerical solutions show that both chaotic and periodic solutions of the driven damped pendulum equation are possible depending on the particular choice of system parameters v, T, and f.

We now consider the question of the required dimensionality for chaos for the case of maps. In this case, we must distinguish between invertible and noninvertible maps. We say the map $\mathbf{M}$ is invertible if, given $\mathbf{x}_{n+1}$, we can solve $\mathbf{x}_{n+1} = \mathbf{M}(\mathbf{x}_n)$ for $\mathbf{x}_n$. If this is so, we denote the solution for $\mathbf{x}_n$ as

$$\mathbf{x}_n = \mathbf{M}^{-1}(\mathbf{x}_{n+1}), \tag{1.7}$$

and we call $\mathbf{M}^{-1}$ the inverse of $\mathbf{M}$. For example, consider the one-dimensional ($N = 1$) map[14],

$$M(x) = rx(1 - x), \tag{1.8}$$

which is commonly called the 'logistic map.' As shown in Figure 1.8, this map is not invertible because for a given x_{n+1} there are two possible values of x_n from which it could have come. On the other hand, consider the two-dimensional map,

$$\begin{array}{l} x_{n+1}^{(1)} = f(x_n^{(1)}) - Jx_n^{(2)}, \\ x_{n+1}^{(2)} = x_n^{(1)}. \end{array} \tag{1.9}$$

This map is clearly invertible as long as $J \neq 0$,

$$\begin{array}{l} x_n^{(1)} = x_{n+1}^{(2)}, \\ x_n^{(2)} = J^{-1}[f(x_{n+1}^{(2)}) - x_{n+1}^{(1)}]. \end{array} \tag{1.10}$$

We can now state the dimensionality requirements on maps. If the map is invertible, then there can be no chaos unless

$$N \geq 2. \tag{1.11}$$

If the map is noninvertible, chaos is possible even in one-dimensional maps. Indeed, the logistic map Eq. (1.8) exhibits chaos for large enough r.

It is often useful to reduce a continuous time system (or 'flow') to a discrete time map by a technique called the Poincaré surface of section method. We consider N first-order autonomous ordinary differential equations (Eq. (1.2)). The 'Poincaré map' represents a reduction of the N-dimensional flow to an $(N - 1)$-dimensional map. For illustrative purposes, we take $N = 3$ and illustrate the construction in Figure 1.9. Consider a solution of (1.2). Now, choose some appropriate $(N - 1)$-dimensional surface (the 'surface of section') in the N-dimensional phase space, and observe the intersections of the orbit with the surface. In Figure 1.9, the surface of section is the plane $x^{(3)} = K$, but we emphasize that in general the choice of the surface can be tailored in a convenient way to the particular problem. Points A and B represent two successive crossings of the surface of section. Point A uniquely determines point B, because A can be used as an initial condition in (1.2) to determine B. Likewise, B uniquely determines A by reversing time in (1.2) and using B as the initial condition. Thus, the Poincaré map in this illustration represents an invertible two-dimensional map transforming the coordinates $(x_n^{(1)}, x_n^{(2)})$ of the nth piercing of the surface of section to the coordinates $(x_{n+1}^{(1)}, x_{n+1}^{(2)})$ at piercing $n + 1$. This equivalence of an N-dimensional flow with an $(N - 1)$-dimensional invertible map shows that the requirement Eq. (1.11) for chaos in a map follows from Eq. (1.5) for chaos in a flow.

Another way to create a map from the flow generated by the system of autonomous differential equations (1.3) is to sample the flow at discrete times $t_n = t_0 + nT$ $(n = 0, 1, 2, \ldots)$, where the sampling interval T can be chosen on the basis of convenience. Thus, a continuous time trajectory $\mathbf{x}(t)$ yields a discrete time trajectory $\mathbf{x}_n \equiv \mathbf{x}(t_n)$. The quantity $\mathbf{x}_{n+1}$ is

Figure 1.8 Noninvertibility of the logistic map.

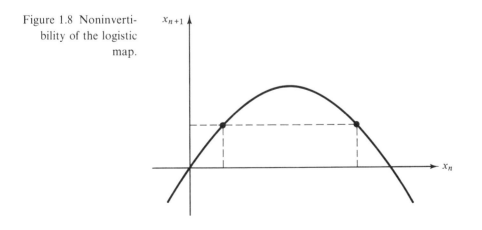

uniquely determined from $\mathbf{x}_n$ since we can use $\mathbf{x}_n$ as an initial condition in Eqs. (1.3) and integrate the equations forward for an amount of time T to determine $\mathbf{x}_{n+1}$. Thus, in principle, we have a map $\mathbf{x}_{n+1} = \mathbf{M}(\mathbf{x}_n)$. We call this map the time T map. The time T map is invertible (like the Poincaré map), since the differential equations (1.3) can be integrated backward in time. Unlike the Poincaré map, the dimensionality of the time T map is the same as that of the flow.

1.4 Attractors

In Hamiltonian systems (cf. Chapter 7) such as arise in Newton's equations for the motion of particles without friction, there are choices of the phase space variables (e.g., the canonically conjugate position and momentum variables) such that phase space volumes are preserved under the time evolution. That is, if we choose an initial ($t = 0$) closed ($N - 1$)-dimensional surface S_0 in the N-dimensional $\mathbf{x}$-phase space, and then evolve each point on the surface S_0 forward in time by using them as initial conditions in Eq. (1.3), then the closed surface S_0 evolves to a closed surface S_t at some later time t, and the N-dimensional volumes $V(0)$ of the region enclosed by S_0 and $V(t)$ of the region enclosed by S_t are the same, $V(t) = V(0)$. We call such a volume preserving system *conservative*. On the other hand, if the flow does not preserve volumes, and cannot be made to do so by a change of variables, then we say that the system is

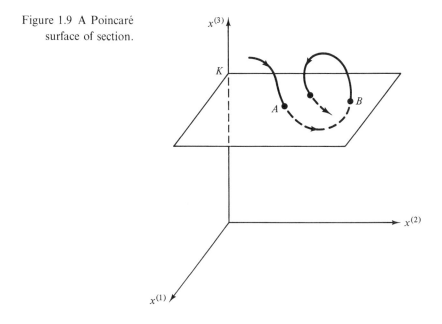

Figure 1.9 A Poincaré surface of section.

nonconservative. By the divergence theorem, we have that

$$dV(t)/dt = \int_{S_t} \nabla \cdot \mathbf{F} d^N x, \qquad (1.12)$$

where $\int_{S_t}$ signifies the integral over the volume interior to the surface S_t, and $\nabla \cdot \mathbf{F} \equiv \Sigma_{i=1}^{N} \partial F_i(x^{(1)}, \ldots, x^{(N)})/\partial x^{(i)}$. For example, for the forced damped pendulum equation written in first-order autonomous form, Eq. (1.6b), we have that $\nabla \cdot \mathbf{F} = -v$, which is independent of the phase space position $\mathbf{x}$ and is negative. From (1.12), we have $dV(t)/dt = -vV(t)$ so that V decreases exponentially with time, $V(t) = \exp(-vt)V(0)$. In general, $\nabla \cdot \mathbf{F}$ will be a function of the phase space position $\mathbf{x}$. If $\nabla \cdot \mathbf{F} < 0$ in some region of phase space (signifying volume contraction in that region), then we shall refer to the system as a *dissipative* system. It is an important concept in dynamics that dissipative systems typically are characterized by the presence of attracting sets or *attractors* in the phase space. These are bounded subsets to which regions of initial conditions of nonzero phase space volume asymptote as time increases. (Conservative dynamical systems do not have attractors; see the discussion of the Poincaré recurrence theorem in Chapter 7).

As an example of an attractor, consider the damped harmonic oscillator, $d^2y/dt^2 + v\,dy/dt + \omega^2 y = 0$. A typical trajectory in the phase space ($x^{(1)} = y$, $x^{(2)} = dy/dt$) is shown in Figure 1.10(a). We see that, as time goes on, the orbit spirals into the origin, and this is true for any initial condition. Thus, in this case the origin, $x^{(1)} = x^{(2)} = 0$, is said to be the 'attractor' of the dynamical system. As a second example, Figure 1.10(b) shows the case of a limit cycle (the dashed curve). The initial condition (labeled α) outside the limit cycle yields an orbit which, with time, spirals into the closed dashed curve on which it circulates in periodic motion in the $t \to +\infty$ limit. Similarly, the initial condition (labeled β) inside the limit cycle yields an orbit which spirals outward, asymptotically

Figure 1.10(a) The attractor is the point at the origin. (b) The attractor is the closed dashed curve.

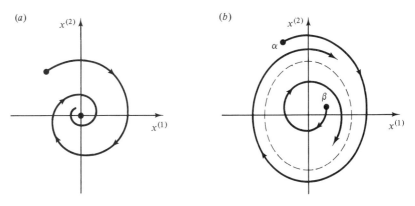

approaching the dashed curve. Thus, in this case, the dashed closed curve is the attractor. An example of an equation displaying a limit cycle attractor as illustrated in Figure 1.10(b) is the van der Pol equation,

$$\frac{d^2x}{dt^2} + (x^2 - \eta)\frac{dx}{dt} + \omega^2 x = 0. \tag{1.13}$$

This equation was introduced in the 1920s as a model for a simple vacuum tube oscillator circuit.

One can speak of conservative and dissipative *maps*. A conservative N-dimensional map is one which preserves N-dimensional phase space volumes on each iterate (or else can be made to do so by a suitable change of variables). A map is volume preserving if the magnitude of the determinant of its Jacobian matrix of partial derivatives is one,

$$J(\mathbf{x}) \equiv |\det[\partial\mathbf{M}(\mathbf{x})/\partial\mathbf{x}]| = 1.$$

For example, for a continuous time Hamiltonian system, a surface of section formed by setting one of the N canonically conjugate variables equal to a constant can be shown to yield a volume preserving map in the remaining $N - 1$ canonically conjugate variables (Chapter 7). On the other hand, if $J(\mathbf{x}) < 1$ in some regions, then we say the map is dissipative, and, as for flows, typically it can have attractors. For example, Figure 1.11 illustrates the Poincaré surface of section map for a three-dimensional

Figure 1.11 Surface of section for a three-dimensional flow with a limit cycle.

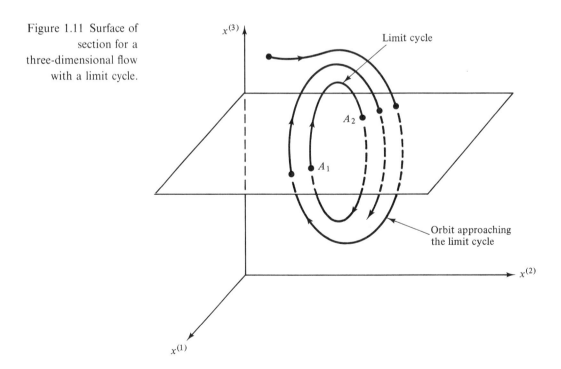

flow with a limit cycle. We see that *for the map*, the two points A_1 and A_2 together constitute the attractor. That is, the orbit of the two-dimensional surface of section map $\mathbf{x}_{n+1} = \mathbf{M}(\mathbf{x}_n)$ yields a sequence $\mathbf{x}_1, \mathbf{x}_2, \ldots$ which converges to the set consisting of the two points A_1 and A_2, between which the map orbit sequentially alternates in the limit $n \to +\infty$.

In Figure 1.10, we have two examples, one in which the attractor of a continuous time system is a set of dimension zero (a single point) and one in which the attractor is a set of dimension one (a closed curve). In Figure 1.11, the attractor of the map has dimension zero (it is the two points, A_1 and A_2). It is a characteristic of chaotic dynamics that the resulting attractors often have a much more intricate geometrical structure in the phase space than do the examples of attractors cited above. In fact, according to a standard definition of dimension (Section 3.1), these attractors commonly have a value for this dimension which is not an integer. In the terminology of Mandelbrot, such geometrical objects are *fractals*. When an attractor is fractal, it is called a *strange attractor*.

As an example of a strange attractor, consider the attractor obtained for the two-dimensional Hénon map,

$$\left.\begin{array}{l} x_{n+1}^{(1)} = A - (x_n^{(1)})^2 + B x_n^{(2)}, \\ x_{n+1}^{(2)} = x_n^{(1)}, \end{array}\right\} \tag{1.14}$$

for $A = 1.4$ and $B = 0.3$. See Hénon (1976). (Note that Eq. (1.14) is in the form of Eq. (1.9).) Figure 1.12(*a*) shows the results of plotting 10^4 successive points obtained by iterating Eqs. (1.14) (with the initial transient before the orbit settles into the attractor deleted). The result is essentially a picture of the attractor. Figure 1.12(*b*) shows that a blow-up of the rectangle in Figure 1.12(*a*) reveals that the attractor apparently has a local small-scale structure consisting of a number of parallel lines. A blow-up of the rectangle in Figure 1.12(*b*) is shown in Figure 1.12(*c*) and reveals more lines. Continuation of this blow-up procedure would show that the attractor has similar structure on *arbitrarily small scale*. In fact, roughly speaking, we can regard the attractor in Figure 1.12(*b*) as consisting of an *uncountable* infinity of lines. Numerical computations show that the fractal dimension D_0 of the attractor in Figure 1.12 is a number between one and two, $D_0 \simeq 1.26$. Hence, this is an example of a strange attractor.

As another example of a strange attractor, consider the forced damped pendulum (Eqs. (1.6) and Figure 1.7) with $v = 0.22$, $T = 2.7$, and $f = 1/2\pi$. Treating $x^{(3)}$ as an angle in phase space, we define

$$\bar{x}^{(3)} = x^{(3)} \text{ modulo } 2\pi$$

and choose a surface of section $\bar{x}^{(3)} = 0$. The modulo operation is defined as

$$y \text{ modulo } K \equiv y + pK.$$

where p is a positive or negative integer chosen to make $0 \leq y + pK < K$. The surface of section $\bar{x}^{(3)} = 0$ is crossed at the times $t = 0, 2\pi, 4\pi, 6\pi, \ldots$. (This type of surface of section for a periodically forced system is often referred to as a *stroboscopic* surface of section, since it shows the system state at successive 'snapshots' of the system at evenly spaced time intervals.) As seen in Figure 1.13(a) and in the blow-up of the rectangle (Figure 1.13(b)), the attractor again apparently consists of a number of parallel curves. The fractal dimension of the intersection of the attractor with the surface of section in this case is approximately 1.38. Correspondingly, if one considers the attracting set in the full three-dimensional phase space, it has a dimension 2.38 (i.e., one greater than its intersection with the surface of section).

Figure 1.12(a) The Hénon attractor. (b) Enlargement of region defined by the rectangle in (a). (c) Enlargement of region defined by the rectangle in (b) (Grebogi *et al.* 1987d).

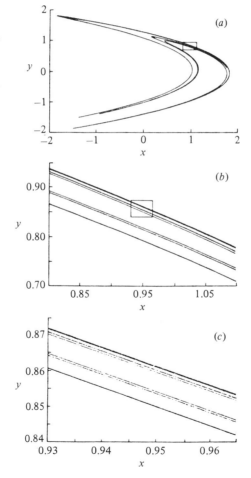

1.5 Sensitive dependence on initial conditions

A defining attribute of an attractor on which the dynamics is *chaotic* is that it displays exponentially sensitive dependence on initial conditions. Consider two nearby initial conditions $\mathbf{x}_1(0)$ and $\mathbf{x}_2(0) = \mathbf{x}_0 + \Delta(0)$, and imagine that they are evolved forward in time by a continuous time dynamical system yielding orbits $\mathbf{x}_1(t)$ and $\mathbf{x}_2(t)$ as shown in Figure 1.14. At time t, the separation between the two orbits is $\Delta(t) = \mathbf{x}_2(t) - \mathbf{x}_1(t)$. If, in the limit $|\Delta(0)| \to 0$, and large t, orbits remain bounded and the difference between the solutions $|\Delta(t)|$ grows exponentially for typical orientation of the vector $\Delta(0)$ (i.e., $|\Delta(t)|/|\Delta(0)| \sim \exp(ht)$, $h > 0$), then we say that the system displays sensitive dependence on initial conditions and

Figure 1.13 The attractor of the forced damped pendulum equation in the surface of section $x^{(3)}$ modulo $2\pi = 0$ (Grebogi *et al.* 1987d).

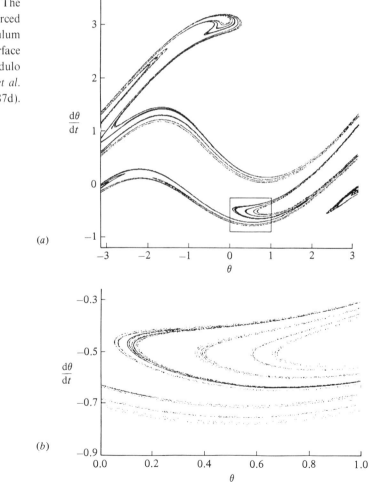

is chaotic. By bounded solutions, we mean that there is some ball in phase space, $|\mathbf{x}| < R < \infty$, which solutions never leave.[15] (Thus, if the motion is on an attractor, then the attractor lies in $|\mathbf{x}| < R$.) The reason we have imposed the restriction that orbits remain bounded is that, if orbits go to infinity, it is relatively simple for their distances to diverge exponentially. An example is the single, autonomous, linear, first-order differential equation $dx/dt = x$. This yields $d[x_2(t) - x_1(t)]/dt = [x_2(t) - x_1(t)]$ and hence $\Delta(t) \sim \exp(t)$. Our requirement of bounded solutions eliminates such trivial cases.[16] For the case of the driven damped pendulum equation, we defined three phase space variables, one of which was $x^{(3)} = 2\pi ft$. As defined, $x^{(3)}$ is unbounded since it is proportional to t. The reason we can speak of the driven damped pendulum as being chaotic is that, as previously mentioned, $x^{(3)}$ only occurs as the argument of a sine, and hence it (as well as $x^{(2)} = \theta$) can be regarded as an angle. Thus, the phase space coordinates can be taken as $x^{(1)}, \bar{x}^{(2)}, \bar{x}^{(3)}$, where $\bar{x}^{(2,3)} \equiv x^{(2,3)}$ modulo 2π. Since the variables $\bar{x}^{(2)}$ and $\bar{x}^{(3)}$ lie between 0 and 2π, they are necessarily bounded.

The exponential sensitivity of chaotic solutions means that, as time goes on, small errors in the solution can grow very rapidly (i.e., exponentially) with time. Hence, after some time, effects such as noise and computer roundoff can totally change the solution from what it would be in the absence of these effects. As an illustration of this, Figure 1.15 shows the results of a computer experiment on the Hénon map, Eq. (1.14), with $A = 1.4$ and $B = 0.3$. In this figure, we show a picture of the attractor (as in Figure 1.12(a)) superposed on which are two computations of iterate numbers 32–36 of an orbit originating from the single initial condition $(x_0^{(1)}, x_0^{(2)}) = (0,0)$ (labeled as an asterisk in the figure). The two computations of the orbits are done identically, but one uses single precision and

Figure 1.14 Evolution of two nearby orbits in phase space.

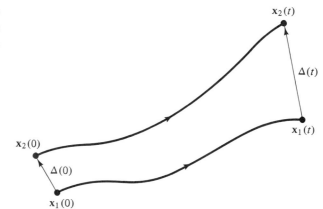

the other double precision. The roundoff error in the single precision computation is about 10^{-14}. The orbit computed using single precision is shown as open diamonds, while the orbit using double precision is shown as asterisks. A straight line joins the two orbit locations at each iterate. We see that the difference in the two computations has become as large as the variables themselves. Thus, we cannot meaningfully compute the orbit on the Hénon attractor using a computer with 10^{-14} roundoff for more than of the order of 30–40 iterates. Hence, given the state of a chaotic system, its future becomes difficult to predict after a certain point. Returning to the Hénon map example, we note that, after the first iterate, the two solutions differ by of the order of 10^{-14} (the roundoff). If the subsequent computations were made *without error*, and the error doubled on each iterate (i.e., an exponential increase of $2^n = \exp(n \ln 2)$), then the orbits would be separated by an amount of the order of the attractor size at a time roughly determined by $2^n 10^{-14} \sim 1$ or $n \sim 45$. If errors double on

Figure 1.15 After a relatively small number of iterates, two trajectories, one computed using single precision, the other computed using double precision, both originating from the same initial condition, are far apart (This figure courtesy of Y. Du).

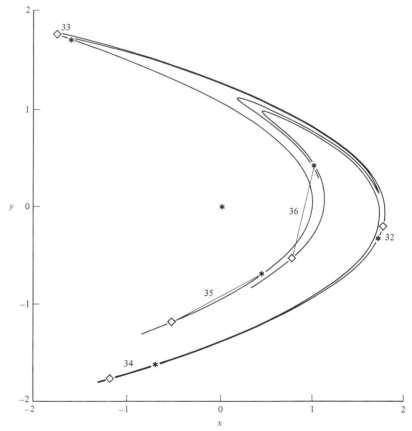

each iterate, it becomes almost impossible to improve prediction. Say, we can compute exactly, but our initial measurement of the system state is only accurate to within 10^{-14}. The above shows that we cannot predict the state of the system past $n \sim 45$. Suppose that we wish to predict to a longer time, say, twice as long, i.e., to $n \sim 90$. Then we must improve the accuracy of our initial measurement from 10^{-14} to 10^{-28}. That is, we must improve our accuracy by a tremendous amount, namely, 14 orders of magnitude! In any practical situation, this is likely to be impossible. Thus, the relatively modest goal of an improvement of prediction time by a factor of two is not feasible.

The fact that chaos may make prediction past a certain time difficult, and essentially impossible in a practical sense, has important consequences. Indeed, the work of Lorenz was motivated by the problem of weather prediction. Lorenz was concerned with whether it is possible to do long-range prediction of atmospheric conditions. His demonstration that thermally driven convection could result in chaos raises the possibility that the atmosphere is chaotic. Thus, even the smallest perturbation, such as a butterfly flapping its wings, *eventually* has a large effect. Long-term prediction becomes impossible.

Given the difficulty of accurate computation, illustrated in Figure 1.15, one might question the validity of pictures such as Figures 1.12 and 1.13 which show thousands of iterates of the Hénon map. Is the figure real, or is it merely an artifact of chaos-amplified computer roundoff? A partial answer to this question comes from rigorous mathematical proofs of the *shadowing* property for certain chaotic systems. Although a numerical trajectory diverges exponentially from the true trajectory with the same initial condition, there exists a true (i.e. errorless) trajectory with a slightly

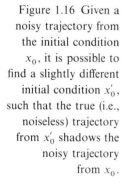

Figure 1.16 Given a noisy trajectory from the initial condition x_0, it is possible to find a slightly different initial condition x_0', such that the true (i.e., noiseless) trajectory from x_0' shadows the noisy trajectory from x_0.

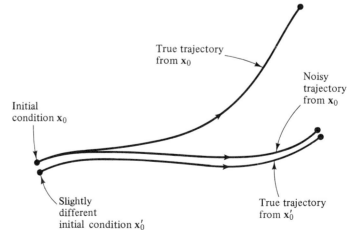

different initial condition (Fig. 1.16) that stays near (shadows) the numerical trajectory (Anosov, 1967; Bowen, 1970; Hammel, Yorke and Grebogi, 1987). Thus, there is good reason to believe that the apparent fractal structure seen in pictures like Figures 1.12 and 1.13 is real.

We emphasize that the nonchaotic cases, shown in Figure 1.10(*a*) and 1.10(*b*), do not yield long-term exponential divergence of solutions. For the damped harmonic oscillator example (Figure 1.10(*a*)), two initially nearby points approach the point attractor and their energies decrease exponentially to zero with time. Hence, orbits *converge* exponentially for large time. For the case of a limit cycle (Figure 1.10(*b*)), orbits initially separated by an amount $\Delta(0)$ typically eventually wind up on the limit cycle attractor separated by an amount of order $|\Delta(0)|$ and maintain a separation of this order forever. Thus, a small initial error leads to small errors *for all time*. As another example, consider the motion of a particle in a one-dimensional anharmonic potential well in the absence of friction (a conservative system). The total particle energy (potential energy plus kinetic energy) is constant with time on an orbit. Each orbit is periodic and the period depends on the particle energy. Two nearby initial conditions, in general, will have slightly different energies and hence slightly different orbit frequencies. This leads to divergence of these orbits, but the divergence is only linear with time rather than exponential; $|\Delta(t)| \sim (\Delta\omega)t$, where $\Delta\omega$ is the difference of the orbital frequencies. Thus, if $|\Delta(0)|$ is reduced by a factor of two (reducing $\Delta\omega$ by a factor of two), then t can be doubled, and the same error will be produced. This is in contrast with our chaotic example above where errors doubled on each iterate. In that case, to increase the time by a factor of two, $|\Delta(0)|$ had to be reduced by a factor of order 10^{14}.

The dynamics on an attractor is said to be chaotic if there is exponential sensitivity to initial conditions. We will say that an attractor is strange if it is fractal (this definition of strange is often used but is not universally accepted). Thus, chaos describes the dynamics on the attractor, while 'strange' refers to the geometry of the attractor. It is possible for chaotic attractors not to be strange (typically the case for one-dimensional maps (see the next chapter)), and it is also possible for attractors to be strange but not chaotic (Grebogi *et al.*, 1984; Romeiras and Ott, 1987). For most cases involving differential equations, strangeness and chaos commonly occur together.

1.6 Delay coordinates

In experiments one cannot always measure all the components of the vector $\mathbf{x}(t)$ giving the state of the system. Let us suppose that we can only measure one component, or, more generally, one scalar function of the state vector,

$$g(t) = G(\mathbf{x}(t)). \tag{1.15}$$

Given such a situation, can we obtain phase space information on the geometry of the attractor? For example, can we somehow make a surface of section revealing fractal structure as in Figures 1.12 and 1.13? The answer is yes. To see that this is so define the so-called delay coordinate vector (Takens, 1980), $\mathbf{y} = (y^{(1)}, y^{(2)}, \ldots, y^{(M)})$, by

$$\left. \begin{aligned} y^{(1)}(t) &= g(t), \\ y^{(2)}(t) &= g(t - \tau), \\ y^{(3)}(t) &= g(t - 2\tau), \\ &\vdots \\ y^{(M)}(t) &= g[t - (M - 1)\tau], \end{aligned} \right\} \tag{1.16}$$

where τ is some fixed time interval, which should be chosen to be of the order of the characteristic time over which $g(t)$ varies. Given $\mathbf{x}$ at a specific time t_0, one could, in principle, obtain $\mathbf{x}(t_0 - m\tau)$ by integrating Eq. (1.3) backwards in time by an amount $m\tau$. Thus, $\mathbf{x}(t_0 - m\tau)$ is uniquely determined by $\mathbf{x}(t_0)$ and can hence be regarded as a function of $\mathbf{x}(t_0)$,

$$\mathbf{x}(t - m\tau) = \mathbf{L}_m(\mathbf{x}(t)).$$

Hence, $g(t - m\tau) = G(\mathbf{L}_m(\mathbf{x}(t)))$, and we may thus regard the vector $\mathbf{y}(t)$ as a function of $\mathbf{x}(t)$

$$\mathbf{y} = \mathbf{H}(\mathbf{x}).$$

We can now imagine making a surface of section in the $\mathbf{y}$-space. It can be shown (Section 3.8) that, if the number of delays M is sufficiently large, then we will typically see a qualitatively similar structure as would be seen had we made our surface of section in the original phase space $\mathbf{x}$.

Figure 1.17 Experimental delay coordinate plot showing a closed curve corresponding to a limit cycle attractor.

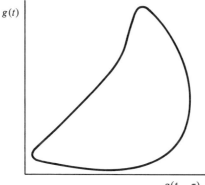

Alternatively, we might simply examine the continuous time trajectory in **y**. For example, Figure 1.17 shows a result for an experiment involving chemical reactions (cf. Section 2.4.3). The vertical axis is the measured concentration $g(t)$ of one chemical constituent at time t and the horizontal axis is the same quantity evaluated at $t - (8.8$ seconds). We see that the delay coordinates $\mathbf{y} = (g(t), g(t - 8.8))$ traces out a closed curve indicating a limit cycle.

Problems

1. Consider the following systems and specify (i) whether chaos can or cannot be ruled out for these systems, and (ii) whether the system is conservative or dissipative. Justify your answer

   (*a*) $\theta_{n+1} = [\theta_n + \Omega + 1.5 \sin \theta_n]$ modulo 2π,

   (*b*) $\theta_{n+1} = [\theta_n + \Omega + 0.5 \sin \theta_n]$ modulo 2π,

   (*c*) $x_{n+1} = [2x_n - x_{n-1} + k \sin x_n]$ modulo 2π,

   (*d*) $x_{n+1} = x_n + k(x_n - y_n)^2$, $y_{n+1} = y_n + k(x_n - y_n)^2$,

   (*e*) $dx/dt = v$, $dv/dt = -\alpha v + C \sin(\omega t - kx)$,

   (*f*) $dx/dt = B \cos y + C \sin z$

   $\qquad dy/dt = C \cos z + A \sin x$

   $\qquad dz/dt = A \cos x + B \sin y$.

2. Consider the one-dimensional motion of a free particle which bounces elastically between a stationary wall located at $x = 0$ and a wall whose position oscillates with time and is given by $x = L \Delta \sin(\omega t)$. Derive a map relating the times T_n of the nth bounce off the oscillating wall and the particle speed v_n between the nth bounce and the $(n + 1)$th bounce off the oscillating wall to T_{n+1} and v_{n+1}. Assume that $L \gg \Delta$ so that $v_n(T_{n+1} - T_n) \approx 2L$. Is the map relating (T_n, v_n) to (T_{n+1}, v_{n+1}) conservative? Show that a new variable can be introduced in place of T_n, such that the new variable is bounded and results in a map which yields the same v_n as for the original map for all n.

3. Write a computer program to take iterates of the Hénon map. Considering the case $A = 1.4$, $B = 0.3$ and starting from an initial condition $(x_0, y_0) = (0, 0)$ iterate the map 20 times and then plot the next 1000 iterates to get a picture of the attractor.

4. Plot the first 25 iterates of the map given by Eq. (1.8) starting from $x_0 = 1/2$; (*a*) for $r = 3.8$ (chaotic attractor), (*b*) for $r = 2.5$ (period one attractor), and (*c*) for $r = 3.1$ (period two attractor).

5. For the map (1.8) with $r = 3.8$ plot the iterates of the two orbits orginating from the initial conditions $x_0 = 0.2$ and $x_0 = 0.2 + 10^{-5}$ versus iterate number. When does the separation between the two orbits first exceed 0.2?

Notes

1. Some review articles giving compact overview of chaotic dynamics are those of Helleman (1980), Ott (1981), Shaw (1981), and Grebogi *et al.* (1987d).
2. Some experiments on chaos in fluids are those of Libchaber and Maurer (1980), Gollub and Benson (1980), Bergé *et al.* (1980), Brandstater *et al.* (1983), and Sreenivasan (1986).
3. Applications of chaos to plasmas as well as many other topics are dealt with in the book by Sagdeev, Usikov and Zaslavsky (1990).
4. For example, Bryant and Jefferies (1984), Iansiti *et al.* (1985), Carroll *et al.* (1987), Roukes and Alerhand (1990), and Ditto *et al.* (1990b).
5. For example, Linsay (1981), Testa *et al.* (1982), and Rollins and Hunt (1984).
6. For example, Arecchi *et al.* (1982), Gioggia and Abraham (1984), and Mork *et al.* (1990).
7. The book by Moon (1987) on chaos contains outlines of results from a number of mechanical applications.
8. Chaotic phenomena and nonlinear dynamics in biology are dealt with in the book by Glass and Mackey (1988).
9. For example, Rössler (1976), Roux *et al.* (1980), Hudson and Mankin (1981), and Simoyi *et al.* (1982).
10. For example, Lauterborn (1981).
11. For example, Wisdom (1987) and Petit and Hénon (1986).
12. A review on the topic of routes to chaos is that of Eckmann (1981).
13. For example, according to the Poincaré–Bendixon theorem (e.g., see Hirsch and Smale (1974)), the only possible attracting solutions of (1.3) for **x** a two-dimensional vector in the plane are periodic solutions, steady states and solutions in which the orbit approaches a figure 8 or one of its lobes. In all these cases, the solution is not chaotic.
14. See May (1976) for an early discussion of the dynamics of this map.
15. Note that, if we take $|\Delta(0)|$ to be a small constant value (rather than examining $|\Delta(t)|/|\Delta(0)|$ in the limit that $|\Delta(0)| \to 0$), then the growth of $|\Delta(t)|$ cannot be exponential forever. In particular, $|\Delta(t)| < 2R$, and hence exponential growth must cease when $|\Delta(t)|$ becomes of the order of the attractor size. Thus later on (in Chapters 2 and 4) we shall be defining sensitive dependence on initial conditions in terms of the exponential growth of *differential* separations between orbits.
16. The definition of chaos given here is for chaotic *attractors*. When dealing with *nonattracting* chaotic sets (treated in Chapter 5) a more general definition of chaos is called for. Such a more general definition, which seems suitable very broadly, equates chaos with the condition of positive topological entropy. Topological entropy is defined in Chapter 4.

CHAPTER TWO

One-dimensional maps

One-dimensional noninvertible maps are the simplest systems capable of chaotic motion.[1] As such, they serve as a convenient starting point for the study of chaos. Indeed, we shall find that a surprisingly large proportion of the phenomena encountered in higher dimensional systems is already present, in some form, in one-dimensional maps.

2.1 Piecewise linear one-dimensional maps

As a first example, we consider the *tent map*,

$$x_{n+1} = 1 - 2|x_n - \tfrac{1}{2}|. \tag{2.1}$$

This map is illustrated in Figure 2.1(a). For $x_n < \tfrac{1}{2}$, Eq. (2.1) is $x_{n+1} = 2x_n$. Hence initial conditions that are negative remain negative and move off to $-\infty$, doubling their distance from the origin on each iterate. For $x_n > \tfrac{1}{2}$, $x_{n+1} = 2(1 - x_n)$. Hence, if $x_0 > 1$, then $x_1 < 0$, and the subsequent orbit points again move off to $-\infty$. For x_n in the interval $[0, 1]$, we have $0 \le 1 - 2|x_n - \tfrac{1}{2}| \le 1$, and so the subsequent iterate, x_{n+1}, is also in $[0, 1]$; hence, if $0 \le x_0 \le 1$, the orbit remains bounded and confined to $[0, 1]$ for all $n \ge 0$. We henceforth focus on the dynamics of orbits in $[0, 1]$. Figure 2.1(b) illustrates the action of the map on the interval $[0, 1]$ as consisting of two steps. In the first step, the interval is uniformly stretched to twice its original length. In the second step, the stretched interval is folded in half, so that the folded line segment is now contained in the original interval. Following a point on the original line segment $[0, 1]$ through this stretching and folding process, its final location is given in terms of its location before stretching and folding by Eq. (2.1). The stretching leads to exponential divergence of nearby trajectories (by a factor of two on each iterate). The folding process keeps the orbit bounded. Note, also, that the folding process causes the map to be noninvertible, since it results in two different values of x_n mapping to the same x_{n+1}. This example illustrates a general result for one-dimensional maps mapping an interval into itself (here the interval is $[0, 1]$). That is, in order for there to be chaos, the map must, on average, be stretching. On the other hand, for the orbit to remain bounded in the presence of stretching, there must also be folding. Hence, for a one-dimensional map to be chaotic, it must be noninvertible.

To further illustrate the sensitive dependence on initial conditions for the tent map, consider composing the map with itself m times to obtain M^m. Here, M^m is defined as

$$M^m(x) = M(M^{m-1}(x)) = \underbrace{M(M(M(\ldots(M(x))\ldots)))}_{m \text{ times}}$$

$$M^1(x) = M(x).$$

Figure 2.1 The tent map.

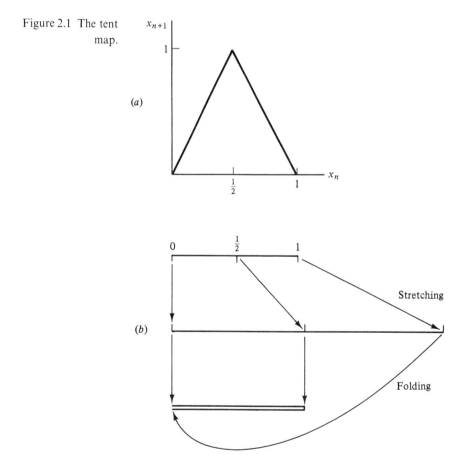

Thus,

$$x_{n+m} = M^m(x_n). \qquad (2.2)$$

Figure 2.2(a) shows $x_{n+2} = M(M(x)) = M^2(x_n)$ versus x_n for the tent map. To obtain Figure 2.2(a), we note that, if x_n is equal to 0, $\frac{1}{2}$, or 1, then two applications of (2.1) yield $x_{n+2} = 0$, while, if x_n is either $\frac{1}{4}$ or $\frac{3}{4}$, then two applications of (2.1) yield $x_{n+2} = 1$. Noting that the variation of x_{n+2} with x_n between these points is linear, Figure 2.2(a) follows. Figure 2.2(b) shows $x_{n+m} = M^m(x_n)$ for arbitrary m. Thus, given the knowledge that an initial condition lies within $\pm 2^{-m}$ of some point, then Figure 2.2(b) shows that x_m can lie anywhere in the interval $[0, 1]$. Hence, the knowledge we have of the small range in which the initial condition falls leads to absolutely no knowledge of the location of orbit points x_n for times $n \geq m$.

Figure 2.2(a) The second and (b) the mth iterate of the tent map.

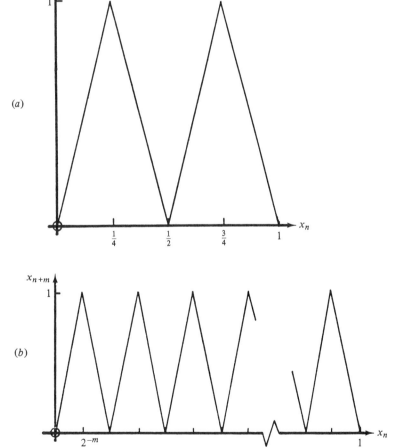

This is a consequence of the exponential sensitivity of chaotic orbits to small changes in initial conditions.

Another simple map, closely related to the tent map, is $M(x) = 2x$ modulo 1,

$$x_{n+1} = 2x_n \text{ modulo } 1. \tag{2.3}$$

Figure 2.3 shows the map and its mth iterate. This map can be regarded as a map on a circle, since the modulo 1 in Eq. (2.3) makes x like an angle variable, where x increasing from 0 to 1 corresponds to one circuit around the circle. Viewed in this way, the action of the $2x$ modulo 1 map on the circle may be thought of as the stretch-twist-fold operation illustrated in Figure 2.4. First, the circle is uniformly stretched so that its circumference is twice its original length. Then it is twisted into a figure 8, the upper and lower lobes of which are circles of the original length. The upper circle is then folded down on to the lower circle, and the two circles are pressed together. By following this operation, a point on the original circle (Figure 2.4(a)) is mapped to a point on the final pressed-together circle in such a way that its x coordinate transforms as in Eq. (2.3). Both Figures 2.3(b) and 2.4 illustrate the chaotic separation of nearby points (by a factor of two) on each iterate of this map.

An alternate way of viewing the $2x$ modulo 1 map is as an example of a

Figure 2.3(a) The $2x$ modulo 1 map, and (b) its mth iterate.

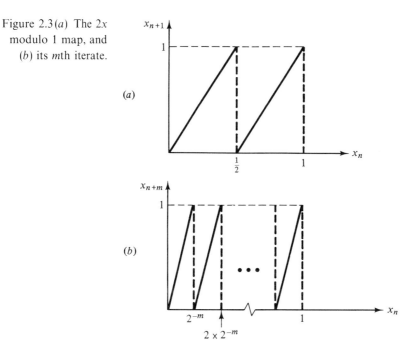

Bernoulli shift. Say we represent the initial condition x_0 as a binary decimal

$$x_0 = 0.a_1 a_2 a_3 \ldots \equiv \sum_{j=1}^{\infty} 2^{-j} a_j, \qquad (2.4)$$

where each of the digits a_j is either 0 or 1. Then, the next iterate is obtained by setting the first digit to zero and then moving the decimal point one space to the right,

$$x_1 = 0.a_2 a_3 a_4 \ldots,$$
$$x_2 = 0.a_3 a_4 a_5 \ldots,$$

and so on. Thus, digits that are initially far to the right of the decimal point, and hence have only a very slight influence on the initial value of x, eventually become the first digit. Thus, a small change of the initial condition, such as changing a_{40} from a zero to a one (a change of x_0 by 2^{-40}), eventually, at time $n = 39$, makes a large change in x_n.

We define a periodic orbit of a map to have period p if the orbit successively cycles through p *distinct* points $\bar{x}_0, \bar{x}_1, \ldots, \bar{x}_{p-1}$. (These points are 'distinct' if $\bar{x}_i \neq \bar{x}_j$ unless $i = j$.) Thus for each such point $\bar{x}_j$ we have $\bar{x}_j = M^p(\bar{x}_j)$ for $j = 0, 1, \ldots, p-1$. We can now use the binary representation (2.4) to construct periodic orbits of the map (2.3). In fact, any infinite sequence of zeros and ones which is made up of identically

Figure 2.4 Stretch-twist-fold operation.

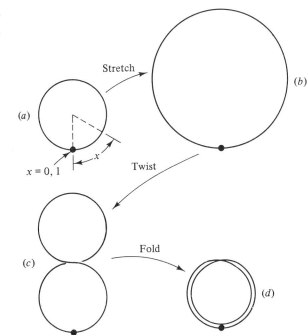

repeating finite sequence segments produces a binary digit expansion of an initial condition for a periodic orbit. For example, the initial condition $x_0 = 0.10101010\ldots = \frac{2}{3}$, which is made by repeating the two-digit sequence 10 *ad infinitum*, is the initial condition for a period 2 orbit. Applying (2.3) to x_0 yields $x_1 = 0.010101\ldots = \frac{1}{3}$. Applying (2.3) to $\frac{1}{3}$ reproduces $\frac{2}{3}$. Thus, we produce a period 2 orbit $(\frac{2}{3}, \frac{1}{3}, \frac{2}{3}, \frac{1}{3}, \ldots)$ which repeats after every second iterate. Similarly, orbits of any arbitrarily large period p arise from initial conditions of the form $0.a_1 a_2 \ldots a_p a_1 a_2 \ldots a_p a_1 a_2 \ldots$. We can ask, how many different initial conditions are there that return to themselves after p iterations? Since there are 2^p distinct sequences $a_1, a_2, \ldots, a_p$, we conclude that there are $2^p - 1$ such initial conditions. The minus one arises because the two sequences $(a_1, a_2, \ldots, a_p) = (0, 0, \ldots, 0)$ and $(a_1, a_2, \ldots, a_p) = (1, 1, \ldots, 1)$ yield the same results, namely, $0 = 0.000\ldots$ and $1 = 0.111\ldots$, which are the same values modulo 1. A point y on a period p orbit is also a fixed point (i.e., a period one point) of the p times composed map,

$$y = M^p(y).$$

We illustrate this for period $p = 2$ and the $2x$ modulo 1 map in Figure 2.5. We note that there are $2^p = 2^2 = 4$ intersections of the diagonal line $x_{n+2} = x_n$ with the two times composed map function $M^2(x_n)$. These intersections are $0, \frac{1}{3}, \frac{2}{3}$, and 1. Since 0 and 1 are equivalent, there are $2^2 - 1 = 3$ distinct initial conditions which repeat after two iterates. The

Figure 2.5 Fixed points of M^2.

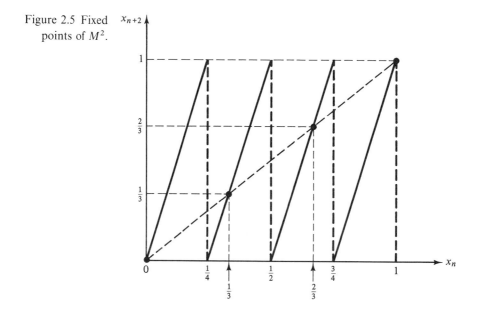

point 0 is a fixed point of the original map. The points $\frac{1}{3}$ and $\frac{2}{3}$ are on the period two orbit $(\frac{2}{3}, \frac{1}{3}, \frac{2}{3}, \frac{1}{3}, \ldots)$.

Example: How many period four orbits are there for the $2x$ modulo 1 map? There are $2^4 - 1 = 15$ fixed points of M^4. Of these 15, one is 0, and two are $\frac{1}{3}$ and $\frac{2}{3}$ (since fixed points of M and M^2 are necessarily fixed points of $M^4 = (M^2)^2$). This leaves 12 fixed points of M^4 which are not fixed points of M^p for any $p < 4$. These must lie on orbits of period four. Thus, there are $12/4 = 3$ distinct period four orbits.

Note that the number of fixed points of M^p for the tent maps is 2^p, since, as is evident from Figure 2.2, the graph of $x_{n+p} = M^p(x_n)$ will have 2^p intersections with $x_{n+p} = x_n$. Unlike the $2x$ modulo 1 map, the tent map has two distinct fixed points, $x = 0$, and $x = \frac{2}{3}$.

For both the tent map and the $2x$ modulo 1 map, the number of fixed points of M^p that are not fixed points of M is $2^p - 2$. If p is a prime number, then all of these $2^p - 2$ fixed points must lie on periodic orbits of period p. (If p is not prime and has integer factors $p_1, p_2, \ldots$ ($p = p_1^{n_1} p_2^{n_2} \ldots$), then some of these $2^p - 2$ points will be on orbits of the lower periods $p_1, p_2, \ldots$.) Hence, if p is prime, the number of periodic orbits of period p is[2]

$$N_p = (2^p - 2)/p \qquad (2.5)$$

for both the tent map and the $2x$ modulo 1 map. This number gets large rapidly; for example, $N_{11} = 186$, $N_{13} = 630$, and $N_{17} = 7710$. For p not prime, the number N_p of periodic orbits satisfies $N_p < (2^p - 2)/p$ and is more difficult to obtain, as our example for $p = 4$ above demonstrates. Nevertheless, for *large p*, we always have that $N_p \simeq 2^p/p$ with a correction which is small compared to N_p.

We now address the question of the stability of periodic orbits of a one-dimensional map. Say we have a periodic orbit of period p: $\bar{x}_0, \bar{x}_1, \ldots, \bar{x}_{p-1}, \bar{x}_p, \ldots$, where $\bar{x}_p = \bar{x}_0$. Then, for each of these x-values $\bar{x}_j = M^p(\bar{x}_j)$, for $j = 0, 1, \ldots, p - 1$. Say we take an initial condition slightly different from $\bar{x}_j$. We denote this initial condition $x_0 = \bar{x}_j + \delta_0$. As a result of the deviation δ_0, the pth iterate of x_0 is slightly different from $\bar{x}_j$, and we denote it $x_p = \bar{x}_j + \delta_p$. Thus,

$$\bar{x}_j + \delta_p = M^p(\bar{x}_j + \delta_0).$$

Since δ_0 is small, we Taylor expand to first order in δ_0, to obtain

$$\delta_p = \lambda_p \delta_0, \qquad (2.6)$$

where

$$\lambda_p = \left. \frac{dM^p(x)}{dx} \right|_{x = \bar{x}_j} = \left. \frac{dx_{n+p}}{dx_n} \right|_{x_n = \bar{x}_j}.$$

Using

$$\frac{dx_{n+p}}{dx_n} = \frac{dx_{n+1}}{dx_n}\frac{dx_{n+2}}{dx_{n+1}}\cdots\frac{dx_{n+p}}{dx_{n+p-1}}$$

$$= M'(x_n)M'(x_{n+1})\ldots M'(x_{n+p-1}),$$

where $M'(x) \equiv dM(x)/dx$, we find that $\lambda_p = dM^p(x)/dx|_{x=\bar{x}_j}$ is the same for all points $\bar{x}_j$ on the periodic orbit,

$$\lambda_p = M'(\bar{x}_0)M'(\bar{x}_1)\ldots M'(\bar{x}_{p-1}). \tag{2.7}$$

(For any $\bar{x}_j = x_n$, the points $x_n, x_{n+1}, \ldots, x_{n+p-1}$ cycle through each point on the given periodic orbit, leading to the above product of terms M' at every point in the cycle.) What happens if we follow the point $\bar{x}_j + \delta_p = \bar{x}_j + \lambda_p\delta_0$ another p iterates around the periodic cycle? If we do this, it maps to $\bar{x}_j + \delta_{2p} = \bar{x}_j + \lambda_p\delta_p = \bar{x}_j + \lambda_p^2\delta_0$, yielding the deviation $\delta_{2p} = \lambda_p^2\delta_0$. In general,

$$\delta_{mp} = \lambda_p^m\delta_0. \tag{2.8}$$

Thus, the deviation from the periodic orbit grows (if $|\lambda_p| > 1$) or shrinks (if $|\lambda_p| < 1$) by a factor $|\lambda_p|$ on each circuit around the periodic cycle. If $|\lambda_p| > 1$, the periodic orbit is said to be *unstable*. This is the case for all the periodic orbits of the tent map and the $2x$ modulo 1 map. This follows from the fact that $|M'(x)| = 2$ for these maps for all x (except at the single point $x = 1/2$, where the derivative is not defined, and which, in any case, does not lie on any periodic orbit). Thus, by Eq. (2.7), for these maps, $|\lambda_p| = 2^p > 1$, and all of the periodic orbits are unstable. On the other hand, in the next section we shall deal with situations where periodic orbits are stable, $|\lambda_p| < 1$. In this case, initial conditions near the periodic orbit asymptote to it, and the periodic orbit is an attractor. We say that the periodic orbit is *stable* if $|\lambda_p| < 1$ and *superstable* if $\lambda_p = 0$. We call λ_p the *stability coefficient* for the periodic orbit.

From our previous discussion of the fixed points of M^p for the tent map and the $2x$ modulo 1 map, it is clear that, for these maps, points on periodic orbits are *dense* in the interval $[0, 1]$. That is, for any x in $[0, 1]$ and any ε, *no matter how small* ε is, there is at least one point on a periodic orbit (actually an infinite number of such points) in $[x - \varepsilon, x + \varepsilon]$. For example, from Figure 2.3(*b*), there is one fixed point of M^p in each interval $[2^{-p}(m - 1), 2^{-p}m]$ for $m = 1, 2, \ldots, 2^p$. Thus, there is at least one fixed point of M^p and hence at least one periodic point of M in $[x - \varepsilon, x + \varepsilon]$ for $p > \ln(1/\varepsilon)/\ln 2$ (i.e., $2^{-p} < \varepsilon$). The fact that periodic points are dense is very significant.[3] It is important to note, however, that the periodic points are a *countably* infinite set, while the set of *all* points in the interval $[0, 1]$ is *uncountable*. In this sense, the periodic points, while dense, are still a much smaller set than all the points in $[0, 1]$. This implies, in particular, that, if one were to choose an initial condition x_0 at random in $[0, 1]$ according to a uniform probability distribution in $[0, 1]$, then the probability that x_0 lies on a periodic point of the map is zero. Hence, randomly chosen initial conditions do not produce periodic orbits for the tent and $2x$ modulo 1 maps. Thus, we say that nonperiodic orbits are *typical* for these maps, and periodic orbits are not typical. If we make a histogram of the fraction of times a finite length orbit originating from a typical initial condition falls in bins of equal size along the x-axis, $[(m-1)/N, m/N]$ for $m = 1, 2, \ldots, N$, then the fraction of time spent in a bin approaches $1/N$ for each bin as the length of the orbit is allowed to increase to infinity. Thus, defining a function $\rho(x)$ such that, for any interval $[a, b]$ in $[0, 1]$, the fraction of the time typical orbits spend in $[a, b]$ is $\int_a^b \rho(x)\,dx$, we have that, for the tent map and the $2x$ modulo 1 map,

$$\rho(x) = 1 \text{ in } [0, 1]. \tag{2.9}$$

We call $\rho(x)$ the *natural invariant density*.

Of course, if x_0 is chosen to lie exactly on an unstable periodic point, the orbit does not generate a uniform density. We emphasize, however, that such points have zero probability when x_0 is chosen randomly. The uniform invariant density generated by orbits from typical initial conditions for the tent and $2x$ modulo 1 maps is an example of a *natural measure*, a concept which we discuss in Section 2.3.3.

2.2 The logistic map

In this section, we consider the logistic map (1.8),

$$x_{n+1} = rx_n(1 - x_n). \tag{2.10}$$

As pointed out by May (1976), this map may be thought of as a simple idealized ecological model for the yearly variations in the population of an insect species. Imagine that every spring these insects hatch out of eggs laid the previous fall; they eat, grow, mature, mate, lay eggs, and then die. Assuming constant conditions each year (same weather, predator population, etc.), the population at year n uniquely determines the population at year $n + 1$. Thus a one-dimensional map applies. Say that the number z_n of insects hatching out of eggs is not too large. Then we can imagine that for each insect, on average, there will be r eggs laid, each of which hatches at year $n + 1$. This yields a population at year $n + 1$ of $z_{n+1} = rz_n$. Assuming $r > 1$ this also yields an exponentially increasing population $z_n = r^n z_0$. However, if the population is too large, the insects may begin to exhaust their food supply as they eat and grow. Thus some insects may die before they reach maturity. Hence the average number of eggs laid per hatched insect will become less than r as z_n is increased. The simplest possible assumption incorporating this overcrowding effect would be to say that the number of eggs laid per insect decreases linearly with the insect population, $r[1 - (z_n/\bar{z})]$, where $\bar{z}$ is the insect population at which the insects exhaust all their food supply such that none of them reach maturity and lay eggs. This yields the one-dimensional map $z_{n+1} = rz_n[1 - (z_n/\bar{z})]$. Dividing through by $\bar{z}$ and letting $x = z/\bar{z}$, we obtain the logistic map Eq. (2.10).

We now examine the dynamics of the logistics map. In particular, we shall be concerned with the question of how the character of the orbits originating from typically chosen initial conditions changes as the parameter r is varied. The map function is shown in Figure 1.8. The maximum of $M(x)$ occurs at $x = \frac{1}{2}$ and is $M(\frac{1}{2}) = r/4$. Thus, for $0 \leq r \leq 4$, if x_n is in $[0, 1]$, then so is x_{n+1}, and the orbit remains in $[0, 1]$ for all subsequent time. If $r > 1$, then $M'(0) > 1$ and the fixed point at $x = 0$ is unstable. Also, if $r > 1$, then $M(x) < x$ for negative x, implying that (as for the tent map) negative initial conditions and initial conditions in $x > 1$ generate orbits which tend to $-\infty$ with increasing time. In this section, we restrict our considerations to $4 \geq r \geq 1$ and x in the interval $[0, 1]$.

First, consider the case $r = 4$,

$$x_{n+1} = 4x_n(1 - x_n). \tag{2.11}$$

In this case, a change of variables transforms the logistic map into the tent map, Eq. (2.1). For x in the interval $[0, 1]$, define y, also in $[0, 1]$, by

$$x = \sin^2\left(\frac{\pi y}{2}\right) = \frac{1}{2}[1 - \cos(\pi y)]. \tag{2.12}$$

(See Figure 2.6.) Substituting in (2.11), we obtain $\sin^2(\pi y_{n+1}/2) = 1 - \cos^2(\pi y_n) = \sin^2(\pi y_n)$. This yields $(\pi y_{n+1}/2) = \pm(\pi y_n) + s\pi$, where s

Figure 2.6 The change of variables (2.12).

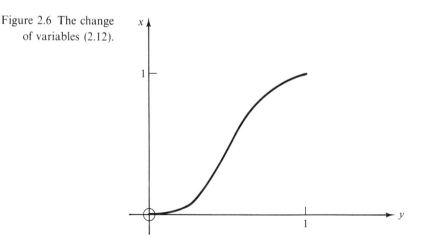

is an integer. Recalling that y is defined to lie in $[0, 1]$ determines the choice of s and the sign of the term πy_n. Thus, we obtain $y_{n+1} = 2y_n$ (i.e., the plus sign and $s = 0$) for $0 \leq y_n \leq \frac{1}{2}$, and $y_{n+1} = 2 - 2y_n$ (i.e., the minus sign and $s = 1$) for $\frac{1}{2} \leq y_n \leq 1$. This is just the tent map. Since the tent map is chaotic, so too must be the logistic map at $r = 4$. Also, the logistic map at $r = 4$ must have the same number of unstable periodic orbits as does the tent map. Since periodic orbits are dense in $[0, 1]$ for the tent map, they are also dense in $[0, 1]$ for the logistic map at $r = 4$.

In addition, for $r = 4$ typical (e.g., randomly chosen) initial conditions x_0 in $[0, 1]$ yield orbits which generate a smooth invariant density of points as for the tent map. Let $\rho(x)$ denote the natural invariant density for the logistic map with $r = 4$, and let $\tilde{\rho}(y)$ be the natural invariant density for the tent map. According to Eq. (2.9), $\tilde{\rho}(y) = 1$ for $1 \geq y \geq 0$. To find $\rho(x)$, we make use of the fact that the interval in y, $[y, y + dy]$, and the corresponding interval in x, $[x, x + dx]$, obtained by applying the change of variables, must be visited by typical orbits from their respective maps with the same frequency. Thus, $\rho(x)|dx| = \tilde{\rho}(y)|dy|$ or

$$\rho(x) = \left| \frac{dy(x)}{dx} \right| \tilde{\rho}(y(x)).$$

Using (2.9) for $\tilde{\rho}(y)$ and (2.12) for $y(x)$, we obtain the natural invariant density for the logistic map at $r = 4$,

$$\rho(x) = \pi^{-1}/[x(1 - x)]^{1/2}. \tag{2.13}$$

This density is graphed in Figure 2.7. Note the singularities at $x = 0$ and $x = 1$; $\rho \sim 1/x^{1/2}$ near $x = 0$ and $\rho \sim 1/(1 - x)^{1/2}$ near $x = 1$.

Having determined that there is chaos at $r = 4$, let us now examine smaller values of r. For $r \neq 1$, the logistic map has two fixed points given by the two solutions of $x = rx(1 - x)$, namely, $x = 0$ and $x = 1 - 1/r$. As already mentioned, the $x = 0$ fixed point is unstable for $r > 1$. Noting that $M'(1 - 1/r) = 2 - r$, we see that $x = 1 - 1/r$ is stable ($|2 - r| < 1$) for $3 > r > 1$. Hence, $x = 1 - 1/r$ is a fixed point attractor in this range. Furthermore, it may be shown that there are no periodic orbits with periods $p > 1$ for $3 > r$, and that, in the range $3 > r > 1$, any initial condition x_0 which satisfies $1 > x_0 > 0$ approaches the attractor at $x = 1 - 1/r$. We say that $[0, 1]$ is the *basin of attraction* of the attractor $x = 1 - 1/r$.

We have seen that there is chaos and an infinite number of unstable periodic orbits which are dense in $[0, 1]$ at $r = 4$. For $1 < r < 3$, there are only two periodic orbits of period one and no chaos. How is the infinite number of periodic orbits at $r = 4$ and the accompanying chaotic dynamics created as r is increased continuously from $r = 3$ to $r = 4$? This is a question we shall be addresing in some detail.

To begin, it is instructive to consider the two times iterated logistic map M^2 as shown in Figure 2.8. The fixed point of M at $x = x_* \equiv 1 - 1/r$ is also a fixed point of M^2, and the slope of $M^2(x)$ at $x = x_*$ is $[M'(x_*)]^2 = (2 - r)^2$. Thus, as r increases through $r = 3$, the orbit $x = x_*$ becomes unstable ($M'(x_*)$ decreases through -1), and, simultaneously, the slope of $M^2(x)$ at x_* increases from below one to above one. As shown in Figure 2.8, this leads to the creation of two new fixed points of M^2. Since these two new fixed points of M^2 are not fixed points of M, they must lie on a period two orbit. Thus, at precisely the point where the period one fixed point $x = x_*$ becomes unstable, a period two orbit is created. Furthermore, this period two orbit is stable when it is created. This can be seen from the fact that for r slightly larger than 3 the slope of $M^2(x)$ at the period two points is necessarily less than 1. The situation is schematically illustrated in Figure 2.8(c) which shows the solutions of $M^2(x) = x$ as a function of r, with the solutions corresponding to stable orbits shown as solid lines, and the unstable solution ($x = x_*$ for $r > 3$) shown as a dashed line. The change in the orbit structure, illustrated in Figure 2.8, is called a *period doubling bifurcation*.

Another way to visualize the occurrence of the period doubling bifurcation is shown in Figure 2.9. First, we note that the orbit starting from x_n can be obtained by the graphical interpretation of the map relation, $x_{n+1} = M(x_n)$, shown in Figure 2.9(a). Starting at some value x_n on the horizontal axis, the vertical dashed line in Figure 2.9(a) locates the

Figure 2.7 The invariant density $\rho(x)$ generated by typical orbits of the logistic map for $r = 4$.

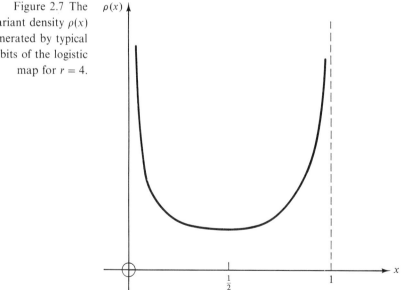

point (x_n, x_{n+1}). Then, going along the horizontal line from the point (x_n, x_{n+1}) to the 45° line $M(x) = x$, we locate the point (x_{n+1}, x_{n+1}). Going vertically from this point to the curve $M(x)$, we locate the point (x_{n+1}, x_{n+2}). Again, going horizontally to the 45° line, we come to the point (x_{n+2}, x_{n+2}). Proceeding in this way we can generate a sequence of orbit points. Using this type of construction, Figure 2.9(b) shows convergence to the period one fixed point $x = x_*$ for $1 < r < 3$, and Figure 2.9(c) shows convergence to the period two orbit for r a little larger than 3. In fact, when the stable period two orbit exists, it attracts typical

Figure 2.8(a) $M^2(x)$ versus x for $r < 3$, and (b) for $r > 3$ (the straight dashed line is the tangent to $M^2(x)$ at $x = x_*$). (c) Bifurcation of the fixed point at $x = 1 - 1/r$ from stability to instability accompanied by the creation of a stable period two as r increases through 3.

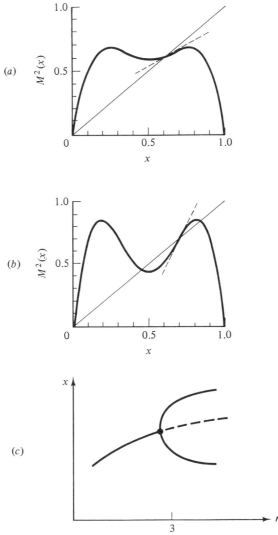

(in the sense of randomly chosen) initial conditions in $[0, 1]$. Another way of saying this is that it attracts all points in $[0, 1]$ except for a set of Lebesgue measure zero (roughly, the nonattracted set has zero length). (A compact set A is of Lebesgue measure zero if, for any $\varepsilon > 0$, we can cover A with a finite number of intervals whose total length is less than ε.) The set of points in $[0,1]$ not attracted to the period two is just the three points $x = 0$ and $x = x_*$ (which are unstable fixed points) and $x = 1$ (which maps to $x = 0$ on one iterate).

We have seen that the period one orbit has a stability coefficient $\lambda_1 = M'(x_*) = 2 - r$ which is 1 at $r = 1$ and decreases to $\lambda_1 = -1$ at $r = 3$, the point of the period doubling bifurcation. The period two orbit has a stability coefficient $\lambda_2 = M'(e)M'(f) = (M^2)'(e) = (M^2)'(f)$ (where $x = e, f$ are the points on the period 2 orbit), and $\lambda_2 = 1$ for $r = 3$. As r increases past $r = r_0 = 3$, the stability coefficient λ_2 decreases from $\lambda_2 = 1$, eventually becoming negative (e.g., the slope of $M^2(x)$ at e and f has decreased to negative values in Figure 2.8(b)). At some value $r = r_1$, the quantity λ_2 becomes -1, and for $r > r_1$, we have $\lambda_2 < -1$; hence, the period two orbit is unstable ($|\lambda_2| > 1$). Thus, $x = x_*$ is stable in a range

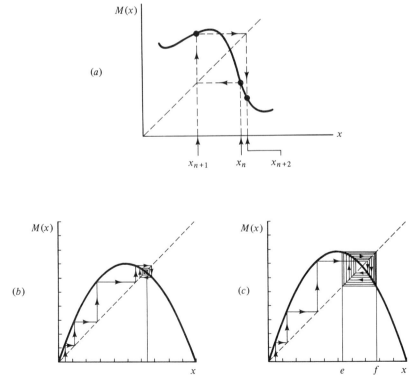

Figure 2.9(a) Graphical construction for following the orbit of a one-dimensional map. (b) Orbit of the logistic map for $r < 3$. (c) Orbit of the logistic map when the period two is stable.

$1 < r \leq r_0 = 3$, while the period two orbit is stable in a range $r_0 < r \leq r_1$. As r increases through r_1, the period two orbit period doubles to a period four orbit. In doing this, the picture is essentially the same as for the period doubling at $r = r_1$, except that the roles of M and M^2 in our interpretation of the original period doubling are now played by M^2 and M^4. Increasing r from r_1 to a value r_2, the stability coefficient of the period four orbit decreases from $\lambda_4 = 1$ to $\lambda_4 = -1$, past which point a stable period eight orbit appears, and remains stable for a range $r_2 < r \leq r_3$. This process of period doublings continues, successively producing an infinite cascade of period doublings with ranges, $r_{m-1} < r \leq r_m$, in which period 2^m orbits are stable. The length in r of the range of stability for an orbit of period 2^m decreases approximately geometrically with m. In particular (Feigenbaum, 1978, 1980a)

$$\frac{r_m - r_{m-1}}{r_{m+1} - r_m} \to 4.669201\ldots \equiv \hat{\delta} \tag{2.14}$$

as $m \to \infty$. Also, there is an accumulation point of an infinite number of period doubling bifurcations at a finite r value denoted r_∞,

$$r_\infty \equiv \lim_{m \to \infty} r_m = 3.57\ldots . \tag{2.15}$$

Equation (2.14) implies that for large m

$$|r_\infty - r_m| \simeq (\text{const.}) \, \hat{\delta}^{-m}. \tag{2.16}$$

Figure 2.10(a) shows a schematic plot of the points on the stable 2^m cycle as a function of r. (Orbits larger than eight are not shown in this figure because their stability ranges become so tiny.) Figure 2.10(b) shows a similar plot, but now the horizontal coordinate is replaced by $-\log(r_\infty - r)$. On this latter plot, the horizontal distance between successive period doublings approaches a constant as r approaches r_∞; namely, it approaches $\log \hat{\delta}$.

There is another scaling ratio in addition to $\hat{\delta}$ that one can define. For this purpose, it is useful to define a *superstable* period 2^m orbit as occurring at that value of r (denoted $\bar{r}_m$) at which the stability coefficient for the period 2^m orbit is zero. Recall that the stability coefficient decreases from 1 to -1 as r goes from r_{m-1} to r_m, so that $\bar{r}_m$ may, in some sense, be regarded as the middle of the range of the stable period 2^m orbit. Since the stability coefficient is zero at $\bar{r}_m$, we see from Eq. (2.7) that it must be the case that the *critical point* (defined as the point at the maximum of M, $M'(x) = 0$), $x = \frac{1}{2}$, is a point on the superstable orbit. Let Δ_m be the distance between this point and the nearest to it of the other $2^m - 1$ points in the cycle. This nearest point turns out to be the point which is one half period displaced from the critical point,

$$\Delta_m = M^{2^{m-1}}(\tfrac{1}{2}) - \tfrac{1}{2}. \tag{2.17}$$

It is found that for large m, the quantity Δ_m decreases geometrically (Feigenbaum, 1978, 1980a)

$$\Delta_m/\Delta_{m+1} \to -2.50280\ldots \equiv -\hat{\alpha}. \qquad (2.18)$$

(The minus sign in Eq. (2.18) signifies that the nearest orbit point to $x = \frac{1}{2}$ switches between below $\frac{1}{2}$ and above $\frac{1}{2}$ each time m is increased by 1.)

As we shall discuss later (Chapter 8), the scaling numbers $\hat{\delta}$ and $\hat{\alpha}$ appearing in Eqs. (2.14) and (2.18) were shown by Feigenbaum to be *universal* in the sense that they apply not only to the logistic map, but to any typical dissipative system which undergoes a period doubling cascade. Furthermore, these scaling numbers have been verified in experiments on a variety of physical systems, including ones where the describing dynamical system is infinite dimensional (e.g., fluid flows). Thus, the result for a one-dimensional map is found to apply to systems with arbitrarily high dimensionality.

Figure 2.10 Period doubling cascade.

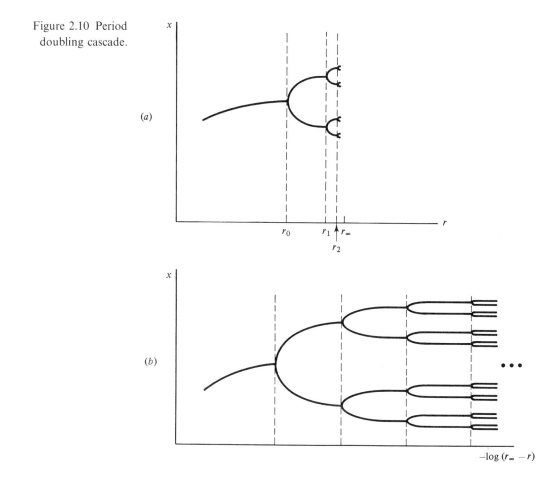

What happens beyond $r = r_\infty$ (i.e., in the range $r_\infty \le r \le 4$)? To answer this question, Figure 2.11(*a*) shows the *bifurcation diagram* for the logistic map for the range $3.5 \le r \le 4$. This diagram is computed in the following way:

(*a*) Set $r = 3.5$.

(*b*) Set $x_0 = \frac{1}{2}$.

(*c*) Iterate the map 500 times.

(*d*) Iterate the map another 1000 times (starting from x_{500}) and plot the resulting 1000 values of x.

(*e*) Increase r by a small amount, $r \to r + 10^{-2}$, and return to step (*b*).

This procedure is followed until $r = 4$ is reached. The reason for not plotting the first 500 iterates is that we wish our plot to show the orbit on (or very close to) the attractor and not the transient motion leading to it. Thus, the figure essentially shows the attracting set in x as a function of the parameter r. For example, at $r = 4$, the orbit fills the entire interval $[0, 1]$. For a value of r slightly less than 4, the attractor is a single interval contained within $[0, 1]$ as shown in Figure 2.11(*b*). As for $r = 4$, this orbit is apparently chaotic, but its motion is restricted to the smaller interval shown in Figure 2.11(*b*). As r is decreased through a value r'_0 labeled in Figure 2.11(*a*), the attractor splits into two bands, as shown in Figure 2.11(*c*). At the value of r corresponding to Figure 2.11(*c*), the orbit on the attractor alternates between the two bands on every iterate. If one were to examine the orbit on every second iterate, then the orbit would always be in the same one of the two bands and would undergo an apparently chaotic sequence restricted to that band, eventually coming arbitrarily close to every point in the band. As r is decreased from the situation shown in Figure 2.11(c), the two band attractor splits into a four band attractor at $r = r'_1$, into an eight band attractor at $r = r'_2$, and so on. The band doublings accumulate on r_∞ from above with the same geometric scaling as for the accumulation of period doublings on r_∞ from below,

$$\frac{r'_{m-1} - r'_m}{r'_m - r'_{m+1}} \to \hat{\delta}, \qquad (2.19)$$

where $\hat{\delta}$ is the same universal number as in Eq. (2.14).

In addition to apparently chaotic orbits, Figure 2.11(*a*) also shows that there are narrow ranges within $r_\infty \le r \le 4$ in which the attracting orbit is periodic. For example, the widest such range is occupied by the period three orbit. A blow-up of the bifurcation diagram in the range where the period three orbit occurs is shown in Figure 2.12(*a*). We see that the period three orbit is born at $r = r_{*3}$; undergoes a period doubling cascade in which orbits of period 3×2^m are successively produced; becomes chaotic and undergoes a cascade of band mergings $(3 \times 2^m$

bands → $3 \times 2^{m-1}$ bands), until a range of r is reached where chaos apparently appears in three bands. Finally, at $r = r_{c3}$, the attractor abruptly widens into a single band similar in size to that before the stable period three orbit came into existence. We call the range of r-values between the point where the period three orbit is born and the point where the three bands widen into one band a *period three window*. There are an infinite number of windows of arbitrarily high period within the chaotic range $4 \geq r \geq r_\infty$. For example, there are $(2^p - 2)/(2p)$ windows of period p if p is prime (the reason for this will be given subsequently). Each period p window essentially contains a replication of the bifurcation diagram for the map over its whole range (e.g., Fig. 2.12(b)). Thus, the windows themselves have windows, which themselves have windows, etc. For example, a window of period $p = 9 = 3 \times 3$ is discernible in Figure 2.12. As shown by Yorke *et al.* (1985) the bifurcation diagram within typical high period windows becomes universal (map independent) as the period

Figure 2.11 (*a*) Bifurcation diagram. (*b*) A single band chaotic attractor. (*c*) A two band chaotic attractor.

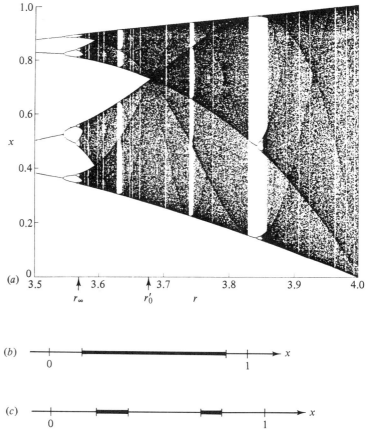

Figure 2.12(a) The period three window. (b) A blow-up of the middle section of the period three window from a value of r before the period doubling from period 12 to period 24 to $r = r_{c3}$. The details reproduce, almost exactly, the features of Figure 2.11(a) (from Yorke *et al.* 1985).

is increased. For example, the ratio of the width in the parameter of a window to the parameter difference between the initiation of the window and the occurrance of the first period doubling in the window universally approaches $\frac{9}{4}$.

It is generally thought that the windows are dense throughout the chaotic range. That is, given a value of r for which the orbit is chaotic, then in any ε neighborhood of that r value $[r - \varepsilon, r + \varepsilon]$, one can always find windows no matter how small ε is. We can give a heuristic argument for why windows should be dense in the chaotic range of r values, as follows. Say we have a chaotic orbit at $r = \bar{r}$, and we wish to argue that there is a stable periodic orbit (and hence a window) in the r interval $[\bar{r} - \varepsilon, \bar{r} + \varepsilon]$ for any ε. We assume that the initial condition $x_0 = \frac{1}{2}$ behaves like a

(a)

(b)

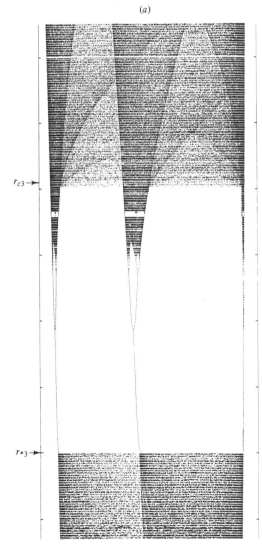
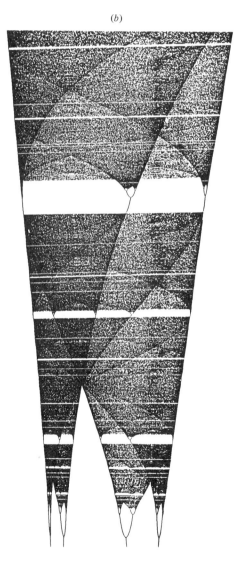

typical initial condition and hence generates an orbit which comes arbitrarily close to every point on the attractor. (This appears to be true except at special r values, one of which is $r = 4$ for which $M(\frac{1}{2}) = 1$, $M(1) = 0$, $M(0) = 0$.) Thus, if we wait long enough, at some time, $m = \tau$, the orbit will fall very close to $\frac{1}{2}$. Since $x_\tau = M^\tau(\frac{1}{2})$, we can regard x_τ as purely a function of r. Since x_τ is close to $x = \frac{1}{2}$, only a small change of r, say Δr, should be required to shift x_τ to $\frac{1}{2}$. At $r = \bar{r} + \Delta r$ we would then necessarily have a stable (in fact, superstable) periodic orbit of period τ. The orbit is stable by Eq. (2.7) because $M'(\frac{1}{2}) = 0$, and $x = \frac{1}{2}$ is a point on the orbit. If $|\Delta r|$ turns out to exceed ε, we can wait longer and find another x_τ that is much closer to $x = \frac{1}{2}$ (indeed, arbitrarily close). Hence, we should be able to make $|\Delta r|$ as small as we wish, and we thus believe that the windows are dense.

Given that stable periodic attractors are dense in r, one might question whether there is any room in r left for chaotic attractors to exist in. That is, if we choose a value of r randomly according to a uniform probability distribution in $[r_\infty, 4]$, is the probability zero that our choice yields chaos? From our bifurcation diagrams it certainly appears that we often see r-values where the orbits are apparently not periodic. Nevertheless, one might argue that these are only periodic orbits of extremely large period. The question has been settled in the proof by Jacobson (1981) which shows that the probability of choosing a chaotic r is not zero. Hence, chaos for this map is said to be 'typical.' Nevertheless, it may still seem strange that r intervals of nonchaotic (i.e., periodic) attractors are dense, yet the set of r-values yielding chaotic orbits is still not probability zero. Thus it may be useful at this point to give a simple example of a set with these characteristics.

The example is as follows. Say we consider the rational numbers in the interval $[0, 1]$. These numbers are dense in $[0, 1]$, since any irrational can be approximated by a rational to arbitrary accuracy. The rationals are also countable, since we can arrange them in a linear ordering (such as, $\frac{1}{2}, \frac{1}{3}, \frac{2}{3}, \frac{1}{4}, \frac{3}{4}, \frac{1}{5}, \frac{2}{5}, \frac{3}{5}, \frac{4}{5}, \frac{1}{6}, \frac{5}{6}, \frac{1}{7}, \ldots$). To the nth rational on this list (denoted s_n), we now associate an interval $I_n = (s_n - (\eta/2)(\frac{1}{2})^n, s_n + (\eta/2)(\frac{1}{2})^n)$ of length $2^{-n}\eta$. We are interested in the set S_* formed by taking the interval $[0, 1]$ and then successively removing $I_1, I_2, I_3, \ldots, I_n$ in the limit $n \to \infty$. Since the total length of all the removed interval sets is $\Sigma_{n=1}^{\infty}(\frac{1}{2})^n\eta = \eta$, we have that the Lebesgue measure ('length') of S_*, denoted $\mu(S_*)$, satisfies

$$\mu(S_*) > 1 - \eta, \tag{2.20}$$

which is positive if $\eta < 1$. (The greater than symbol, rather than an equals sign, appears in (2.20) because some of the removed intervals overlap.) Note that the Lebesgue measure of S_* is also the probability of randomly choosing a point in S_* from points in $[0, 1]$. Thus, for $\eta < 1$, for any point in the set S_*, an ε neighborhood always contains intervals I_n (which by definition are not in S_*), yet S_* has positive Lebesgue measure. The set S_* is an example of a *Cantor set* of positive (as opposed to zero) Lebesgue measure. Cantor sets both of zero Lebesgue measure and of positive Lebesgue measure, are common in chaotic dynamics. The appendix, to this chapter, reviews some elementary material on sets, including Lebesgue measure and Cantor sets.

We now ask, what is the mechanism by which the stable period p orbit initiating the period p window arises as r is increased? To answer this question, we consider the example of the period three window. Figure 2.13(a) shows the third iterate of the map $M^3(x)$ as a function of x for r below r_{*3} (solid curve) and for r above r_{*3} (dashed curve). At $r = r_{*3}$, the graph of $M^3(x)$ becomes tangent to the line $x_{n+3} = x_n$ at three points near

Figure 2.13 (a) x_{n+3} versus x_n from $r = 3.7 < r_{*3}$ (solid curve) and for $r = 3.9 > r_{*3}$ (dashed curve). (b) Schematic of $M^3(x)$ versus x near $x = 0.5$.

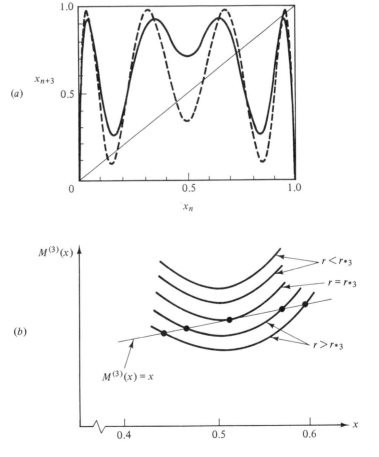

the first and second minimum and the fourth maximum of $M^3(x)$. For slightly larger r, the graph of $M^3(x)$ intersects $x_{n+3} = x_n$ at two points near each of the three tangencies that occurred for $r = r_{*3}$. The slope at three of the intersections is less than 1 and hence represents the stable period three orbit. The slope at the other three intersections is greater than 1 and represents an unstable period three orbit. Thus, as r increases through r_{*3}, we simultaneously create a period three stable attracting orbit and an unstable period three orbit. Figure 2.13(b) schematically shows the graph of $M^3(x)$ near the middle minimum at $x = \frac{1}{2}$ for five successively larger values of r. The situation is illustrated in Figure 2.14 (only the x coordinates of the period three orbit points near $x = \frac{1}{2}$ are plotted). This type of phenomenon is called a *tangent bifurcation*. As we shall see in Sections 2.3.1 and 2.3.2, the occurrence of windows as r is increased proceeds in a very regular and general order.

2.3   General discussion of smooth one-dimensional maps

2.3.1   General bifurcations of smooth one-dimensional maps

A qualitative change in the dynamics which occurs as a system parameter varies is called a *bifurcation*. In this subsection we shall be concerned with bifurcations of *smooth* one-dimensional maps which depend smoothly on a single parameter r. (We say a function is smooth if it is continuous and several times differentiable for all values of its argument.) To emphasize the parameter dependence we write the map function as $M(x, r)$, with the logistic map, Eq. (1.8), as a specific example. Without loss of generality,

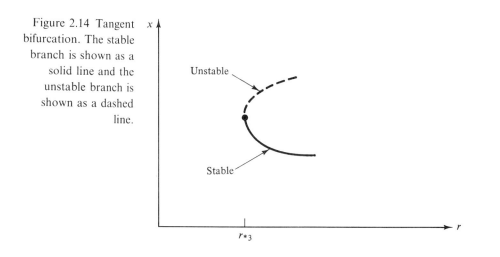

Figure 2.14 Tangent bifurcation. The stable branch is shown as a solid line and the unstable branch is shown as a dashed line.

we shall consider bifurcations of period one orbits (i.e., fixed points). (Bifurcations of a period p orbit can be reduced to consideration of a period one orbit by shifting attention to the p times iterated map M^p for which each point on the period p orbit is a fixed point.) We say a bifurcation is *generic* if the basic character of the bifurcation cannot be altered by arbitrarily small perturbations that are smooth in x and r. That is, if $M(x, r)$ is replaced by $M(x, r) + \varepsilon g(x, r)$, where g is smooth, then, if ε is small enough, the qualitative bifurcation behavior is unchanged. There are three generic types of bifurcations of smooth one-dimensional maps: the period doubling bifurcation, the tangent bifurcation, and the inverse period doubling bifurcation. These are illustrated in Figure 2.15, where the parameter r is taken as increasing to the right, and we have defined forward and backward senses for each bifurcation. Dashed lines are used for unstable orbits and solid lines for stable orbits. Thus, for example, in the forward inverse period doubling bifurcation, an initially unstable period one orbit bifurcates into an unstable period two and a stable period one orbit. (The forward period doubling bifurcation and forward tangent bifurcations have already been discussed in the previous section in the context of the logistic map.) Figure 2.16 shows how the three forward bifurcations can occur as the shape of the map[4] changes with increasing r. Note that in all three cases the magnitude of the stability coefficient of the period one orbit $|M'|$ is 1 at the bifurcation point.

Figure 2.15 Generic bifurcations of differentiable one-dimensional maps. The system parameter r increases toward the right. The vertical scale represents the value of the map variable. Dashed lines represent unstable orbits. Solid lines represent stable orbits.

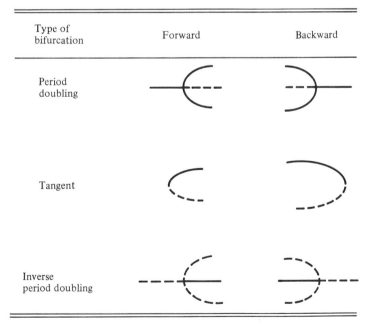

In order to get a better idea of the meaning of the word generic as applied to bifurcations, we now give an example of a nongeneric bifurcation. We consider the bifurcation of the logistic map at $r = 1$. The logistic map has two fixed points $x = 0$ and $x = x_* = 1 - 1/r$. These fixed points coincide at $r = 1$. For r slightly less than 1, $x = 0$ is stable and $x = x_*$ is unstable. For r slightly greater than 1, the stability characteristics of the two fixed points are interchanged. Thus the bifurcation diagram near $r = 1$ is as shown in Figure 2.17(a). Now say we perturb the logistic map by adding to it a small number ε

$$M(x; r) = rx(1 - x) + \varepsilon. \tag{2.21}$$

For ε positive, we obtain the picture in Figure 2.17(b), in which there is no bifurcation at all, but only a continuous change of the two fixed points with increasing r. For ε negative, we obtain two tangent bifurcations as shown in Figure 2.17(c). Nongeneric bifurcations require 'special' conditions on the map function, and one therefore expects that they would be unlikely to occur in applications. For example, the nongeneric

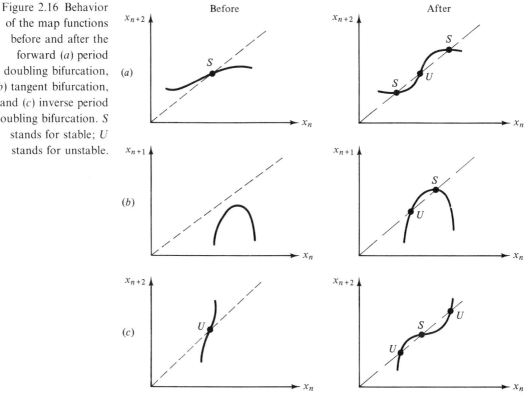

Figure 2.16 Behavior of the map functions before and after the forward (a) period doubling bifurcation, (b) tangent bifurcation, and (c) inverse period doubling bifurcation. S stands for stable; U stands for unstable.

bifurcation of the logistic map at $r = 1$ occurs because the map satisfies the special condition $M(0, r) = 0$ *for all r*.

We now ask how the infinite number of unstable periodic orbits for the logistic map at $r = 4$ is created as r is increased to $r = 4$ from a value of r in the range $3 > r > 1$ (where the only periodic orbits are the fixed points (period one orbits) $x = 0$ and $x = x_*$). The key point (which we will not demonstrate) is that the logistic map can be shown to have no backward bifurcations and no inverse period doublings.[5] Hence the answer to our question lies in the character of the forward period doubling and tangent bifurcations (the only bifurcations of the logistic map in $r > 1$). Namely, as r is increased a tangent bifurcation *creates* two new orbits, one stable and one unstable. The period doubling bifurcation takes an original stable period p orbit and replaces it by an unstable period p and a stable period $2p$ orbit. Thus the period doubling also creates a new orbit, one of period $2p$, as r is increased. Also since every stable period p orbit period doubles, every stable orbit that is created eventually yields an unstable orbit of the same period. The unstable orbits remain unstable and are not destroyed as r is increased, because they can only be destroyed or rendered stable by backward bifurcations (Figure 2.15) or by the forward inverse period doubling bifurcation, and these do not occur for the logistic map. Thus every periodic orbit created at lower r must be present at $r = 4$.

As an application of this discussion we can use it to deduce the number of windows that exist for a given period for the logistic map. Assuming that all tangent bifurcations initiate the start of a window (as illustrated, for example, in Figures 2.12 and 2.13 for period three), we can utilize our knowledge of the value of N_p, the number of periodic orbits at $r = 4$, to

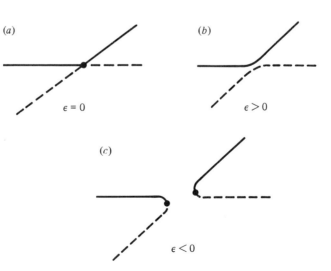

Figure 2.17 (*a*) Bifurcation of the logistic map at $r = 1$. (*b*) There is no bifurcation when ε is a small positive number. (*c*) There are two tangent bifurcations (denoted by dots) when ε is a small negative number.

deduce the number of tangent bifurcations needed to create them, which then gives the number of windows. For example, in Section 2.1 we found that the tent map has three period four orbits. Hence the logistic map at $r = 4$ must also have three period four orbits. Of these three, one is created when the period two period doubles. The remaining two must have been created by a tangent bifurcation. Hence we conclude that as r is increased from $r = 1$ to $r = 4$ there must be one period four tangent bifurcation (each tangent bifurcation creates two orbits), and hence there is one period four window. If p is an odd number, then orbits of period p can only be produced by tangent bifurcations, and the number of windows is $N_p/2$. In particular, if $p > 1$ is prime, we have by Eq. (2.5) that the number of period p windows of the logistic map is $(2^{p-1} - 1)/p$.

To conclude our discussion of generic bifurcations of one-dimensional maps, we emphasize that these bifurcations have their counterparts in higher dimensional maps and flows. Figures 2.18(a) and (b) illustrate the occurrence of a period doubling bifurcation of a flow in phase space. Before the bifurcation (Figure 2.18(a)) there is a single periodic attracting orbit manifested as a single point in the Poincaré surface of a section shown in the figure. After the bifurcation there are two periodic orbits (Figure 2.18(b)). One of these is an unstable 'saddle' periodic orbit (the dashed curve); i.e., it repels orbits in one direction and attracts them in the other direction. This saddle is essentially the continuation of the stable periodic orbit that existed before the bifurcation and manifests itself as a

Figure 2.18(a) and (b) A period doubling bifurcation. (c) A saddle node-bifurcated situation. In (c) the arrows on the surface of section indicate directions of attraction and repulsion from the saddle and the node fixed points of the surface of section map.

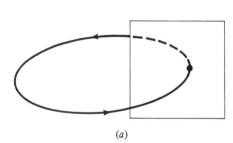

(a)

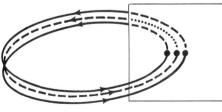

(b)

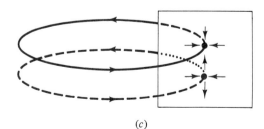

(c)

single point in the surface of the section. The other, period doubled, orbit shown in Figure 2.18(b) is stable and passes twice through the surface of the section. We may view the period doubled orbit as forming the edges of a Möbius strip in which the dashed unstable orbit lies. Points in the Möbius strip are repelled by the unstable orbit and pushed toward the edge of the strip (i.e., they approach the period doubled orbit). Figure 2.18(c) illustrates the counterpart of the tangent bifurcation in a flow. (In the flow context this bifurcation is called a *saddle-node* bifurcation.) At the bifurcation, coincident saddle periodic and attracting periodic orbits are created. As the parameter is increased the two orbits separate as shown.

2.3.2 Organization of the appearance of periodic orbits

One-dimensional maps are relatively more constrained in their possible dynamics than higher dimensional systems. One consequence of this is the remarkable theorem of Šarkovskii (1964). Consider the following ordering of all the positive integers

$$3, 5, 7, \ldots, 2 \times 3, 2 \times 5, 2 \times 7, \ldots, 2^2 \times 3, 2^2 \times 5, 2^2 \times 7, \ldots,$$
$$2^3 \times 3, 2^3 \times 5, 2^3 \times 7, \ldots, 2^5, 2^4, 2^3, 2^2, 2, 1 \qquad (2.22)$$

That is, first we list all the odd numbers except one. Then we list two times all the odd numbers except 1. Then we list 2^2 times all the odd numbers except 1, and so on. Having done this, we have accounted for all the positive integers except for those which are a power of 2, which we add to the list in decreasing order.

Theorem. Suppose a continuous map $M(x)$ of the real line (i.e., the set of points $-\infty \le x \le +\infty$) has a periodic orbit of period p (i.e., a point $\bar{x}$ on the orbit returns to itself after p iterates and not before). If, in Šarkovskii's ordering (2.22), p occurs before another integer l in the list, then the map M also has a periodic orbit of period l.

Note, that, if M has a periodic orbit of period p, and p is not a power of 2, then the theorem implies that M must have an infinite number of periodic orbits (in particular all orbits of period 2^m for $m = 0, 1, 2, 3, \ldots$). In addition, if M has a period three orbit, it must also have orbits of all other periods. Thus, for example, in the range of the period three windows for which there is an attracting period three orbit there must be an infinite number of periodic orbits of all periods. In this range of r, however, all periodic orbits except the period three are unstable, and hence do not attract typical initial conditions. In addition to period three implying the existence of all other periods, Li and Yorke (1975) show that the existence of a period three orbit also implies the existence of an uncountable set of orbits which never settle into a periodic cycle and remain nonperiodic forever. (They introduced the term 'chaos' to describe this situation.) These nonperiodic orbits are also nonattracting when the stable period three exists. Thus, for a given value of r for which the period three orbit is attracting, there are an uncountably infinite number of initial conditions in $[0,1]$ for which nonperiodic, very complicated orbits result. However, these initial conditions are of zero Lebesgue measure, and the orbits they yield are unstable. Thus, starting at a typical initial condition very near to one that follows a nonperiodic orbit, the orbit initially follows the nonperiodic orbit, but eventually diverges from it and is then attracted to the period three orbit. The same situation with respect to the existence of nonperiodic orbits applies for all ranges of r, above r_∞, for which stable periodic orbits occur. Note the fundamental difference between this situation, and the situation that holds in $3 > r > 1$. In that range there is an attracting period one orbit, and the only initial conditions in $[0, 1]$ not attracted to it are $x = 0$ and $x = 1$.

Still more can be said concerning the organization of periodic orbits. Metropolis, Stein and Stein (1973) consider maps of the form

$$M(x,r) = rf(x),$$

where $f(x)$ has a single maximum. (Actually their considerations are somewhat more general.) At the value of r at which a periodic orbit of period p is superstable (i.e., has stability coefficient $\lambda_p \equiv 0$), the orbit is labeled by a string of $p - 1$ symbols. Since the orbit is superstable, it goes through the critical point. If the nth iterate from the critical point falls to the right of the critical point, the nth symbol in the string is an R; if it falls to the left, the nth symbol in the string is an L. For example, the symbol string corresponding to the period four orbit shown in Figure 2.19 is RLL

Figure 2.19 The symbol sequence for this period four superstable orbit is *RLL*.

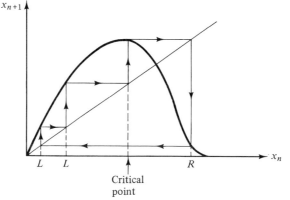

or RL^2 (no character label is used for the initial condition at the critical point). Metropolis *et al.* show that, as r is increased, orbits with particular symbol strings appear in accord with a well-defined rule. In particular, they show how to produce an ordered list of orbits such that if, as r is increased, two different orbits on the list occur, then all orbits in between them on the list must also have occurred. For example, if a period nine with symbol sequence RL^2RLR^2L appears at some value of r, and a period four with symbol sequence RL^2 appears at some larger value of r, then there must appear at some r-value in between a period five with symbol sequence RL^2R.

2.3.3 Measure, ergodicity and Lyapunov exponents for one-dimensional maps

We have seen that typical initial conditions for the tent map and logistic map at $r = 4$ generate orbits whose frequency of visits to any given interval is described by a function $\rho(x)$ which we have called the *natural invariant density*. Another view of the invariant density is the following. Imagine that we start off with an infinite number of initial conditions sprinkled along the x-axis with a smooth density $\rho_0(x)$, such that the fraction of those initial conditions in an interval $[a, b]$ is $\int_a^b \rho_0(x)\,dx$. Now imagine applying the map $M(x)$ to each initial condition. Thus a new density $\rho_1(x)$ is generated. Similarly, applying the map again, we obtain a density $\rho_2(x)$, and so on. The relation evolving a density forward in time is

$$\rho_{n+1}(x) = \int \rho_n(y)\delta[x - M(y)]\,dy, \tag{2.23a}$$

where $\delta(x)$ is a delta function. Equation (2.23a) is called the *Frobenius–Peron equation*. The *invariant* density previously discussed satisfies the equation obtained by setting $\rho_{n+1}(x) = \rho_n(x) = \rho(x)$,

$$\rho(x) = \int \rho(y)\delta[x - M(y)]\,dy. \tag{2.23b}$$

In order to see how Eq. (2.23a) comes about, consider Figure 2.20. Orbit points in the range x to $x + dx$ at time $n + 1$ came from the ranges $y^{(i)}$ to $y^{(i)} + dy^{(i)}$ at time n, where $y^{(i)}$ denote the solutions of the equation $M(y) = x$ (for Figure 2.20 there are three such solutions). Thus, the number of orbit points in the interval x to $x + dx$ at time $n + 1$ is the sum of the number of orbit points in the intervals $y^{(i)}$ to $y^{(i)} + dy^{(i)}$ at time n,

$$\rho_{n+1}(x) = \sum_i \rho_n(y^{(i)})\left|\frac{dx}{dy^{(i)}}\right|^{-1} = \sum_i \rho_n(y^i)|M'(y^{(i)})|^{-1},$$

which is Eq. (2.23a) by virtue of the delta function identity

$$\delta(x - M(y)) = \sum_i \delta(y - y^{(i)})|M'(y^{(i)})|^{-1}$$

(This can be shown by expanding $x - M(y)$ to first order around each of its zeros.)

For the logistic map at $r = 4$ we have found that the natural invariant density, given by Eq. (2.13), has singular behavior at the two points $x = 0$ and $x = 1$. The reason for this is as follows. If we examine the fraction of time a typical orbit spends within a small interval $I_\varepsilon(\bar{x}) = [\bar{x} - \varepsilon, \bar{x} + \varepsilon]$, then this quantity, which we call the natural measure of $I_\varepsilon(\bar{x})$, is

$$\mu(I_\varepsilon(\bar{x})) = \int_{\bar{x} - \varepsilon}^{\bar{x} + \varepsilon} \rho(x)\,dx. \tag{2.24}$$

If $\rho(x)$ is smooth and bounded, then $\mu(I_\varepsilon(\bar{x})) \sim \varepsilon$ for small ε. Now consider the small ε interval, centered at the critical point, $I_\varepsilon(\frac{1}{2})$, for the logistic map at $r = 4$. Since $\rho(x)$ varies smoothly at $x = \frac{1}{2}$, we have $\mu(I_\varepsilon(\frac{1}{2})) \sim \varepsilon$. Mapping the interval $I_\varepsilon(\frac{1}{2})$ forward in time by one iterate, it maps to $[1 - 4\varepsilon^2, 1]$. Since every point in $I_\varepsilon(\frac{1}{2})$ maps to $[1 - 4\varepsilon^2, 1]$, we have that

$$\mu(I_\varepsilon(\tfrac{1}{2})) = \mu(I_{4\varepsilon^2}(1))$$

(because $\rho(x) = 0$ for $x > 1$, we may add the interval $[1, 1 + 4\varepsilon^2]$ to the interval $[1 - 4\varepsilon^2, 1]$ without changing the natural measure). Hence $\mu(I_{4\varepsilon^2}(1)) \sim \varepsilon$, implying that

$$\mu(I_\varepsilon(1)) \sim \varepsilon^{1/2}. \tag{2.25}$$

Figure 2.20 Illustration of the derivation of the Frobenius–Peron equation.

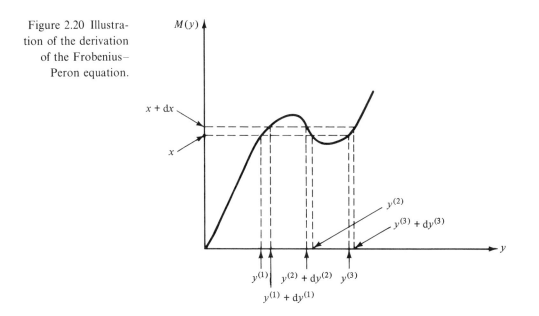

Thus for small ε, we see that $\mu(I_\varepsilon(1))$ is much larger than would be the case if $\rho(x)$ were smooth and bounded, at $x = 1$. Utilizing (2.24), we see that (2.25) implies an inverse square root singularity of $\rho(x)$ at $x = 1$. Now applying the map to the interval $I_\varepsilon(1)$ and noting that $M(1) = 0$, we find that

$$\mu(I_\varepsilon(0)) \sim \varepsilon^{1/2},$$

and there is thus also an inverse square root singularity of $\rho(x)$ at $x = 0$. The main point is that singularities are produced at iterates of the critical point $x = \frac{1}{2}$.

For $r = 4$ the logistic map maps the critical point $x = \frac{1}{2}$ to $x = 1$ and then maps $x = 1$ to the *fixed point* $x = 0$, and subsequent iterates remain at $x = 0$. In contrast, at lower values of r, where chaotic behavior occurs, we observe numerically that the orbit starting at the critical point apparently does not fall on an unstable fixed point or periodic orbit (except for very specially chosen[6] values of r). Rather, the orbit from the critical point appears to wander throughout the attracting interval eventually coming arbitrarily close to every point in the interval. That is, the initial condition at the critical point acts like a typically chosen initial condition. However, we have seen before that, for $r = 4$, the invariant density $\rho(x)$ was singular at the locations of iterates of the critical point. Now these iterates are dense in the attracting interval. Thus we might expect that $\rho(x)$ will typically exhibit very 'nasty' behavior. In particular, the function $\rho(x)$ is expected to be discontinuous everywhere and probably has a dense countable set of x-values (the iterates of the critical point) at which $\rho(x)$ is infinite. We can imagine that a numerical histogram approximation of $\rho(x)$ would reveal more and more bins of unusually large values as the bin size is reduced and the orbit length increased. See Figure 2.21. This type of behavior is to be expected for a typical chaotic one-dimensional map with a smooth maximum.

Rather than deal with the density $\rho(x)$, one can equivalently deal with the corresponding measure μ. (In general, even if a density cannot be sensibly defined, a suitable measure can be. Hence measure is a more generally applicable concept.[7]) More specifically we shall deal with *probability measures*. A probability measure μ for a bounded region R assigns nonnegative numbers to any set in R, is countably additive, and assigns the number 1 to R, $\mu(R) = 1$. By countably additive, we mean that, given any countable family of disjoint (i.e., nonoverlapping) sets S_i in R, then the measure of the union of these sets is the sum of the measures of the sets,

$$\mu\left(\bigcup_i S_i\right) = \sum_i \mu(S_i).$$

Given a set S we define $M^{-1}(S)$ as the set of points which map to S on one iterate. Thus $M^{-1}(S)$ is defined even if the map M is not invertible (e.g., if M is the $2x$ modulo 1 map and $S = (\frac{3}{4}, 1)$ then $M^{-1}(S) = (\frac{3}{8}, \frac{1}{2}) \cup (\frac{7}{8}, 1)$). We say a measure μ is *invariant* if

$$\mu(S) = \mu(M^{-1}(S)).$$

(If the map is invertible this is the same as saying $\mu(S) = \mu(M(S))$.)

Say that we have a chaotic attractor of a one-dimensional map and that this attractor has a *basin of attraction B*. We define the basin of attraction as the closure of the set of initial conditions that are attracted to the attractor. (For example, the logistic map for $0 < r < 4$ may be thought of as having two attractors. One is the point at $x = -\infty$ with basin of attraction $[-\infty, 0] \cup [1, +\infty]$. The other is the attractor in $[0, 1]$ with basin of attraction $[0, 1]$.) Given an interval S, let $\mu(S, x_0)$ denote the fraction of time an orbit originating from an initial condition x_0 in B spends in the interval S in the limit that the orbit length goes to infinity. If $\mu(S, x_0)$ is the same value for every x_0 in the basin of attraction except for a set of x_0-values of Lebesgue measure zero, then we say that $\mu(S, x_0)$ is the *natural measure* of S. We denote the natural measure of S by $\mu(S)$ and say that its value is the common value assumed by $\mu(S, x_0)$ for all x_0 in B except for a set of Lebesgue measure zero. (In our discussion above of the logistic map for $r < 4$ we have implicitly assumed the existence of a

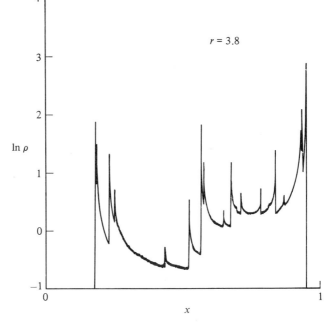

Figure 2.21 $\rho(x)$ versus x for the logistic map at $r = 3.8$. The figure is computed from a histogram. With finer and finer resolution and longer and longer orbit length the sharp peaks in the figure become more numerous and their heights increase without bound (after Shaw 1981).

natural measure.) For smooth one-dimensional maps with chaotic attractors natural measures and densities can be proven to exist under fairly general conditions. We note, however, that proving the existence of a natural measure in cases of higher dimensional systems encountered in applications is an open problem,[8] although numerically the existence of a natural measure in such cases seems fairly clear. (An example of a point x_0 on the zero Lebesgue measure set of which $\mu(S, x_0) \neq \mu(S)$ is the case where x_0 is chosen to lie on an unstable periodic orbit in B.)

The above discussion implies that given a smooth function $f(x)$ and an attractor with a natural measure μ, the time average of $f(x)$ over an orbit originating from a *typical* initial condition in the basin of the attractor is the same as its natural-measure-weighted average over x,

$$\lim_{T \to \infty} \frac{1}{T} \sum_{n=0}^{T} f(M^n(x_0)) = \int f(x) \, d\mu(x), \tag{2.26}$$

where $d\mu(x) = \rho(x) \, dx$ when a density exists[9] (as before, by typical we mean all except for a set of Lebesgue measure zero).

More generally, we say that an invariant probability measure μ (not necessarily the natural measure) is *ergodic* if it cannot be decomposed such that

$$\mu = p\mu_1 + (1 - p)\mu_2, 1 > p > 0,$$

where $\mu_1 \neq \mu_2$ are invariant probability measures. The *ergodic theorem* states if $f(x)$ is an integrable function and μ is an ergodic probability measure (in the sense defined above), then the set A of x_0-values for which the limit on the left-hand side of Eq. (2.26) exists and Eq. (2.26) holds has μ measure 1, $\mu(A) = 1$. Alternatively, the set of x_0-values for which Eq. (2.26) does not hold has μ measure 0. (Note that the result (2.26) for an attractor applies not only for x_0 chosen as a typical point with respect to the natural measure but also for x_0 chosen as a typical point with respect to Lebesgue measure in the basin of attraction.)

A convenient indicator of the sensitivity to small orbit perturbations characteristic of chaotic attractors is the Lyapunov exponent (or, in higher dimensionality, exponents). The Lyapunov exponent h of a one-dimensional map gives the average exponential rate of divergence of infinitesimally nearby initial conditions. That is, on average, the separation between two infinitesimally displaced initial points x_0 and $x_0 + dx_0$ typically grows exponentially as the two points are evolved by the map M. Thus, for large n

$$dx_n \sim \exp(hn) \, dx_0,$$

where dx_n denotes the infinitesimal separation between the two points after they have both been iterated by the map n times. We define the

Lyapunov exponent h as

$$h = \lim_{T \to \infty} \frac{1}{T} \ln \left| \frac{dx_T}{dx_0} \right|$$

Noting that

$$dx_T/dx_0 = (dx_T/dx_{T-1})(dx_{T-1}/dx_{T-2}) \ldots (dx_2/dx_1)(dx_1/dx_0)$$
$$= M'(x_{T-1})M'(x_{T-2}) \ldots M'(x_1)M'(x_0),$$

we have

$$h = \lim_{T \to \infty} \frac{1}{T} \sum_{n=0}^{T-1} \ln |M'(x_n)|. \tag{2.27}$$

The existence of a natural measure implies that the time average on the right-hand side of (2.27) will be the same for all orbits x_n in the basin, except for those starting at a set of initial conditions of Lebesgue measure zero. Roughly, two nearby points initially separated by a distance Δ_0 typically diverge from each other with time as $\Delta_n \sim \Delta_0 \exp(hn)$ (cf. Figure 1.14). Thus, a positive Lyapunov exponent $h > 0$ indicates chaos. Using $\ln |M'(x)|$ for $f(x)$ in (2.26), we have

$$h = \int \ln |M'(x)| \, d\mu(x). \tag{2.28}$$

Given a map $x_{n+1} = M(x_n)$ we might wish to make a change of variables $y = g(x)$, where we assume the function g is continuous and invertible. For this new variable the map becomes $y_{n+1} = \tilde{M}(y_n)$ where $\tilde{M} = g \circ M \circ g^{-1}$, where the symbol $\circ$ denotes functional composition [i.e., $\tilde{M}(z) = g^{-1}(M(g(z)))$]. We say M and $\tilde{M}$ are *conjugate*. An example is the conjugacy between the logistic map at $r = 4$ and the tent map (cf. Figure 2.6). If g is smooth, then M and $\tilde{M}$ have the same Lyapunov exponent. (See Problem 13.)

2.4   Examples of applications of one-dimensional maps to chaotic systems of higher dimensionality

In Chapter 1 we saw that systems of differential equations necessarily yield invertible maps when the Poincaré surface of section technique is applied. Since the one-dimensional maps studied in the present chapter are all noninvertible, it may be somewhat puzzling as to how noninvertible one-dimensional maps might be relevant to situations involving differential equations. To clarify this point, consider the two-dimensional invertible map (1.9). Eliminating the variable $x_n^{(2)}$, we obtain

$$x_{n+1}^{(1)} = f(x_n^{(1)}) - Jx_{n-1}^{(1)}, \tag{2.29}$$

where J is the Jacobian determinant of the original two-dimensional map. If we set $J = 0$, Eq. (2.29) becomes a one-dimensional map which can be noninvertible and chaotic. If J is very small, but not zero, then (1.9) is always invertible. However, for small J the first term in (2.29) is much larger than the second, and we have $x_{n+1}^{(1)} \simeq f(x_n^{(1)})$. Thus for small J (implying rapid shrinking of areas as the map is iterated), the one-dimensional map yields an approximation to the dynamics of the invertible system with J nonzero but small. Say we generate a chaotic orbit from some particular initial condition $(x_0^{(1)}, x_0^{(2)})$ of the two-dimensional map (1.9). We then record the values obtained for $x_n^{(1)}$ for $n = 0, 1, 2, \ldots$ for our particular orbit. Now say we plot $x_{n+1}^{(1)}$ versus $x_n^{(1)}$ using these data. From Eq. (2.29) the plotted points should fall approximately on the one-dimensional curve $x_{n+1}^{(1)} = f(x_n^{(1)})$. Actually, due to the term $-Jx_{n-1}^{(1)}$, there will be some spread about this curve. To see the character of this spread we note that the Hénon map, Eq. (1.14), is in the same form as Eq. (1.9). Thus, if the Jacobian for the situation plotted in Figure 1.12(a) were small (it is 0.3 which is not small), then we should expect to see that the points fall on the curve $x^{(1)} = A - (x^{(2)})^2$. This is a parabola turned on its side. Alternatively, since for the Hénon map (and Eq. (1.9)) $x_{n+1}^{(2)} = x_n^{(1)}$, it is also the approximate (for small Jacobian) one-dimensional map function turned on its side. Indeed, we see that the attractor in Figure 1.12(a) roughly follows the parabola $x^{(1)} = A - (x^{(2)})^2$ ($A = 1.4$), but has appreciable 'width' about the parabola. Within this width is the fractal-like structure seen in the blow-ups in Figures 1.12(b) and (c). As J is made smaller this width decreases, and eventually the attractor may look like a one-dimensional curve. Nevertheless, as long as J is not exactly zero, magnification of such an *apparent* curve will always reveal fractal structure. Although our discussion above has been in the context of Eq. (1.9), we expect that, in general, when very strong phase space contraction is present in higher dimensional systems, an approximate one-dimensional map may apply. Basically, what can happen is that the attractor becomes highly elongated along a one-dimensional unstable direction, and other directions transverse to it are so highly contracted that the dynamics in the transverse directions becomes difficult to discern. We now give some examples in which chaos in a one-dimensional map has provided a key to obtaining an understanding of a specific physical model or experiment.

2.4.1 The Lorenz system

Lorenz (1963) considered the Rayleigh–Benard instability discussed in Chapter 1. Assuming variations of the fluid to occur in only two spatial dimensions as shown in Figure 1.4, Saltzman (1962) had previously derived a set of first-order differential equations by expanding a suitable set of fluid variables in a double spatial Fourier series with coefficients depending on time. Substituting the expansion into the fluid equations results in an infinite set of coupled first-order ordinary differential equations. Truncation of this system by setting Fourier terms beyond a certain order to zero (the 'Galerkin approximation') results in a finite-dimensional system which presumably yields an adequate approximation to the infinite-dimensional dynamics if the truncation is at sufficiently high order. To gain insight into the types of dynamics that are possible, Lorenz considered a truncation to just three variables. While this truncation is not of high enough order to model the real fluid behavior faithfully, it was assumed that the resulting solutions would give an indication of the type of qualitative behavior of which the actual physical system was capable. The equations Lorenz considered were the following:

$$dX/dt = -\tilde{\sigma}X + \tilde{\sigma}Y, \tag{2.30a}$$

$$dY/dt = -XZ + \tilde{r}X - Y, \tag{2.30b}$$

$$dZ/dt = XY - \tilde{b}Z, \tag{2.30c}$$

where $\tilde{\sigma}$, $\tilde{r}$, and $\tilde{b}$ are dimensionless parameters. Referring to Figure 1.4, the quantity X is proportional to the circulatory fluid flow velocity, Y characterizes the temperature difference between rising and falling fluid regions, and Z characterizes the distortion of the vertical temperature profile from its linear-with-height equilibrium variation. Lorenz numerically considered the case $\tilde{\sigma} = 10$, $\tilde{b} = 8/3$ and $\tilde{r} = 28$. Taking the divergence of the phase space flow, we find from Eq. (1.12) that phase space volumes contract at an exponential rate of $(1 + \tilde{\sigma} + \tilde{b}) = 41/3$, $V(t) = V(0)\exp[-(41/3)t]$. It is this relatively rapid volume contraction which leads to the applicability of one-dimensional map dynamics to this problem.

Figure 2.22 shows a projection of the phase space orbit obtained by Lorenz onto the YZ plane. The points labelled C and C' represent *steady* convective equilibria (i.e., solutions of Eqs. (2.30) with $dX/dt = dY/dt = dZ/dt = 0$) which are unstable for the parameter values investigated by Lorenz. We see that the solution spirals outward from one of the equilibria C or C' for some time, then switches to spiraling outward from the other equilibrium point. This pattern repeats forever with the number of circuits around an equilibrium before switching appearing to vary in an erratic manner.

As one of the ways of analyzing this motion, Lorenz obtained the sequence m_n giving the nth maximum of the function $Z(t)$. He then plotted m_{n+1} versus m_n. The resulting data are shown in Figure 2.23, and they clearly tend to fall on an approximate one-dimensional map function.

Furthermore, we note that the magnitude of the slope $|dm_{n+1}/dm_n|$ for this function is greater than 1 throughout the range visited by the orbit. This is very similar to the situation for the tent map (cf. Figure 2.1(*a*)). Since $|dm_{n+1}/dm_n| > 1$ we have, by Eq. (2.28), that the Lyapunov exponent h is positive, indicating chaos. Note that the maxima of Z may be regarded as lying on the surface of section $bZ = XY$ obtained by setting $dZ/dt = 0$ in Eq. (2.30c).

2.4.2 Instability saturation by quadratically nonlinear three wave coupling

Consider a small amplitude wave propagating in a homogeneous medium with the wave field represented as

$$\{C_1 \exp(-i\omega_1 t + i\mathbf{k}_1 \cdot \mathbf{x}) + (\text{complex conjugate})\},$$

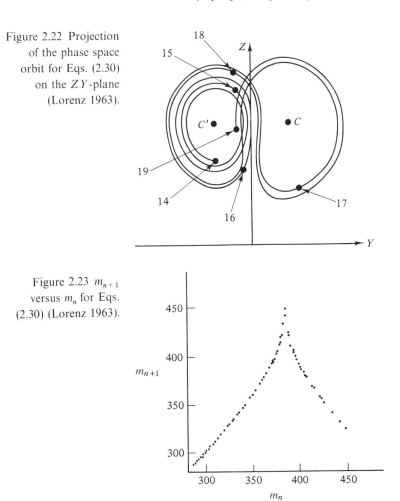

Figure 2.22 Projection of the phase space orbit for Eqs. (2.30) on the ZY-plane (Lorenz 1963).

Figure 2.23 m_{n+1} versus m_n for Eqs. (2.30) (Lorenz 1963).

where C_1 is a complex number. The quantities ω_1 and $\mathbf{k}_1$ are the (real) frequency and wavevector of the wave and are, in general, related by some dispersion relation, $\omega_1 = \omega_1(\mathbf{k})$. Due to nonlinearities in the medium, this wave can couple strongly to two other linear waves, represented as

$$\{C_{2,3} \exp(-i\omega_{2,3}t + i\mathbf{k}_{2,3} \cdot \mathbf{x}) + (\text{complex conjugate})\},$$

if the three waves are nearly in resonance, that is, if

$$\mathbf{k}_1 = \mathbf{k}_2 + \mathbf{k}_3, \tag{2.31a}$$

$$\omega_1 = \omega_2 + \omega_3 + \delta, \tag{2.31b}$$

where δ is small compared to $\omega_{1,2,3}$. We assume $\omega_{1,2,3} > 0$ so that wave 1, by convention, is the wave of largest frequency. In this case, the complex wave amplitude $C_{1,2,3}$ (which were constant in the absence of nonlinear interaction) become slow functions of time satisfying the three-wave mode coupling equations,

$$\mathrm{d}C_1/\mathrm{d}t = C_3 C_2 \exp(i\delta t), \tag{2.32a}$$

$$\mathrm{d}C_{2,3}/\mathrm{d}t = -C_1 C_{3,2}^* \exp(i\delta t), \tag{2.32b}$$

where C^* denotes the conjugate of C, and $C_{1,2,3}$ have been normalized to make the coefficient of the nonlinear term on the right-hand side equal to 1. These equations apply quite generally to the case where the medium in which the waves propagate is conservative (in the sense that there is no net exchange of energy between the waves and the medium).

Now let us say that we consider a case where wave 1 is unstable such that the medium is capable of transferring some of its internal energy to the wave. This can happen in systems that are not in thermodynamic equilibrium; examples include, pumped lasing media, many situations in plasma physics, and stratified fluids in shear flow. In such a situation the net effect is for the amplitude C_1 to increase exponentially in time, $|C_1| \sim \exp(\gamma_1 t)$, provided the wave amplitude is small enough that nonlinearity can be neglected. As the wave amplitude $|C_1| \equiv a_1$ grows, nonlinear coupling to nearly resonant waves, as in Eqs. (2.32), can become a significant effect. Assuming that the two lower frequency waves, waves 2 and 3, are damped, so that, in the absence of nonlinearity, $a_{2,3} = |C_{2,3}| \sim \exp(-\gamma_{2,3}t)$, Eqs. (2.32) are modified by the linear growth and damping and become

$$\mathrm{d}C_1/\mathrm{d}t = \gamma_1 C_1 + C_2 C_3 \exp(i\delta t), \tag{2.33a}$$

$$\mathrm{d}C_{2,3}/\mathrm{d}t = -\gamma_{2,3} C_{2,3} + C_1 C_{3,2}^* \exp(i\delta t). \tag{2.33b}$$

Thus there is the possibility that the linear exponential growth of the unstable wave can be arrested by nonlinearity coupling its energy to damped waves. Introducing $a_1 \exp(i\phi_1) = C_1, a_{2,3} \exp(i\phi_{2,3}) = C_{2,3} \exp(i\delta t/2)$, $\phi = \phi_1 - \phi_2 - \phi_3$, (where $a_{1,2,3}$ and $\phi_{1,2,3}$ are real variables) and restricting consideration to the important case $\gamma_2 = \gamma_3 \equiv \gamma$ and $a_2 = a_3$, Eqs. (2.33) reduce to three real first-order equations,

$$da_1/dt = a_1 + a_2^2 \cos \phi, \qquad (2.34a)$$

$$da_2/dt = -a_2(\gamma + a_1 \cos \phi), \qquad (2.24b)$$

$$d\phi/dt = -\gamma + a_1^{-1}(2a_1^2 - a_2^2)\sin \phi. \qquad (2.34c)$$

In Eq. (2.34a) we have set $\gamma_1 = 1$ (this can be accomplished by proper normalization).

This system has been solved numerically by Vyshkind and Rabinovich (1976) and Wersinger *et al.* (1980). Figure 2.24 shows numerical solutions of Wersinger *et al.* for a_1 versus t for $\delta = 2$ and several values of γ. We see that at $\gamma = 3$ (Figure 2.24(a)) the orbit settles into a simple periodic motion (a limit cycle attractor). Utilizing a surface of section at $\phi = \pi/2$, this periodic motion is manifested as a single fixed point. Increasing γ to $\gamma = 9$, the orbit shown in Figure 2.24(b) is obtained. The solution (after the initial transient) is still periodic, but the single fixed point in the surface of section that previously manifested the periodic attractor splits into two points that are visited alternately. Correspondingly, the single peak per period function of Figure 2.24(a) becomes a function with two alternating maxima (Figure 2.24(b)), thus doubling the period. As γ is further increased, more period doublings are observed, and the time evolution eventually becomes apparently chaotic. Figure 2.24(c) shows the time evolution for such an apparently chaotic case at $\gamma = 15$. Figure 2.25(a) shows that points in the surface of section *appear* to fall on an arc. Since this arc has no *apparent* thickness, it is natural to try to obtain an approximate reduction to a one-dimensional map.

This is done in Figure 2.25(b) which shows $x_{n+1} = a_2(t_{n+1})$ versus $x_n = a_2(t_n)$, where t_n denotes the time at the nth piercing of the surface of section. By a change of variables, $\bar{x} = (\text{const.}) - x$, the apparently

Figure 2.24 a_1 versus t for $\delta = 2$ and three different values of γ: (a) $\gamma = 3$, (b) $\gamma = 9$, and (c) $\gamma = 15$ (Wersinger *et al.*, 1980).

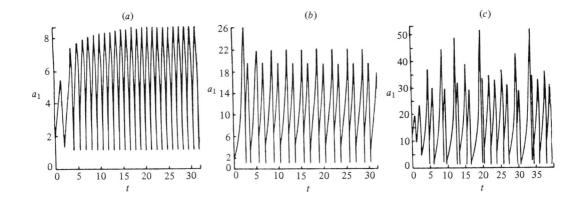

one-dimensional map in Figure 2.25 (*b*) can be turned upside down so that it becomes a map with smooth rounded maximum. Thus the map is similar in character to the logistic map, and, correspondingly, the observed phenomenology of the solutions as γ is increased (i.e., period doubling followed by chaos) is similar to that for the logistic map as r is increased.

Figure 2.25(*a*) Attractor in the surface of section for $\gamma = 15$. (*b*) x_{n+1} versus x_n at $\gamma = 15$ (Wersinger *et al.*, 1980).

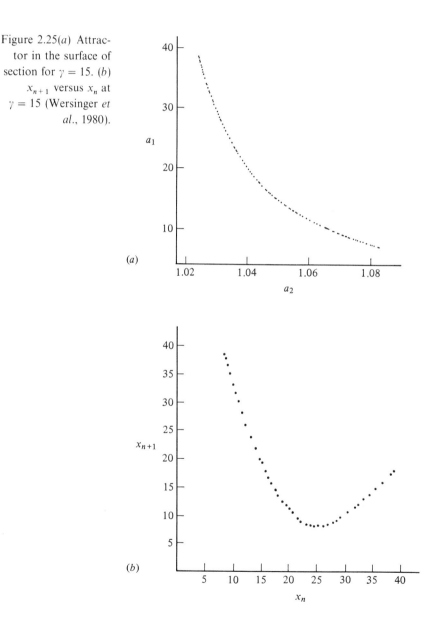

2.4.3 Experiments: Chemical chaos and the dripping faucet

An experimental system whose chaotic dynamics has been studied by several research groups is the Belousov–Zhabotinskii reaction in a well-stirred chemical reactor. A reactor consists of a tank into which chemicals are pumped and an output pipe out of which the reaction product is taken. In the experiment the fluid in the tank is stirred rapidly enough that the medium within the tank may be considered homogeneous. The reactants in the Belousov–Zhabotinskii reaction are $Ce_2(SO_4)_3$, $NaBrO_3$, $CH_2(COOH)_2$, and H_2SO_4. These reactants undergo a complex sequence of reactions involving about 25 chemical species. Presumably, a description of this experiment is provided by the solution of the coupled system of rate equations giving the evolution of the concentration of each chemical species (one first-order equation for each species). All the reaction rates and intermediate reactions are, however, not accurately known. Experimentally it is observed that the time dependence of the chemical concentrations in the reactor can be periodic or chaotic.

The experiment of Simoyi *et al.* (1982) examined this reaction and demonstrated that the observed chaotic dynamics was well described by a one-dimensional map in the parameter regime that they considered. They also examined in detail the sequence of the appearance of periodic orbits in windows as a flow rate parameter was varied and verified that the Metropolis–Stein–Stein sequence was followed.

In their experiment they measured the time dependence of the concentration of one of the chemicals in the reactor, namely the bromide ion. They then used delay coordinates (Section 1.6) to deduce the presence of approximately one-dimensional dynamics. In particular, for a chaotic case, they plot $B(t + T)$ versus $B(t)$ where $T = 53$ seconds and $B(t)$ denotes the concentration of the bromide ion. The result is shown in Figure 2.26(a) and may be regarded as a particular projection of the attractor onto a plane. They then consider the value of $B(t_n) \equiv x_n$ at successive crossings of the dashed line shown in Figure 2.26(a). Plotting

Figure 2.26(*a*) Attractor projection and (*b*) map function for the experiment of Simoyi *et al.* (1982).

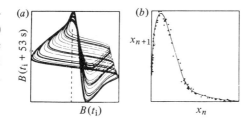

x_{n+1} versus x_n for this data, they observe that it appears to lie on a smooth one-dimensional map function with a rounded maximum. Figure 2.26(b) shows the experimental data as dots and a fitted curve as a solid line. (The map function has a single maximum as required for applicability of the theory of Metropolis, Stein and Stein.)

Another experiment which is apparently described by chaotic one-dimensional map dynamics is the dripping water faucet experiment of Shaw discussed in Section 1.2. In particular when the water flow rate is such that the behavior is chaotic, a plot of experimental data for the time between drops Δt_{n+1} versus Δt_n appears to lie on a one-dimensional map function (although there is appreciable spread of the experimental data about a fitted curve). See Figure 2.27.

Appendix: Some elementary definitions and theorems concerning sets

For simplicity we consider sets of the real numbers.

An *open interval*, denoted (a, b), is the set of all x such that $a < x < b$. A *closed interval*, denoted $[a, b]$, is the set of all x such that $a \le x \le b$ with $b > a$.

An *interior point* P of a set S is a point such that there exists an ε-neighborhood $(P - \varepsilon, P + \varepsilon)$ contained entirely in S. A point P is a *boundary point* of S if any ε-neighborhood of P possesses points that are in S as well as points not in S.

A point P is a *limit point* of the set S if every ε-neighborhood of P contains at least one point in S (distinct from P). This can be shown to be

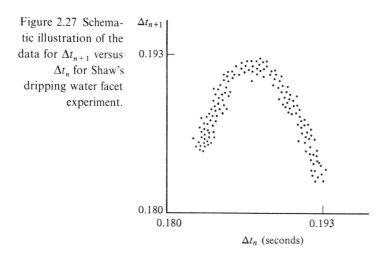

Figure 2.27 Schematic illustration of the data for Δt_{n+1} versus Δt_n for Shaw's dripping water facet experiment.

equivalent to the statement that there exists an infinite sequence of distinct points $x_1, x_2, \ldots$ all in S such that $\lim_{n \to \infty} x_n = P$.

If a set contains all its limit points it is called a *closed* set. The *closure* $\bar{S}$ of a set S is the set S plus its limit points.

A set S is *open* if all its points are interior points. The following two theorems concerning the structure of open and closed sets will be important for our further discussions.

Theorem. Every nonempty bounded open set S can be represented as the sum of a finite or a countably infinite number of disjoint open intervals whose end points do not belong to S.

That is,

$$S = \sum_k (a_k, b_k).$$

The *Lebesgue measure of the open set S* is

$$\mu(S) = \sum_k (b_k - a_k).$$

Theorem. A nonempty closed set S is either a closed interval or else can be obtained from a closed interval by removing a finite or countably infinite family of disjoint open intervals whose end points belong to S.

Thus a closed set S can be expressed as

$$S = [a, b] - \sum_k (a_k, b_k).$$

The *Lebesgue measure of the closed set S* is

$$\mu_L(S) = (b - a) - \sum_k (b_k - a_k).$$

(We shall not give the more general definition of the Lebesgue measure applicable to sets that are neither open nor closed, since, for the most part, we shall be dealing with open or closed sets.)

In particular, a set has Lebesgue measure zero if for any $\varepsilon > 0$ the set can be covered by a countable union of intervals such that the sum of the length of the intervals is smaller than ε. For sets in an N-dimensional Cartesian space, we have the analogous definition of a zero Lebesgue measure set: for any $\varepsilon > 0$ the set can be covered by a countable union of N-dimensional cubes whose total volume is less than ε.

A *Cantor set* is a closed set which consists entirely of boundary points each of which is a limit point of the set. Examples of Cantor sets are given in the text. In general, Cantor sets can have either zero Lebesgue measure or else positive Lebesgue measure. They are also uncountable.

Problems

1. For the $2x$ modulo 1 map and the tent map find the number of periodic orbits N_p for periods $p = 2, 3, \ldots, 10$. How many distinct period four orbits are there for the map $x_{n+1} = 3x_n$ modulo 1?

2. Consider the $2x$ modulo 1 map with noise, $y_{n+1} = (2y_n$ modulo 1) + (noise). Assume that the form of the noise is such as to change randomly all the digits a_j in the binary representation of y for $j \geq 50$. (Thus the noise is of the order of $2^{-50} \sim 10^{-5}$.) Assume that you are given exact observations of the noisy orbit for a time T: $y_0, y_1, \ldots, y_T$ (where T is much larger than 50). Show that there is an initial condition x_0 such that the exact 'true' orbit, $x_0, x_1, \ldots, x_T$, followed by the noiseless map, $x_{n+1} = 2x_n$ modulo 1, shadows the noisy orbit, $y_0, y_1, \ldots, y_T$. In particular, show that x_0 can be chosen so that $|y_n - x_n| \leq 2^{-49}$ for all n from $n = 0$ to $n = T$.

3. Consider the one-dimensional map
$$x_{n+1} = \begin{cases} 3x_n & \text{if } 0 \leq x_n \leq 1/3 \\ \frac{3}{2}(1 - x_n) & \text{if } 1/3 \leq x_n \leq 1 \end{cases}$$
   (a) Find the locations and stability coefficients for the fixed points.
   (b) Find the location of the orbit points and the stability coefficient for the period two orbit.

4. For the logistic map find the value of r at which the superstable period one orbit exists. Find the value of r at which the superstable period two exists.

5. Show for the logistic map at the value of r at the merging of the two band attractor to form a one band attractor (i.e., at $r = r_0'$) that the third iterate of $x = \frac{1}{2}$ lands on the unstable fixed point $x = 1 - 1/r$.

6. Consider the map $M(x; r) = r - x^2$. Show that it has a forward tangent bifurcation at some value $r = r_0$ at which a stable and an unstable fixed point are created. Find r_0 and the locations of the stable and unstable fixed points. Find r_1, the value of r at which the stable fixed point created at r_0 becomes unstable.

7. Consider the map $x_{n+1} = x_n^3 + \alpha x_n$.
   (a) For some range of α values $\alpha_- < \alpha < \alpha_+$ the fixed point at the origin is stable. What are α_+ and α_-?
   (b) Describe the bifurcation that takes place as α increased through α_-. Is it generic?
   (c) Describe the bifurcation that takes place as α increases through α_+. Is it generic?

8. How many period six windows of the logistic map are there?

9. Write a computer program to make a bifurcation diagram for the cubic map, $M(x) = rx(1 - x^2)$, $r > 0$, and plot the bifurcation diagram using as your initial condition $x_0 = 1/\sqrt{3}$. Repeat using $x_0 = -1/\sqrt{3}$. Comment on the results.

10. Consider the map

$$x_{n+1} = \begin{cases} 2x_n & \text{for } 0 \leq x_n \leq \frac{1}{2} \\ (x_n - \frac{1}{2}) & \text{for } \frac{1}{2} < x_n \leq 1 \end{cases}$$

(a) Find the invariant density generated by orbits orginating from typical initial conditions x_0 in $(0, 1)$. Assume that the density is of the form $\rho(x) = \rho_a$ for $0 \leq x \leq \frac{1}{2}$ and $\rho(x) = \rho_b$ for $\frac{1}{2} \leq x \leq 1$, where ρ_a and ρ_b are constants.

(b) Find the Lyapunov exponent.

11. Consider the map

$$x_{n+1} = \begin{cases} (\frac{3}{2})x_n & \text{for } 0 \leq x_n \leq \frac{2}{3} \\ 2(x_n - \frac{2}{3}) & \text{for } \frac{2}{3} < x_n \leq 1 \end{cases}$$

(a) What is the natural invariant density? (Assume $\rho(x)$ to be constant in the interval $(0, \frac{2}{3})$ and in the interval $(\frac{2}{3}, 1)$.)

(b) What is the fraction of time a typical orbit spends in the region $0 \leq x \leq \frac{1}{2}$?

12. Show that the invariant measure for the tent map is stable in the sense that a small smooth perturbation of the density from the natural invariant density decays to zero as the map is iterated. (Use the Frobenius–Peron equation to do this.)

13. Show that if M and $\tilde{M}$ are conjugate by a smooth change of variables g then they have the same Lyapunov exponent. Use Eq. (2.28). If you wish, you may assume that a natural invariant density $\rho(x)$ exists so that $d\mu(x) = \rho(x)\,dx$.

14. (a) Find the natural invariant density for the map pictured in Figure 2.28. (Hint: assume that $\rho(x)$ is constant in each of the three intervals $(0, \frac{1}{3})$, $(\frac{1}{3}, \frac{2}{3})$, $(\frac{2}{3}, 1)$.) (b) Find the Lyapunov exponent for this map. (c) What is the fraction of time a typical orbit spends in the interval $\frac{1}{2} \leq x \leq \frac{3}{4}$?

Figure 2.28 Plot of the map for Problem 14.

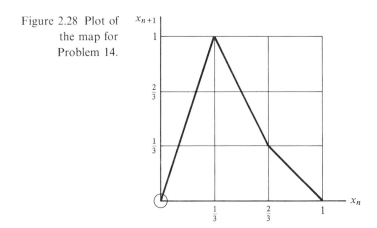

Notes

1. Much material on one-dimensional maps is contained in the monograph by Collet and Eckmann (1980) and the textbook by Devaney (1986).
2. This exponential increase of the number of unstable periodic orbits with increasing period is typical for chaotic systems, including maps of dimensionality larger than one, as well as flows. More precisely, the quantity

$$\lim_{p \to \infty} \frac{1}{p} \ln N_p,$$

which defines the rate of exponential increase of N_p, is positive for these chaotic systems.
3. It is probably the case that chaotic attractors for systems commonly encountered in practice (e.g., the forced damped pendulum) have embedded within them a dense set of periodic orbits.
4. The occurrence of the tangent bifurcation illustrated in Figure 2.16(b) shows the map function as concave down with the curve approaching tangency from below the $45°$ line. In contrast, Figure 2.13(b) shows the map function as concave up and approaching tangency from above. These two pictures are equivalent as a change of variables $y = (\text{const.}) - x$ readily shows. Similar comments apply to Figure 2.16(a) and (c).
5. This is shown in Milnor and Thurston (1987) and is a special property of the logistic map.
6. For example, those at the upper boundary of a window.
7. A particular example of a measure μ is when there is a density $\rho(x)$, and the measure of a set A is defined as $\mu(A) = \int_A \rho(x) \, dx$. In other naturally occurring cases that will be of interest to us, the measure will be concentrated on a Cantor set that has Lebesgue measure zero. In such a case a density function $\rho(x)$ does not exist.
8. Sinai (1972) and Bowen and Ruelle (1975) introduced the concept of natural measure and show that it exists for certain types of attractors (Axiom A attractors; cf. Chapter 4).
9. For a definition of the integral $\int \ldots d\mu(x)$ and further discussion of measure and ergodic theory in dynamics see Ruelle (1989).

CHAPTER THREE

Strange attractors and fractal dimension

Perhaps the most basic aspect of a set is its dimension. In Figures 1.10(*a*) and (*b*) we have given two examples of attractors; one is a steady state of a flow represented by a single point in the phase space, while the other is a limit cycle, represented by a simple closed curve. While it is clear what the dimensions of these attracting sets are (zero for the point and one for the curve), it is also the case that invariant sets arising in dynamical systems (such as chaotic attractors) often have structure on arbitrarily fine scale, and the determination of the dimension of such sets is nontrivial. Also the frequency with which orbits visit different regions of a chaotic attractor can have its own arbitrarily fine scaled structure. In such cases the assignment of a dimension value gives a much needed quantitative characterization of the geometrical structure of a complicated object. Furthermore, experimental determination of a dimension value from data for an experimental dynamical process can provide information on the dimensionality of the phase space required of a mathematical dynamical system used to model the observations. These issues are the subjects of this chapter.

3.1 The box-counting dimension

The *box-counting dimension*[1] (also called the 'capacity' of the set) provides a relatively simple and appealing way of assigning a dimension to a set in such a way that certain kinds of sets are assigned a dimension which is not an integer. Such sets are called fractals by Mandelbrot, while, in the context of dynamics, attracting sets with fractal properties have been called strange attractors. (The latter term was introduced by Ruelle and Takens (1971).)

Assume that we have a set which lies in an N-dimensional Cartesian space. We then imagine covering the space by a grid of N-dimensional cubes of edge length ε. (If $N = 2$ then the 'cubes' are squares, while if $N = 1$ the 'cubes' are intervals of length ε.) We then count the number of cubes $\tilde{N}(\varepsilon)$ needed to cover the set. We do this for successively smaller ε values. The box-counting dimension is then given by[2]

$$D_0 = \lim_{\varepsilon \to 0} \frac{\ln \tilde{N}(\varepsilon)}{\ln(1/\varepsilon)}. \tag{3.1}$$

As an example, consider the case of some simple sets lying in a two-dimensional Cartesian space, Figure 3.1. The three geometrical sets shown are (*a*) a set consisting of two points, (*b*) a curve segment, and (*c*) the area inside a closed curve. The squares required to cover the sets are shown cross-hatched in the figure. In the case of Figure 3.1(*a*), we see that $\tilde{N}(\varepsilon) = 2$ independent of ε; thus Eq. (3.1) yields $D_0 = 0$. In the case of Figure 3.1(*b*), we have $\tilde{N}(\varepsilon) \sim l/\varepsilon$ for small ε, where l is the length of the

Figure 3.1 Illustration of $\tilde{N}(\varepsilon)$ for sets consisting of (*a*) two points, (*b*) a curve segment, and (*c*) the area inside a closed curve.

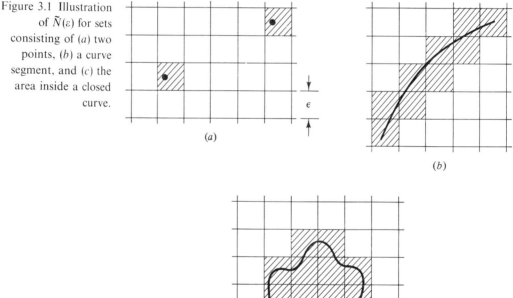

curve; thus Eq. (3.1) yields $D_0 = 1$. Similarly for the area (Figure 3.1(c)), $\tilde{N}(\varepsilon) \sim A/\varepsilon^2$ where A is the area, and $D_0 = 2$. Hence we see that the box-counting dimension yields, as it should, correct dimension values for simple nonfractal sets: 0, 1, and 2 for a set of a finite number of points, a simple smooth curve, and an area. (Note that to obtain D_0 we only require a rather crude estimate of the dependence of $\tilde{N}(\varepsilon)$ on ε. For example, plugging $\tilde{N}(\varepsilon) = K\varepsilon^{-d}$ in Eq. (3.1) yields $D_0 = d$ *independent* of the constant of proportionality K. In this regard see also Problem 2 of Chapter 5.)

Now let us consider a somewhat more interesting set, the middle third Cantor set. This set is defined as follows. Take the closed interval $[0, 1]$, and remove the open middle third interval $(\frac{1}{3}, \frac{2}{3})$, leaving the two intervals $[0, \frac{1}{3}]$ and $[\frac{2}{3}, 1]$. Now remove the open middle thirds of each of these two intervals leaving four closed intervals of length $\frac{1}{9}$ each, namely the intervals $[0, \frac{1}{9}]$, $[\frac{2}{9}, \frac{1}{3}]$, $[\frac{2}{3}, \frac{7}{9}]$ and $[\frac{8}{9}, 1]$. Continuing in this way *ad infinitum*, the set of remaining points is the middle third Cantor set. The construction is illustrated in Figure 3.2. This set has zero Lebesgue measure since at the nth stage of the construction the total length of the remaining intervals is $(\frac{2}{3})^n$, and this length goes to zero as n goes to infinity. Although of zero length, the set is also uncountable. To see that this is so, we make a one to one correspondence of points in the Cantor set with all the numbers in the interval $[0, 1]$. Each point in the Cantor set can be specified by giving its location at successive stages of the construction of the set. For example, 'at the first stage it is to the right of the removed interval (which has length $\frac{1}{3}$); at the second stage it is to the left of the interval of length $\frac{1}{9}$ that is removed from the center of the interval of length $\frac{1}{3}$ in which it fell in the first stage; at the third stage it is to the left of the removed interval; at the fourth stage it is to the right of the removed interval', etc. Associating right with 1 and left with 0, yields a representation of an element of the Cantor set as an infinite string of zeros and ones. For the example above, the string is 1001 . . . . All combinations of zeros and ones are possible and each string represents a different element of the

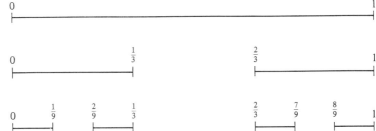

Cantor set. Infinite strings of zeros and ones can also be used to represent all the numbers between 0 and 1 via the binary decimal representation, Eq. (2.4). Hence, by identifying similar sequences of zeros and ones in the two representations, we have a one to one correspondence between points in the Cantor set and points in the set [0, 1], and the Cantor set is therefore uncountable.

To calculate the box-counting dimension of the middle third Cantor set, let us consider a sequence ε_n of ε-values converging to zero as n approaches infinity, $\lim_{n \to \infty} \varepsilon_n = 0$. Then by Eq. (3.1) we have $D_0 = \lim_{n \to \infty} [\ln \tilde{N}(\varepsilon_n)]/\ln(1/\varepsilon_n)$. The most convenient choice for ε_n is $\varepsilon_n = (\frac{1}{3})^n$. By the construction of the Cantor set (Figure 3.2), we then have $\tilde{N}(\varepsilon_n) = 2^n$ and

$$D_0 = \ln 2/\ln 3 = 0.63\ldots.$$

Hence we obtain for the dimension a number between zero and one, indicating that the set is a fractal.

In the examples we have given (i.e., the sets in Figure 3.1 and the middle third Cantor set), we have that

$$\tilde{N}(\varepsilon) \sim \varepsilon^{-D_0}. \tag{3.2}$$

That is, the number of 'cubes' needed to cover the set increases with ε in a power law fashion with exponent D_0. If one assumes that one only needs to resolve the location of points in the set to within an accuracy of ε, then $\tilde{N}(\varepsilon)$ tells us roughly how much information we need to do this (the information is the locations of the $\tilde{N}(\varepsilon)$ cubes), and D_0 tells us how rapidly the required information increases as the required accuracy increases.

The middle third Cantor set is *self-similar* in the sense that smaller pieces of it reproduce the entire set upon magnification. For example, magnifying the interval $[\frac{2}{3}, \frac{7}{9}]$ by a factor of 9 yields a picture which is identical to that for the set in the interval [0, 1]. In the case of fractal sets arising in typical dynamical systems, such as the forced damped pendulum (Figure 1.13), self-similarity rarely holds.[3] In such cases the fractal nature reveals itself upon successive magnifications as structure on all scales. That is, as successive magnifications are made about any point in the set, we do not arrive at a situation where, at some sufficiently large magnification and all magnifications beyond that, we see only a single point, a line, or a flat surface.

We now give a simple example of a dynamical system yielding a fractal set. We consider the one-dimensional map

$$M(x) = \begin{cases} 2\eta x, & \text{if } x < \frac{1}{2}, \\ 2\eta(x-1) + 1, & \text{if } x > \frac{1}{2}, \end{cases} \tag{3.3}$$

illustrated in Figure 3.3. For $\eta = 1$ and x restricted to [0, 1] this is the $2x$ modulo 1 map, Eq. (2.3), discussed in Section 2.1. Here we consider the case $\eta > 1$. Note that any point in $x > 1$ gets mapped toward $x = +\infty$ on successive iterates, while any point in $x < 0$ gets mapped toward $x = -\infty$ on successive iterates. Thus we focus on the x interval $[0, 1]$. We note that $M(x) > 1$ for x in $[1/2\eta, \frac{1}{2}]$, so that whenever x falls in this interval it is mapped to $x > 1$ and then toward $x = +\infty$ on subsequent iterates. Likewise, $M(x) < 0$ for x in the interval $[\frac{1}{2}, 1 - 1/2\eta]$, and whenever x falls in this interval it is consequently mapped to $x < 0$ and then toward $x = -\infty$. Say we consider an initial uniform distribution of points in the interval $[0, 1]$, $\rho_0(x) = 1$ in $[0, 1]$ and $\rho_0(x) = 0$ outside $[0, 1]$. Then a fraction, $\Delta = 1 - 1/\eta$, of these points will be mapped to $x > 1$ or $x < 0$ on one iterate. By the Frobenius–Peron equation, Eq. (2.23a), the density $\rho_1(x)$ is again uniform in $[0, 1]$, $\rho_1(x) = 1 - \Delta$ for x in $[0, 1]$. Repeating this process we obtain

$$\rho_n(x) = (1 - \Delta)^n, \text{ for } x \text{ in } [0, 1]. \tag{3.4}$$

We see that the fraction of points remaining in $[0, 1]$ decreases exponentially with time

$$\int_0^1 \rho_n(x)\,\mathrm{d}x = \exp(-\gamma n), \tag{3.5}$$

Figure 3.3 The map Eq. (3.3) with $\eta > 1$.

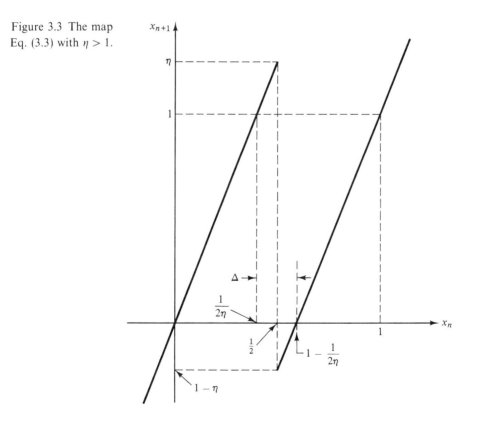

where $\gamma = \ln(1 - \Delta)^{-1} = \ln[\eta/(\eta - 1)]$. We now ask what is the character of the set of initial conditions which never leaves the interval $[0, 1]$. Since the fraction of a uniformly distributed initial distribution remaining in $[0, 1]$ decreases exponentially with time, Eq. (3.5), the set of points which remain forever must have zero Lebesgue measure. To see what the character of this set is, consider those initial conditions which remain for at least one iterate. This is clearly the set of two intervals $[0, 1/2\eta]$ and $[1 - 1/2\eta, 1]$. Now we ask, what is the set which remains for at least two iterates. The action of the map on both the interval $[0, 1/2\eta]$ and the interval $[1 - 1/2\eta, 1]$ is to stretch each interval uniformly to a length of one and map it to the interval $[0, 1]$. Hence there is an interval of length $(\Delta/2)(1 - \Delta)$ in the middle of the intervals $[0, 1/2\eta]$ and $[1 - 1/2\eta, 1]$ which leaves on the second iterate. Thus there are four intervals of initial conditions which remain for at least two iterates. For example, for $\eta = \frac{3}{2}$ (corresponding to $\Delta = 1 - 1/\eta = 1/3$), the sets of points that remain for at least one iterate and for at least two iterates are the sets of intervals seen in Figure 3.2 at the first and second stage of construction of the middle third Cantor set. Thus we see that for $\eta = \frac{3}{2}$ the set of points which remains forever in $[0, 1]$ is just the middle third Cantor set. For arbitrary $\eta > 1$, the set which remains forever is also a Cantor set, but its dimension is a function of η (cf. Problem 2),

$$D_0 = (\ln 2)/(\ln 2\eta). \tag{3.6}$$

It is interesting to consider the evolution of points on the Cantor set which remains forever in the interval $[0, 1]$. This set is invariant under the map, since a point which remains in $[0, 1]$ forever is necessarily mapped to another point which remains forever. Previously we made a correspondence between points x on the middle third Cantor set and points on the interval $[0, 1]$ expressed as a binary decimal. The same can be done for the invariant Cantor set of the map, Eq. (3.3), for arbitrary $\eta > 1$. That is, assign a 1 or a 0 according to whether the point is in the right or left interval at each stage of the construction. Alternatively, if we iterate the point x with time and let $a_n = 1$ if $M^n(x) > \frac{1}{2}$ and let $a_n = 0$ if $M^n(x) < \frac{1}{2}$, then the point x has the symbol sequence representation

$$a_0 a_1 a_2 \ldots .$$

Furthermore, operation of the map on the point x in the Cantor set transforms it to the location corresponding to the point with symbol sequence

$$a_1 a_2 a_3 \ldots .$$

Thus, as for the case $\eta = 1$ (Section 2.1), the dynamics of those points which do not go to $x = \pm\infty$ is fully specified by the *symbolic dynamics* of the Bernoulli shift operation. The difference between $\eta = 1$ and $\eta > 1$ is that, in the former case, the invariant set is the entire interval $[0, 1]$, while for the latter case it is a zero Lebesgue measure Cantor set in $[0, 1]$.

Since for $\eta > 1$ all points in $[0, 1]$ except for a set of Lebesgue measure zero (the Cantor set) eventually leave $[0, 1]$, the Cantor set we have found for $\eta > 1$ is not an attractor. However, if Δ is very small (η close to 1), many initial conditions will remain in $[0, 1]$ for a long time before leaving. During the time that they remain in $[0, 1]$ they undergo orbits which have all the hallmarks of chaos. We call such orbits chaotic transients, and we identify $1/\gamma$ in Eq. (3.5) as the typical duration of such a chaotic transient. Furthermore, since the dynamics of points on the Cantor set is equivalent to the Bernoulli shift, which also describes the dynamics of the $2x$ modulo 1 map, and since the latter is chaotic (its Lyapunov exponent is $h = \ln 2 > 1$), we therefore also call the dynamics of Eq. (3.3) on the invariant Cantor set chaotic.

3.2 The generalized baker's map

In this section we introduce the generalized baker's map[4] and discuss some of its properties. It will become evident that this map is an extremely useful tool for conceptualizing many of the basic properties of strange attractors. We define the generalized baker's map as a transformation of the unit square $[0, 1] \times [0, 1]$,

$$x_{n+1} = \begin{cases} \lambda_a x_n, & \text{if } y_n < \alpha, \\ (1 - \lambda_b) + \lambda_b x_n, & \text{if } y_n > \alpha, \end{cases} \tag{3.7a}$$

$$y_{n+1} = \begin{cases} y_n/\alpha, & \text{if } y_n < \alpha, \\ (y_n - \alpha)/\beta, & \text{if } y_n > \alpha, \end{cases} \tag{3.7b}$$

where $\beta = 1 - \alpha$ and $\lambda_a + \lambda_b \leq 1$. The deformations and rearrangements of the unit square stipulated by (3.7) are illustrated in Figure 3.4(a)–(d). We first divide the unit square into two pieces, $y > \alpha$ and $y < \alpha$ (Figure 3.4(a)). We then compress the two pieces in the horizontal direction by different factors, λ_a for the piece in $y < \alpha$ and λ_b for the piece in $y > \alpha$ (Figure 3.4(b)). Then we vertically stretch the lower piece by a factor $1/\alpha$ and the upper piece by a factor $1/\beta$, so that both are of unit length (Figure 3.4(c)). We then take the upper piece and place it back in the unit square with its right vertical edge coincident with the right vertical edge of the unit square (Figure 3.4(d)). Thus the map Eq. (3.7) maps the unit square into two strips within the square, one in $0 \leq x \leq \lambda_a$ and one in $1 - \lambda_b \leq x \leq 1$.

Applying the map a second time, maps the two strips of Figure 3.4(d) into four strips (Figure 3.5), one of width λ_a^2, one of width λ_b^2, and two of width $\lambda_a \lambda_b$. Application of the map more times results in more strips of narrower width, and the widths approach zero as n approaches infinity. After n applications there will be 2^n strips of varying widths, $\lambda_a^m \lambda_b^{n-m}$ for $m = 0, 1, 2, \ldots, n$. The number of strips $Z(n, m)$ of width $\lambda_a^m \lambda_b^{n-m}$ at the nth stage is given by the binomial coefficient(Problem 11)

$$Z(n, m) = \frac{n!}{m!(n-m)!}. \tag{3.8}$$

Assuming that $\lambda_a + \lambda_b < 1$ (rather than $\lambda_a + \lambda_b = 1$), computer generated orbits (Problem 9) show that the attractor for the generalized baker's map appears to consist of many parallel vertical lines. In fact, as we shall see, there is a Cantor set of these vertical lines. That is, the intersection of the attractor with a horizontal line is a Cantor set. (Note the apparent qualitative similarity of this structure with the blow-ups of the Hénon attractor seen in Figures 1.12(b) and (c).) Let $\hat{D}_0$ denote the box-counting dimension of the intersection of the strange attractor with a horizontal line. Then the dimension D_0 of the attractor is

$$D_0 = 1 + \hat{D}_0. \tag{3.9}$$

Figure 3.4(a)–(d) Action of the generalized baker's map on the unit square.

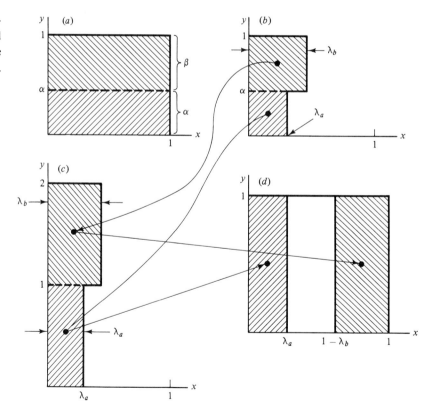

(This follows from the definition, Eq. (3.1).) To find $\hat{D}_0$ we note a self-similarity of the attractor. Namely, if we take the strip in the x interval $[0, \lambda_n]$ in Figure 3.5 and magnify it horizontally by a factor λ_a^{-1}, then Figure 3.4 (d) is reproduced. Likewise, if we horizontally magnify the strip of Figure 3.5 in the x interval $[(1 - \lambda_b), \lambda_b]$ by the factor λ_b^{-1}, then Figure 3.4 (d) is again reproduced. Let us express $\hat{N}(\varepsilon)$, the number of ε length intervals needed to cover the intersection of the attractor with a horizontal line, as

$$\hat{N}(\varepsilon) = \hat{N}_a(\varepsilon) + \hat{N}_b(\varepsilon), \tag{3.10}$$

where $\hat{N}_a(\varepsilon)$ is the number of intervals needed to cover that part of the attractor that lies in $[0, \lambda_a]$, and $\hat{N}_b(\varepsilon)$ is the number needed for that part of the attractor in $[(1 - \lambda_b), 1]$. By the self-similarity of the attractor we have

$$\hat{N}_a(\varepsilon) = \hat{N}(\varepsilon/\lambda_a), \hat{N}_b(\varepsilon) = \hat{N}(\varepsilon/\lambda_b). \tag{3.11}$$

Assuming that $\hat{N}(\varepsilon)$ scales like $\hat{N}(\varepsilon) \simeq K\varepsilon^{-\hat{D}_0}$ (Eq. (3.2)) and substituting

Figure 3.5 Two applications of the generalized baker's map map the square into the four shaded strips shown.

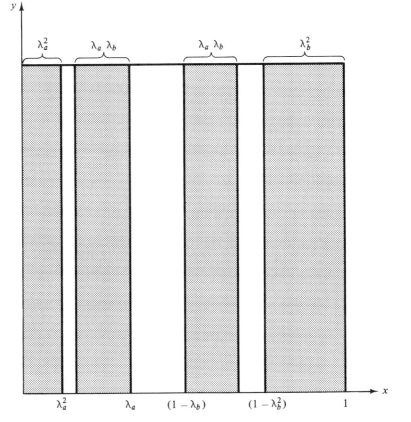

this and Eq. (3.11) into Eq. (3.10), we obtain a transcendental equation for $\hat{D}_0$,

$$\lambda_a^{\hat{D}_0} + \lambda_b^{\hat{D}_0} = 1. \tag{3.12}$$

For $\lambda_a + \lambda_b < 1$, the solution for $\hat{D}_0$ is between zero and one. Hence by Eq. (3.9) the attractor has a dimension between 1 and 2. For example, if $\lambda_a = \lambda_b = \frac{1}{3}$, we obtain from Eq. (3.12) the result $\hat{D}_0 = (\ln 2)/(\ln 3)$, and the intersection of the attractor with a horizontal line is just the middle third Cantor set. In this case $D_0 = 1.63\ldots$. For $\lambda_a + \lambda_b = 1$, there are no gaps between the vertical strips in Figure 3.4(*d*) and the solution of Eq. (3.12) is $\hat{D}_0 = 1$ corresponding to $D_0 = 2$ (the attractor in this case is the entire unit square).

3.3   Measure and the spectrum of D_q dimensions

Say we cover a chaotic attractor with a grid of cubes (as we would do if we were interested in computing D_0), and then we look at the frequency with which typical orbits visit the various cubes covering the attractor in the limit that the orbit length goes to infinity. If these frequencies are the same for all initial conditions in the basin of attraction of the attractor except for a set of Lebesgue measure zero, then we say that these frequencies are the natural measures of the cubes. That is, for a typical $\mathbf{x}_0$ in the basin of the attractor, the natural measure of a typical cube C_i is

$$\mu_i = \lim_{T \to \infty} \frac{\eta(C_i, \mathbf{x}_0, T)}{T}, \tag{3.13}$$

where $\eta(C_i, \mathbf{x}_0, T)$ is the amount of time the orbit originating from $\mathbf{x}_0$ spends in C_i in the time interval $0 \leqslant t \leqslant T$.

In cases where a property holds for all points in a set except for a subset whose measure is zero, we say that the property holds for *almost every* point in the set with respect to the particular measure. For example, assuming the existence of a natural measure, we say that the limit in (3.13) yields the same value μ_i for 'almost every point in the basin with respect to Lebesgue measure,' and we call such points *typical*.

The box-counting dimension gives the scaling of the number of cubes needed to cover the attractor. For strange attractors, however, it is commonly the case that the frequency with which different cubes are visited can be vastly different from cube to cube. In fact, for very small ε it is common that only a very small percentage of the cubes needed to cover the chaotic attractor contain the vast majority of the natural measure on the attractor. That is, typical orbits will spend most of their time in a small minority of those cubes that are needed to cover the attractor. The box-counting dimension definition counts all cubes needed to cover the attractor equally, without regard to the fact that, in some sense, some cubes are much more important (i.e., much more frequently visited) than others. To take into account the different natural measures of the cubes it is possible to introduce another definition of dimension which generalizes the box-counting dimension. This definition of dimension was formulated in the context of chaotic dynamics by Grassberger (1983) and Hentschel and Procaccia (1983). These authors define a dimension D_q which depends on a continuous index q,

$$D_q = \frac{1}{1-q} \lim_{\varepsilon \to 0} \frac{\ln I(q,\varepsilon)}{\ln(1/\varepsilon)}, \qquad (3.14)$$

where

$$I(q,\varepsilon) = \sum_{i=1}^{\tilde{N}(\varepsilon)} \mu_i^q,$$

and the sum is over all the $\tilde{N}(\varepsilon)$ cubes in a grid of unit size ε needed to cover the attractor (see also Renyi (1970)). The point is that for $q > 0$ cubes with larger μ_i have a greater influence in determining the value of D_q. Note that for $q = 0$ we have $I(0,\varepsilon) = \tilde{N}(\varepsilon)$, and we recover the box-counting dimension definition. In the special case where all the μ_i are equal, we have $\mu_i = 1/\tilde{N}(\varepsilon)$, $\ln I(q,\varepsilon) = (1-q)\ln\tilde{N}(\varepsilon)$, and we recover the box-counting dimension independent of q. Defining D_1 by $D_1 = \lim_{q \to 1} D_q$, we have from L'Hospital's rule and Eq. (3.14)

$$D_1 = \lim_{\varepsilon \to 0} \frac{\displaystyle\sum_{i=1}^{\tilde{N}(\varepsilon)} \mu_i \ln \mu_i}{\ln \varepsilon}. \qquad (3.15)$$

The quantity D_1 is, as we shall see, of particular interest (Balatoni and Renyi, 1956) and is called the *information dimension*. Another property of the D_q is that they generally decrease with increasing q (except for the exceptional case where the measure is fairly homogeneously spread through the attractor so that $\mu_i \sim 1/\tilde{N}(\varepsilon)$ for all boxes, in which case all the D_q are equal, $D_q = D_0$). In general, it can be shown that

$$D_{q_1} \le D_{q_2} \text{ if } q_1 > q_2. \qquad (3.16)$$

Thus, for example, D_2 provides a lower bound for D_1, and D_1 provides a lower bound for D_0.

Although, in this chapter, we are primarily interested in the D_q for the natural measure of chaotic attractors, we emphasize that (3.14) can be applied to any measure.

Determinations of the fractal dimensions of strange attractors occurring in numerical experiments have been done for large number of systems; the earliest example is the paper by Russell *et al.* (1980) who examined the box-counting dimension for a number of different systems which yield strange attractors (one of which was the Hénon attractor, Figure 1.12). In doing such numerical experiments to determine D_q one typically generates a long orbit of length T on the attractor and examines the fraction of time the orbit spends in cubes of an ε-grid. This gives an approximation to μ_i for each cube from which an approximation $I_T(q, \varepsilon)$ to $I(q, \varepsilon)$ is obtained. An approximation to the dimensions D_q can then be obtained by plotting $\ln I_T(q, \varepsilon)$ versus $\ln \varepsilon$. If one is not too unlucky, this plot will yield points that appear to fall approximately on a straight line for some appreciable range of $\ln \varepsilon$. One can then fit a straight line to these points and determine the slope of the line. The approximate D_q is then $(q - 1)^{-1}$ times this slope (see Eq. (3.14)). The range of ε over which such a fitting can be meaningful is limited at large ε by the requirement that ε be sufficiently small compared to the attractor size and at small ε by statistical fluctuations in determining the μ_i (due to the necessarily finite amount of data). Statistical problems at small ε can be less severe at larger q (e.g., D_2 is, in general, easier to calculate than D_0), since D_q for larger q values is determined by higher probability cubes for which the statistics is necessarily better.

As an example of a use of a fractal dimension, consider generating a chaotic orbit on a digital computer. Computers represent numbers as binary decimals of a certain limited length (the computer 'roundoff'). Thus there is only a finite amount of represented numbers. Hence, any computer generated orbit (on the Hénon map, for example) must eventually repeat exactly, thus artificially producing a periodic orbit when, in fact, the orbit should be nonperiodic. This is not necessarily a problem if orbits are run for a time less than the typical computer-roundoff-induced period. However, in situations where very long orbits are examined this can be a problem, and an estimate of the period is consequently desirable. This problem was studied in a paper by Grebogi et al. (1988c). They found that the period scaled as a power of the roundoff with the exponent given by the dimension D_2 of the attractor. Specifically, they found that, if the roundoff level is δ, then the typical roundoff-induced periodicity length scales as $\delta^{-D_2/2}$.

To show how nonuniform the natural measure on an attractor can be, consider the following example due to Sinai (1972),

$$\begin{aligned} x_{n+1} &= (x_n + y_n + \Delta \cos 2\pi y_n) \text{ modulo } 1, \\ y_{n+1} &= (x_n + 2y_n) \text{ modulo } 1. \end{aligned} \tag{3.17}$$

For small Δ, Sinai shows that the attractor is the entire square $[0, 1] \times [0, 1]$. Thus a typical orbit comes arbitrarily close to any point in the square if we wait long enough. Hence, for any grid of boxes of edge length ε, all boxes are visited with some nonzero frequency, and

$\tilde{N}(\varepsilon) = \varepsilon^{-2}$. Now consider the orbit points of a typical trajectory of the map Eq. (3.17) shown in Figure 3.6(*a*). We see that the density of points (natural measure) is highly concentrated along diagonal bands, and, if a small piece of the attractor is magnified (Figure 3.6(*b*)), similar structures of high concentration are evident. In fact, Sinai shows that, for Δ small enough, for any $\xi > 0$ there is a collection of small squares whose total area is less than ξ such that the collection of squares contains a natural measure of $1 - \xi$. Thus it apparently takes an arbitrarily small area to cover most of the natural measure (e.g., take $\xi = 10^{-3}$). This extreme type of behavior is not a property peculiar to Sinai's example. Indeed it is typical of chaotic attractors and is, for example, present in the generalized baker's map as we shall show in the next section.

3.4 Dimension spectrum for the generalized baker's map

From the action of the generalized baker's map as illustrated in Figure 3.4, it is evident that an initial density distribution which has no dependence on y is mapped to one which also has no dependence on y, and this holds for all subsequent iterates. In fact the natural invariant measure can be shown to be uniform in y. Thus the natural measure of the attractor in $0 \leq y \leq \alpha$ (the lower portion of the unit square in Figure 3.4(*a*)) is just α, while the natural measure in $\alpha \leq y \leq 1$ is $\beta = 1 - \alpha$. Mapping these regions forward in time and noting that the natural measure is invariant to application of the map, we then find that the natural measure of the strip $0 \leq x \leq \lambda_a$ is α, and the natural measure of the strip $(1 - \lambda_b) \leq x \leq 1$ is β

Figure 3.6(*a*) 80 000 iterates of the map Eq. (3.17) starting from $x_0 = y_0 = 0.5$ with $\Delta = 0.1$. (*b*) A blow-up of the strip marked in (*a*) (Farmer *et al.*, 1983).

Region blown up

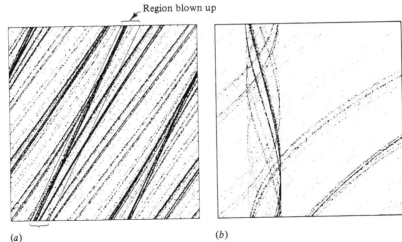

(*a*)   (*b*)

(cf. Figure 3.4). Since the natural measure of the attractor is uniform in y, we can express D_q as

$$D_q = 1 + \hat{D}_q \qquad (3.18)$$

(analogous to Eq. (3.9)), where $\hat{D}_q$ is the dimension in the horizontal direction and is defined as in Eq. (3.14), but with $I(q,\varepsilon)$ replaced by

$$\hat{I}(q,\varepsilon) = \sum_{i=1}^{\hat{N}(\varepsilon)} \hat{\mu}_i^q. \qquad (3.19)$$

Here we assume a uniform ε spacing along the x-axis. For each interval, we determine the total attractor measure in the vertical strip, $0 \le y \le 1$, in that interval. Counting only those strips for which the measure is not zero, we perform the sum Eq. (3.19), where $\hat{\mu}_i$ is the measure in strip i.

We now express $\hat{I}(q,\varepsilon)$ as

$$\hat{I}(q,\varepsilon) = \hat{I}_a(q,\varepsilon) + \hat{I}_b(q,\varepsilon), \qquad (3.20)$$

where $\hat{I}_a$ is the contribution to the sum in (3.19) from $0 \le x \le \lambda_a$, and $\hat{I}_b$ is the contribution from $(1 - \lambda_b) \le x \le 1$. If we magnify the interval $0 \le x \le \lambda_a$ and its ε grid of small intervals in x by the factor $1/\lambda_a$, we get a picture similar to the whole attractor in $0 \le x \le 1$, with the x-axis partitioned by a uniform grid of intervals of lengths ε/λ_a. In addition, since the measure in $0 \le x \le \lambda_a$ is α, we have

$$\hat{I}_a(q,\varepsilon) = \alpha^q \hat{I}(q,\varepsilon/\lambda_a). \qquad (3.21a)$$

Similarly, we obtain from consideration of the interval $(1 - \lambda_b) \le x \le 1$,

$$\hat{I}_b(q,\varepsilon) = \beta^q \hat{I}(q,\varepsilon/\lambda_b). \qquad (3.21b)$$

From Eq. (3.14), we take $\hat{I}(q,\varepsilon)$ to have the small ε dependence

$$\hat{I}(q,\varepsilon) \simeq K\varepsilon^{(q-1)\hat{D}_q}. \qquad (3.22)$$

Putting Eqs. (3.21) and (3.22) in (3.20) we obtain a transcendental equation for $\hat{D}_q$

$$\alpha^q \lambda_a^{(1-q)\hat{D}_q} + \beta^q \lambda_b^{(1-q)\hat{D}_q} = 1. \qquad (3.23)$$

For $q = 0$ this equation reduces to the box-counting dimension result Eq. (3.12). Expanding Eq. (3.23) for small $(q-1)$, we obtain an explicit expression for the information dimension $D_1 = 1 + \hat{D}_1$,

$$D_1 = 1 + \frac{\alpha \ln(1/\alpha) + \beta \ln(1/\beta)}{\alpha \ln(1/\lambda_a) + \beta \ln(1/\lambda_b)}. \qquad (3.24)$$

Also the transcendental equation, Eq. (3.23), can be explicitly solved for the case $\lambda_a = \lambda_b$,

$$D_q = 1 + \frac{1}{q-1} \frac{\ln(\alpha^q + \beta^q)}{\ln \lambda_a}. \qquad (3.25)$$

The information dimension D_1 plays a key role. In the next section we verify using the generalized baker's map that D_1 has the following remarkable property. Consider a subset of the attractor which has a fraction $0 < \theta \leq 1$ of the natural measure of the attractor. We can, in principle, calculate the box-counting dimension of this set. In fact there will be many ways of choosing sets which cover a given fraction θ of the attractor measure. We choose from all these the set with the smallest box-counting dimension and denote its dimension $D_0(\theta)$. For $0 < \theta < 1$ this set is one which is on those regions of the attractor with the greatest concentration of orbit points (e.g., the dark bands in Figure 3.6(*a*)). The result from the next section is that

$$D_0(\theta) = D_1 \tag{3.26}$$

for any $0 < \theta < 1$ (e.g., $\theta = 0.99$). (For $\theta = 1$ the entire attractor must be covered by the set, and $D_0(1) = D_0$.) Thus D_1 is essentially the dimension of the core region of high natural measure of the attractor. Sinai's result for the attractor of Eq. (3.17) arises because $D_0 = 2$, while $D_1 < 2$ for this attractor. That is, the core is fractal while the attractor itself is simply the area (nonfractal) $0 \leq y \leq 1$, $0 \leq x \leq 1$.

3.5 Character of the natural measure for the generalized baker's map

We have seen in Section 3.4 that the natural measure of the strip of width λ_a is α and that of the strip of width λ_b is β. Applying the map to this situation, we find the natural measures of the four strips in Figure 3.5 are as follows: the natural measure of the strip of width λ_a^2 is α^2; the natural measure of the strip of width λ_b^2 is β^2; and the natural measures of the two strips of width $\lambda_a \lambda_b$ are both $\alpha\beta$. As noted in Section 3.2, applying the map n times, we generate 2^n strips, of which $Z(n,m) = n!/(n-m)!m!$ have width $\lambda_a^m \lambda_b^{n-m}$ for $m = 0, 1, 2, \ldots, n$. From the above, we see that each strip of width $\lambda_a^m \lambda_b^{n-m}$ has a natural measure equal to $\alpha^m \beta^{n-m}$. Thus the natural measure contained in all strips of width $\lambda_a^m \lambda_b^{n-m}$ is

$$W(n,m) = \alpha^m \beta^{n-m} Z(n,m). \tag{3.27}$$

(Note that, as it should, the sum $\Sigma_{m=0}^n W(n,m)$ is 1, since, by virtue of $Z(n,m)$ being the binomial coefficient, $(\alpha + \beta)^n = \Sigma \alpha^m \beta^{n-m} Z(n,m)$, and $\alpha + \beta \equiv 1$.)

Using Stirling's approximation,

$$\ln \rho! = (\rho + \tfrac{1}{2}) \ln(\rho + 1) - (\rho + 1) + \ln(2\pi)^{1/2} + O(\rho^{-1}), \tag{3.28}$$

we obtain from (3.8)

$$\ln Z \simeq (n + \tfrac{1}{2}) \ln(n + 1) - (m + \tfrac{1}{2}) \ln(m + 1)$$
$$- (n - m + \tfrac{1}{2}) \ln(n - m + 1) - \ln(2\pi)^{1/2} + 1. \tag{3.29}$$

Expanding this expression in a Taylor series around its maximum value, $m = n/2$, yields

$$Z(n, m) \simeq \frac{2^n}{(2\pi)^{1/2}} \left(\frac{4}{n}\right)^{1/2} \exp\left\{-\frac{1}{2}\left[4n\left(\frac{m}{n} - \frac{1}{2}\right)^2\right]\right\} \quad (3.30)$$

Similarly, from Eq. (3.27)

$$W(n, m) \simeq \frac{1}{(2\pi n \alpha \beta)^{1/2}} \exp\left[-\frac{n(m/n - \alpha)^2}{2\alpha\beta}\right] \quad (3.31)$$

Note that, since these expressions for Z and W are obtained by Taylor series expansions of $\ln Z$ and $\ln W$ about their maxima, they are only valid for $|m/n - \frac{1}{2}| \ll 1$ and $|m/n - \alpha| \ll 1$, respectively. However, since the widths in m/n of these Gaussians are $O(n^{-1/2})$, we see that for large n Eq. (3.30) is valid for most of the strips and Eq. (3.31) is valid for most of the natural measure.

Figure 3.7 shows schematic plots of Z and W. It is clear from this figure that, for large n, almost all of the natural measure is contained in a very small fraction of the total number of strips (i.e., a value k can be chosen so that $\int_{\alpha-k}^{\alpha+k} nW \mathrm{d}(m/n)$ can be close to 1, while $\int_{\alpha-k}^{\alpha+k} nZ \mathrm{d}(m/n)$ can be very compared to 2^n, the number of strips). Furthermore, this situation becomes more and more accentuated as n gets larger, since the widths of the Gaussians decrease as $n^{-1/2}$ (Z and W become delta functions for $n \to \infty$). These properties seem to be typical of chaotic attractors.

To proceed we now take $\lambda_a = \lambda_b \leq \frac{1}{2}$ and examine coverings of the projection of the attractor onto the x-axis by small intervals of length $\varepsilon_n = \lambda_a^n$. In this case $\lambda_a^m \lambda_b^{n-m} = \lambda_a^n$ independent of m. (There is still a distinction to be made for different m, however, since (although all the widths are all the same) the strips have different natural measures $\alpha^m \beta^{n-m}$.)

As an example, let us use the result Eq. (3.31) to calculate the information dimension D_1 in the case $\lambda_a = \lambda_b$ and $\varepsilon = \lambda_a^n$. We convert the sum over i in Eq. (3.15) to a sum over m by noting that there are $Z(n, m)$ intervals of length $\varepsilon = \lambda_a^n$ which each have the measure $\alpha^m \beta^{n-m}$. Thus, $\Sigma_i \hat{\mu}_i \ln \hat{\mu}_i = \Sigma_m \alpha^m \beta^{n-m} Z(n, m) \ln(\alpha^m \beta^{n-m}) = \Sigma_m W(n, m) \ln(\alpha^m \beta^{n-m}) = n\Sigma_m W(n, m)[(m/n)\ln\alpha + (1 - m/n)\ln\beta]$. For large n, we see from Figure 3.7 that W becomes sharply peaked about $m/n = \alpha$. Hence, in the limit $n \to \infty$ we obtain $\Sigma_i \hat{\mu}_i \ln \hat{\mu}_i = n[\alpha \ln \alpha + \beta \ln \beta]$, and from Eq. (3.15) the information dimension projected onto the x-axis is

$$\hat{D}_1 = (\alpha \ln \alpha + \beta \ln \beta)/(\ln \lambda_a), \quad (3.32)$$

in agreement with Eq. (3.24) for $\lambda_a = \lambda_b$.

We now wish to calculate $\hat{D}_0(\theta)$, the dimension of the smallest set containing a natural measure θ, for the generalized baker's map with

$\lambda_a = \lambda_b \leq \frac{1}{2}$. We will find the important result that $D_0(\theta) = D_1$ for all θ in $0 < \theta < 1$. Assuming $\beta > \alpha$, the larger measures (i.e., larger $\alpha^m \beta^{n-m}$) correspond to smaller m. Thus the smallest number of intervals of length $\varepsilon = \lambda_a^n$ needed to cover a fraction θ of the measure (projected to the x axis) is

$$\hat{N}(\varepsilon, \theta) = \sum_{m=0}^{m_\theta} Z(n, m), \tag{3.33}$$

where m_θ is the largest integer such that

$$\sum_{m=0}^{m_\theta - 1} W(n, m) \leq \theta. \tag{3.34}$$

Using (3.31) and approximating the sum by an integral, we have

$$\theta \approx \frac{1}{(2\pi\alpha\beta n)^{1/2}} \int_0^{m_\theta} \exp\left[-\frac{(m - \alpha n)^2}{2n\alpha\beta}\right] dm, \tag{3.35}$$

from which we obtain

$$\frac{m_\theta}{n} \approx \alpha + \left(\frac{\alpha\beta}{n}\right)^{1/2} \mathrm{erfc}^{-1}(\theta), \tag{3.36}$$

where $\mathrm{erfc}(x) = (2\pi)^{-1/2} \int_{-\infty}^x \exp(-x^2/2)\,dx$ and we have assumed $\theta < 1$. (Because the width of the maximum of W is small for large n we can replace the lower limit of integration in (3.35) by $-\infty$.) Now consider Eq. (3.33). For large n, the principal contribution to the sum will come from

Figure 3.7 *W* and *Z* versus m/n.

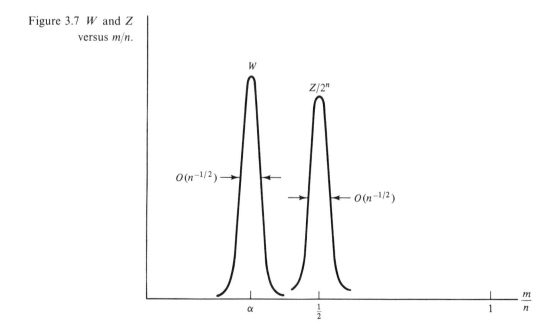

m-values very close to m_θ, since for large n the quantity Z decreases rapidly as m decreases through m_θ (cf. Fig. 3.8). Since $|m_\theta/n - \alpha| \sim O(1/n^{1/2})$, we cannot use the approximation Eq. (3.30) in (3.33). Rather, we divide Eq. (3.31) by $\alpha^m \beta^{n-m}$ to approximate Z near $m = m_\theta$,

$$Z(n, m) \simeq \frac{\beta^{-n}(\beta/\alpha)^m}{(2\pi n \alpha \beta)^{1/2}} \exp\left[-\frac{(m - \alpha n)^2}{2n\alpha\beta}\right]$$

The term $(\beta/\alpha)^m$ decreases as m decreases away from m_θ, and this decrease is much more rapid than the variation of the term $\exp[-(m - \alpha n)^2/2n\alpha\beta]$. Thus, in performing the sum in Eq. (3.33) we replace m by m_θ in this term. Hence the only significant m dependence in the sum is from the term $(\beta/\alpha)^m$. Using $\Sigma_{m=0}^{m_\theta} (\beta/\alpha)^m = (\beta/\alpha)^{m_\theta} \beta/(\beta - \alpha)$, we obtain

$$\hat{N}(\varepsilon, \theta) \sim \beta^{-(n-m_\theta)} \alpha^{-m_\theta} n^{-1/2}.$$

From Eq. (3.1) with $(m_\theta/n) = \alpha + O(n^{-1/2})$ (Eq. (3.36)) and $\varepsilon = \lambda_a^n$ we obtain

$$\hat{D}_0(\theta) = (\alpha \ln \alpha + \beta \ln \beta)/(\ln \lambda_a),$$

which is the same as Eq. (3.32). Hence we see that the information dimension may be thought of as the box-counting dimension of the smallest set which contains most of the attractor measure. Furthermore, since D_1 is in general less than D_0 (at most they can be equal and this is not typical), we see that, on any covering of the attractor by small cubes, the vast majority of them taken together have only a small part of the measure[5] ($\tilde{N}(\varepsilon, \theta) \sim \varepsilon^{-D_1} \ll \varepsilon^{-D_0} \sim \tilde{N}(\varepsilon, 1)$ for $\theta < 1$).

3.6   The pointwise dimension

Another concept of dimension which is useful for the study of strange attractors and other invariant sets is the *pointwise dimension* $D_p(\mathbf{x})$. If

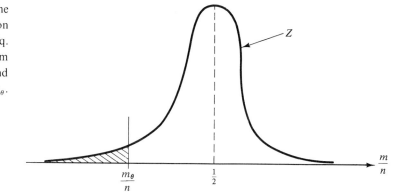

Figure 3.8 The principal contribution to the sum in Eq. (3.33) comes from m-values near and slightly below m_θ.

$B_\varepsilon(\mathbf{x})$ denotes an N-dimensional ball of radius ε centered at a point $\mathbf{x}$ in an N-dimensional phase space, then the pointwise dimension of a probability measure μ at $\mathbf{x}$ is defined as (Young, 1982)

$$D_p(\mathbf{x}) = \lim_{\varepsilon \to 0} \frac{\ln \mu(B_\varepsilon(\mathbf{x}))}{\ln \varepsilon}. \qquad (3.37)$$

We argue below that, if the measure μ is *ergodic*, then $D_p(\mathbf{x})$ assumes a single common value $\bar{D}_p$ for all locations $\mathbf{x}$ except possibly for a set of $\mathbf{x}$ containing zero μ measure. Basically, the $\mathbf{x}$-values yielding this common value constitute the 'core region' of the measure whose box-counting dimension is $D_0(\theta) = D_1(0 < \theta < 1)$. Results of Young (1982) imply that the value $\bar{D}_p$ of $D_p(\mathbf{x})$ assumed for 'almost every' $\mathbf{x}$ with respect to the measure μ (i.e., all $\mathbf{x}$ except for a set of μ measure zero) is $D_0(\theta)(0 < \theta < 1)$,

$$\bar{D}_p = D_0(\theta) = D_1. \qquad (3.38)$$

(This is relatively easy to show for the generalized baker's map example.[6]) To obtain the result that $D_p(\mathbf{x})$ is the same value for almost every $\mathbf{x}$ with respect to μ, recall from Section 2.3.3 that an ergodic measure is an invariant probability measure which cannot be decomposed such that

$$\mu = p\mu_1 + (1 - p)\mu_2, \quad 1 > p > 0,$$

with $\mu_1 \neq \mu_2$ being two other invariant probability measures. The natural measure on a chaotic attractor is of particular interest here, and we note that if it exists (and we assume it does), it is necessarily ergodic by virtue of the fact that it can be constructed from the long-time limit of the frequency that a single typical orbit visits regions of phase space.

To show that $D_p(\mathbf{x})$ assumes a single common value for almost every $\mathbf{x}$ with respect to the ergodic measure μ we first argue that

$$D_p(\mathbf{x}) = D_p(\mathbf{x}'); \qquad (3.39)$$

where $\mathbf{x}' = \mathbf{M}(\mathbf{x})$; that is, the pointwise dimension at $\mathbf{x}$ and at its first iterate under the map are the same. Since the measure is invariant, we have for invertable $\mathbf{M}$ that $\mu(B_\varepsilon(\mathbf{x})) = \mu(\mathbf{M}(B_\varepsilon(\mathbf{x})))$. Assuming the map to be smooth and ε to be small, the region $B_\varepsilon(\mathbf{x})$ is mapped by $\mathbf{M}$ to an ellipsoidal region about $\mathbf{x}' = \mathbf{M}(\mathbf{x})$. Thus, as shown in Figure 3.9, we can define constants $r_1 > r_2$ such that the ball $B_{r_1\varepsilon}(\mathbf{x}')$ contains $\mathbf{M}(B_\varepsilon(\mathbf{x}))$ which contains $B_{r_2\varepsilon}(\mathbf{x}')$. Thus

$$\mu(B_{r_1\varepsilon}(\mathbf{x}')) \geq \mu(\mathbf{M}(B_\varepsilon(\mathbf{x}))) = \mu(B_\varepsilon(\mathbf{x})) \geq \mu(B_{r_2\varepsilon}(\mathbf{x}')). \qquad (3.40)$$

Since (3.37) yields

$$D_p(\mathbf{x}') = \lim_{\varepsilon \to 0} \frac{\ln[\mu(B_{r_{1,2}\varepsilon}(\mathbf{x}'))]}{\ln(r_{1,2}\varepsilon)} = \lim_{\varepsilon \to 0} \frac{\ln[\mu(B_{r_{1,2}\varepsilon}(\mathbf{x}'))]}{\ln \varepsilon},$$

we immediately obtain from (3.40), $D_p(\mathbf{x}') \geq D_p(\mathbf{x}) \geq D_p(\mathbf{x}')$ or $D_p(\mathbf{x}) = D_p(\mathbf{x}')$.

To show that $D_p(\mathbf{x})$ is the same for almost every $\mathbf{x}$ with respect to the ergodic measure μ, first assume that it is not. Then there is some value d_p such that there is a set S_- such that $D_p(\mathbf{x}) \leq d_p$ for $\mathbf{x}$ in S_-, and there is another disjoint set S_+ such that $D_p(\mathbf{x}) > d_p$ for $\mathbf{x}$ in S_+, and further the natural measures of S_+ and S_- are not zero, $\mu(S_\pm) > 0$. By Eq. (3.39) this implies that the sets S_+ and S_- are invariant. Hence the measure is divided into two parts, orbits on one part never visiting the other part. This, however, is not possible because we assume the measure is ergodic. Hence $D_p(\mathbf{x})$ is the same for almost every $\mathbf{x}$ with respect to the measure. Henceforth, we consider μ to be the natural measure on a chaotic attractor.

As we have stated above, $D_p(\mathbf{x})$ assumes the value $D_1 = D_0(\theta)$ $(0 < \theta < 1)$ for $\mathbf{x}$-values in the core region of the attractor. Since, however, D_0 is typically greater than D_1, there is a set on the attractor which is relatively large (in the sense of having a larger box counting dimension $D_0 > D_0(\theta)$ $(0 < \theta < 1)$) which is not on the core, and for which we consequently expect $D_p(\mathbf{x}) \neq D_1$.

In our discussion of fractal dimension we started by defining the box-counting dimension, which gives the dimension of a *set*. We then introduced the spectrum of dimensions D_q which assigns 'dimension' values to a *measure* (for each value of q). Measures for which D_q is not a constant with q are often called *multifractal measures*. Our statement above that there are points on the attractor for which $D_p(\mathbf{x}) \neq D_1$ is

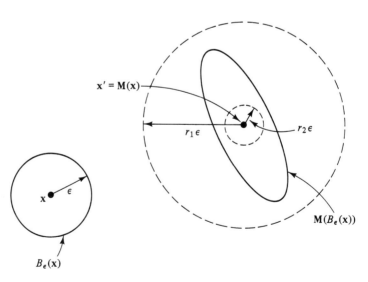

Figure 3.9 Map of a small ball $B_\varepsilon(\mathbf{x})$.

$\mathbf{x}' = \mathbf{M}(\mathbf{x})$

$r_1\epsilon$ $r_2\epsilon$

$\mathbf{M}(B_\varepsilon(\mathbf{x}))$

$\mathbf{x}$ ϵ

$B_\varepsilon(\mathbf{x})$

another consequence of multifractality. Indeed there are a number of intriguing further aspects of multifractal measures that arise. However, these are of a somewhat more advanced nature. For now we drop the discussion of multifractals, but will take it up again in Chapter 9 which will be devoted entirely to the discussion of this interesting topic.

3.7 Implications and determination of fractal dimension in experiments

One of the important issues confronting someone who is examining an experimental dynamical process is the question of how many scalar variables are necessary to model the process. For example, if we model the process using differential equations of the form $d\mathbf{x}/dt = \mathbf{F}(\mathbf{x})$, how large does the dimensionality of $\mathbf{x}$ have to be? Clearly a lower bound to this dimensionality is the fractal dimension of the attractor D_0. If the dimension of the vector $\mathbf{x}$ is less than D_0, the structure of the attractor cannot be reproduced by the model, and one anticipates that important features of the dynamics will be lost. For this reason considerable interest has attached to the problem of determining the dimension of experimental strange attractors. Typically experiments determine D_2 or $\bar{D}_p = D_1$ which are useful since they are lower bounds on D_0 and hence are also lower bounds on the system dimensionality.

Guckenheimer and Buzyna (1983) present measurements of the pointwise dimension of a presumed chaotic attractor in a rotating differentially heated annulus of fluid. In their technique they first choose a number of variables which they regard as their phase space. For these variables they use the temperature readings of 27 thermistors each located at a different point in the fluid. The number 27 was arbitrarily chosen to be large enough to represent the dimensions of the attractors they expected to find. The justification for using thermistor readings at different locations as phase space variables is that these readings are determined by the system state, and thus, like the delay coordinates discussed in Section 1.6, they may be viewed as smooth functions of any other vector variable $\mathbf{x}$ specifying the state. The calculation of the pointwise dimension proceeds as follows. Consider the vector $\mathbf{z}(t) = (\xi_1(t), \xi_2(t), \ldots, \xi_{27}(t))$, where $\xi_j(t)$ is the temperature reading on the jth thermistor. Then a large number of points on the attractor are obtained by sampling $\mathbf{z}(t)$ at discrete time intervals T; $\mathbf{z}_0 = \mathbf{z}(t_0)$, $\mathbf{z}_1 = \mathbf{z}(t_0 + T)$, $\mathbf{z}_2 = \mathbf{z}(t_0 + 2T), \ldots, \mathbf{z}_K = \mathbf{z}(t_0 + KT)$. One then selects one of the $\mathbf{z}_j$s as a reference point, call it $\mathbf{z}_*$, and calculates the distances $d_k = |\mathbf{z}_k - \mathbf{z}_*|$ from $\mathbf{z}_*$ to the K other points $\mathbf{z}_k$. The K distances d_k are then ordered according to their size in a list with the smallest d_k first. The ith distance value on the list gives a value of ε

(namely $\varepsilon = d_k$) such that

$$\mu(B_\varepsilon(\mathbf{z}_*)) \simeq i/K.$$

The quantity $\ln{(i/K)}$ is then plotted as a function of $\ln\varepsilon$. The points are observed to lie approximately on a straight line in some range of ε-values, and the dimension is estimated as the slope of a straight line fitted to the data. Problems can occur due to lack of sufficient data, noise, etc., but useful results were nevertheless obtained. For the experiment, the authors found that as the fluid was driven more strongly into the unstable regime, the dimension of the attractor rose, indicating the excitation of more and more active modes of motion and a consequent transition toward turbulence. Similar results were obtained by Brandstater and Swinney (1987) for an experiment on Couette–Taylor flow. In Couette–Taylor flow one has a fluid contained between two vertical coaxial cylinders (Figure 3.10(a)) and the cylinders are rotated at different angular velocities. In Brandstater and Swinney's experiment the outer cylinder was stationary and the behavior of the system was examined as a function of the rotation rate Ω of the inner cylinder. Figure 3.10(b) shows the computed dimension for this experiment as a function of the rotation rate. In this case improved statistics for the pointwise dimension were obtained by averaging the quantity $\ln\mu(B_\varepsilon(\mathbf{z}_*))$ over different reference points $\mathbf{z}_*$ taken from points on the orbit on the attractor. In performing the dimension computation these authors used delay coordinates (Section 1.6), $V(t)$, $V(t - \tau)$, $V(t - 2\tau), \ldots$, as their phase space variables, where $V(t)$ is the radial component of the fluid velocity measured at a particular point midway between the inner and outer cylinders. Referring to Figure 3.10(b), we see that the measured dimension is apparently close to 2 in the range Ω/Ω_c between 10 and 11.8. (The quantity Ω_c is the theoretical critical rotation rate at which the fluid first develops spatial structure in the vertical direction.) In this range the authors verify that the dynamics is nonchaotic and lies on a two-dimensional toroidal surface (as discussed in Chapter 6 this corresponds to two frequency quasiperiodic motion). As Ω/Ω_c is increased past this range, the motion becomes chaotic and the dimension of the attractor steadily rises.

In addition to the pointwise dimension, it has been emphasized by Grassberger and Procaccia (1983) that the 'correlation dimension' D_2 is particularly suited for relatively easy experimental determination. To calculate the correlation dimension, one must estimate the quantity (cf. Eq. (3.14)),

$$I(2, \varepsilon) = \sum_{i=1}^{\tilde{N}(\varepsilon)} \mu_i^2, \tag{3.41}$$

for different values of ε. Say we have a set of orbit points on the attractor $\mathbf{z}_k$

($k = 0, 1, 2, \ldots, K$). Then we compute the 'correlation integral'

$$C(\varepsilon) = \lim_{K \to \infty} \frac{1}{K^2} \sum_{i,j}^{K} U(\varepsilon - |\mathbf{z}_i - \mathbf{z}_j|), \tag{3.42}$$

where $U(\cdot)$ is the unit step function. (The sum in (3.42) gives the number of point pairs that are separated by a distance less than ε.) The quantity $C(\varepsilon)$ may be shown to scale with ε in the same way as $I(2, \varepsilon)$ scales with ε. Thus

$$D_2 = \lim_{\varepsilon \to 0} \frac{\ln C(\varepsilon)}{\ln \varepsilon}. \tag{3.43}$$

Figure 3.10(a) Configuration of a Couette–Taylor experiment. (b) Dimension of attractor as obtained from experimental data as a function of Ω (Brandstater and Swinney, 1987).

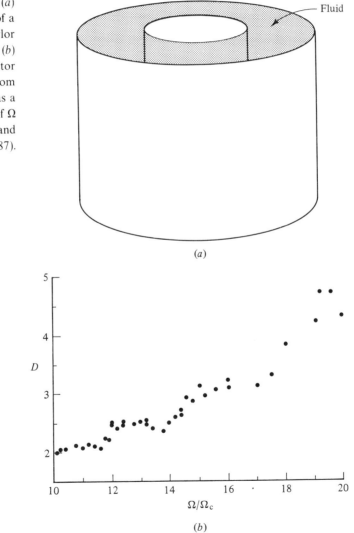

(a)

(b)

To see why $C(\varepsilon)$ and $I(2,\varepsilon)$ have the same scaling we write (3.42) as

$$C(\varepsilon) = \lim_{K \to \infty} \frac{1}{K} \sum_j^K \left[\frac{1}{K} \sum_i^K U(\varepsilon - |\mathbf{z}_i - \mathbf{z}_j|) \right].$$

For large K the sum over i is the natural measure in the ball $B_\varepsilon(\mathbf{z}_j)$, so that

$$C(\varepsilon) = \lim_{K \to \infty} \frac{1}{K} \sum_j^K \mu(B_\varepsilon(\mathbf{z}_j)).$$

This is an orbit average of the quantity $\mu(B_\varepsilon(\mathbf{z}))$. That is, it is an average over the natural measure of a function of the phase space position $\mathbf{z}$. In the notation of Eq. (2.26),

$$C(\varepsilon) = \int \mu(B_\varepsilon(\mathbf{z})) \, d\mu(\mathbf{z}). \tag{3.44}$$

Now refer to Eq. (3.41) for $I(2,\varepsilon)$. Noting that $\mu(B_\varepsilon(\mathbf{z})) \sim \mu_i$ if $\mathbf{z}$ is in cube i, and replacing one of the μ_i in Eq. (3.41) by $\mu(B_\varepsilon(\mathbf{z}))$, we see that (3.41) is also roughly an average of $\mu(B_\varepsilon(\mathbf{z}))$ over the natural measure. Hence

$$I(2,\varepsilon) \sim C(\varepsilon),$$

and Eq. (3.43) follows.

Equation (3.43) provides a useful means of estimating D_2 since $C(\varepsilon)$ from (3.42) can be estimated by using a finite but large K value.[7] This method for calculating D_2 and the method of calculating $\bar{D}_p$ by averaging experimentally determined values of $D_p(\mathbf{z}_*)$ over many reference points $\mathbf{z}_*$ on an orbit on the attractor require similar computational power and data quality, although one might expect better statistics for D_2 since it more heavily weights higher measure regions. Brandstater and Swinney in their paper report that they have used both methods. Since D_p typically is equal to D_1 which typically exceeds D_2 (Eq. (3.16)), one would expect that the pointwise dimension values might be larger than the correlation dimension values. Brandstater and Swinney find, however, that the accuracy of their measurements is insufficient to distinguish the two. Thus Figure 3.10(b) can be regarded as applying to both D_p and D_2.

In the measurement of fractal dimension in experiments it is often important to consider the effect of noise. If we assume the noise is white (i.e., it has a flat frequency power spectrum), then we can regard it as essentially fattening (or 'fuzzing') the attractor by an amount of order η, where η represents the typical noise amplitude (see, for example, Ott and Hanson (1981), Ott *et al.* (1985), Jung and Hänggi (1990)). Thus, for observations of the attractor characteristics on scales ε greater than the noise level η, the attractor appears to be fractal, while for scales $\varepsilon < \eta$ the attractor appears to be an N-dimensional volume, where N is the dimension of the space in which the attractor lies.[8] Figure 3.11 shows an illustration of this effect from numerical experiments by Ben-Mizrachi *et al.* (1984). The figure shows $\log_2 C(\varepsilon)$ versus $\log_2 \varepsilon$ for numerical experiments on the Hénon map embedded in three dimensions (a delay coordinate vector $y_n = (x_i^{(1)}, x_{n-1}^{(1)}, x_{n-2}^{(1)})$ was used; see Eq. (1.13)). Curve 1 is for the map without noise and yields $D_2 \simeq 1.25$ (the slope of the fitted line). Curve 2 is for the map with a random amount of noise added at each iterate. Curve 3 is similar but with larger noise. For $\varepsilon \gtrsim \eta$ the results for all three agree. For $\varepsilon \gtrsim \eta$ the noise causes the slope of the fitted line to be 3, the dimension of the embedding space. Thus white noise effectively limits the smallest size ε that can be used in dimension determinations.

3.8 Embedding

Say we use delay coordinates (as in Section 1.6) to construct a d-dimensional vector $\mathbf{y} = (g(t), g(t - \tau), g(t - 2\tau), \ldots, g[t - (d-1)\tau])$; see Eq. (1.16). In principle, if d is large enough, and the attractor is finite dimensional, then there exists some dynamical system describing the evolution of the vector $\mathbf{y}$. The question we now wish to address is how large must d be for this to be so. We assume that an actual smooth low-dimensional system which describes the dynamics exists,[9]

$$d\mathbf{x}/dt = \mathbf{F}(\mathbf{x}),$$

where $\mathbf{x}$ has some dimensionality d'. The observed quantity $g(t)$ may (as discussed in Section 1.6) be regarded as a smooth function of the state variable $\mathbf{x}$. Hence $\mathbf{y}$ is a function of $\mathbf{x}$,

$$\mathbf{y} = \mathbf{H}(\mathbf{x}).$$

The state of the system is given by $\mathbf{x}$, and knowledge of $\mathbf{x}$ at any time is sufficient to evolve the system into the future by $d\mathbf{x}/dt = \mathbf{F}(\mathbf{x})$. For there to be a dynamical system that evolves delay coordinate vectors $\mathbf{y} = \mathbf{H}(\mathbf{x})$ forward in time, it is sufficient that the function $\mathbf{y} = \mathbf{H}(\mathbf{x})$ be such that if $\mathbf{x}_0$ denotes a system state and $\mathbf{y}_0 = \mathbf{H}(\mathbf{x}_0)$, then there is no other possible

Figure 3.11
$\log_2 C(\varepsilon)$ versus $\log_2 \varepsilon$ for the Hénon map embedded in three dimensions without (curve 1) and with (curves 2 and 3) noise. Noise for curve 3 is greater than noise for curve 2 (Ben-Mizrachi *et al.*, 1984).

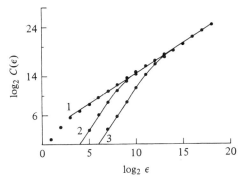

system state $\mathbf{x}_0' \neq \mathbf{x}_0$ satisfying $\mathbf{y}_0 = \mathbf{H}(\mathbf{x}_0')$. Hence, $\mathbf{x}_0$ determines $\mathbf{y}_0$ and vice versa. Thus given delay coordinates $\mathbf{y}_0 = \mathbf{H}(\mathbf{x}_0)$, the state $\mathbf{x}_0$ is uniquely determined and can be evolved forward any amount in time by $\mathrm{d}\mathbf{x}/\mathrm{d}t = \mathbf{F}(\mathbf{x})$ to a new state, which can then be transformed to the $\mathbf{y}$ variable by the function $\mathbf{H}$. This, in principle, defines a dynamical system evolving $\mathbf{y}$ forward in time. The key point is that the function $\mathbf{H}$ must satisfy the condition that $\mathbf{x} \neq \mathbf{x}'$ implies

$$\mathbf{H}(\mathbf{x}) \neq \mathbf{H}(\mathbf{x}').$$

If this is so, then we say that $\mathbf{H}$ is an *embedding* of the d'-dimensional $\mathbf{x}$-space into the d-dimensional $\mathbf{y}$-space. As an example, say our dynamical system in $\mathbf{x}$ is the simple one-dimensional dynamical system $\mathrm{d}x/\mathrm{d}t = \omega$ and x is an angle variable (i.e., in our previous notation, $x = \theta$, and x and $x + 2\pi$ are identified as equivalent). If we were to use a three-dimensional embedding ($d = 3$), then we can write $\mathbf{y}$ as $\mathbf{y} = [G(x(t)), G(x(t - \tau)), G(x(t - 2\tau))]$, and $G(x)$ is 2π periodic in x. Thus, the orbit would be expected to be a limit cycle lying on a closed curve in $\mathbf{y}$-space as shown in Figure 3.12(*a*). Now say we use a two-dimensional embedding, $\mathbf{y} = [G(x(t)), G(x(t - \tau))]$. The picture might look something like that shown in Figure 3.12(*b*). That is, the mapping of x-values to $\mathbf{y}$-space might produce a curve which intersects itself. We cannot now have a dynamical system in $\mathbf{y}$, because at $\mathbf{y}$-values at these intersections there are two possible x-values. Hence, specification of $\mathbf{y}$ does not, in principle, allow us to determine the future system evolution. Thus the question is how large does d typically have to be to ensure that we avoid self-intersections of the d'-dimensional $\mathbf{x}$-space when we attempt to embed it in a d-dimensional delay coordinate $\mathbf{y}$-space. Takens (1980) addresses this question and

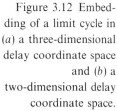

Figure 3.12 Embedding of a limit cycle in (*a*) a three-dimensional delay coordinate space and (*b*) a two-dimensional delay coordinate space.

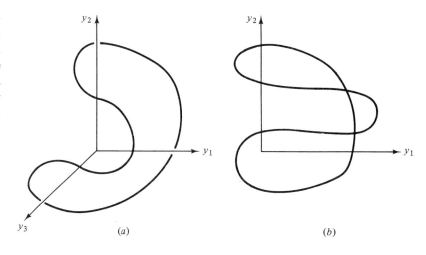

obtained the result that generically

$$d \geq 2d' + 1 \tag{3.45}$$

is sufficient. For our example above $d' = 1$, and Eq. (3.45) says that intersections are generically absent if $d = 3$ (as in Figure 3.12(a)) or larger.

We now give a heuristic discussion and justification for (3.45), following which we discuss some applications of embedding. Say we have a smooth surface of dimension d_1 and another of dimension d_2, both lying in an N-dimensional Cartesian space. If these surfaces intersect in a *generic intersection*, then the dimension d_0 of the intersection set is

$$d_0 = d_1 + d_2 - N. \tag{3.46}$$

If Eq. (3.46) yields $d_0 < 0$, then the two sets do not generically intersect. Figure 3.13 illustrates this equation for several cases. Figure 3.13(a) shows a generic intersection of two curves in a two-dimensional space. In this case $d_1 = d_2 = 1$, $N = 2$ and hence by Eq. (3.46) the dimension of the intersection is zero; the intersection set is two points. Figure 3.13(b) shows a nongeneric intersection of two curves in a two-dimensional space. Here the intersection is one-dimensional. It is nongeneric because it can be destroyed by an arbitrarily small smooth perturbation. For example, rigidly shifting one of the curves by an arbitrarily small amount can either convert the one-dimensional intersection to a zero-dimensional generic intersection (Figure 3.13(c)), or, if the curve is shifted in the other direction, then there is no intersection at all. In contrast, the generic intersections of Figure 3.13(a) cannot be qualitatively altered by small smooth changes in the curves. Figure 3.13(d) shows a case where Eq. (3.46) yields a negative value ($d_1 = d_2 = 1$, $N = 3$) indicating that two one-dimensional curves do not generically intersect in spaces of dimension $N \geq 3$. Figure 3.13(e) shows a curve ($d_1 = 1$) and a two-dimensional surface ($d_0 = 2$) in a three-dimensional space ($N = 3$) generically intersecting at a point with $d_0 = 0$ as predicted by Eq. (3.46). Figure 3.13(f) shows two two-dimensional surfaces ($d_1 = d_2 = 2$) in a three-dimensional space generically intersecting in a curve so that $d_0 = 1$, again as predicted by Eq. (3.46). In Takens' result Eq. (3.45) we were concerned with self-intersections (e.g., Figure 3.12). Thus Eq. (3.45) follows from (3.46) by requiring $d_0 < 0$ and setting $d_1 = d_2 = d'$ and $N = d$ with the delay coordinate transformation function **H** regarded as being a typical function (so that only generic intersections are expected).

Recently, methods have been proposed whereby one can determine a map from experimental data which may be noisy. In principle, this can be done without any knowledge of the physical processes determining the evolution. The only knowledge necessary is that the experimentally observed time evolution results from a low-dimensional dynamical system. (Information on the system dimensionality can be obtained by measurement of the fractal dimension as discussed in Section 3.7.) One way to proceed is as follows. First one forms a delay coordinate vector **y** of sufficient dimensionality. Then a surface of section is used to obtain a large amount of data giving a discrete trajectory $\chi_1, \chi_2, \chi_3, \ldots$, for points in the surface of section. (Alternatively **y**(t) can be sampled at finite time

Figure 3.13 Illustrations of Eq. (3.46).

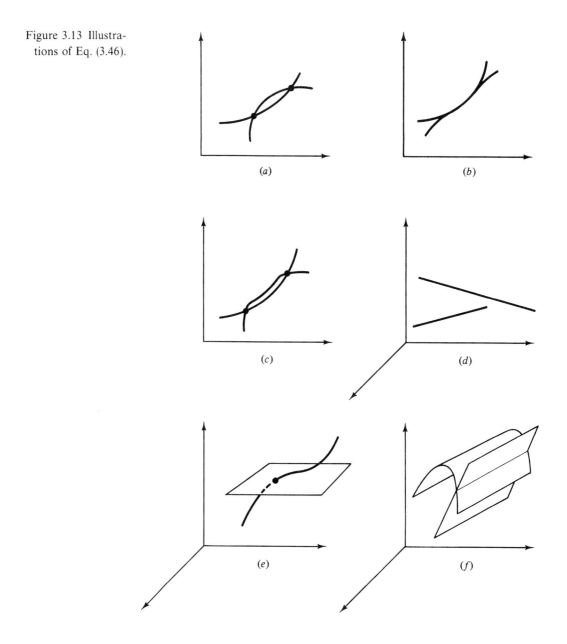

intervals to give an orbit for a time T map as discussed at the end of Section 1.3.) Next one attempts to fit this data to a map, $\chi_{n+1} = \mathbf{F}(\chi_n)$. That is, one attempts to find the function $\mathbf{F}$. This may not be possible unless the dimensionality d of the original delay coordinate vector $\mathbf{y}$ is large enough (e.g., Eq. (3.45) is a sufficient condition for this to be so). One method is to approximate $\mathbf{F}$ as a locally linear function. That is, in a small region of χ-space one can approximate $\mathbf{F}$ as $\chi_{n+1} = \mathbf{A} \cdot \chi_n + \mathbf{b}$ where $\mathbf{A}$ and $\mathbf{b}$ are a matrix and a vector. The matrix $\mathbf{A}$ and the vector $\mathbf{b}$ are obtained by least squares fitting to all the experimental observations χ_i and χ_{i+1} such that χ_i falls near χ_n. By using many data pairs, χ_i and χ_{i+1}, the least squares fitting has the effect of averaging out random noise contamination of the data. By piecing together results from many such small regions, we get a global approximation to $\mathbf{F}$. Hence, we obtain a dynamical system for the experimental process. Possible uses for such an approach include the prediction of the future evolution of the system (Farmer and Sidorowich, 1987; Casdagli, 1989; Abarbanel et al., 1990; Poggio and Girosi, 1990; Linsay, 1991), removing noise from chaotic data (Kostelich and Yorke, 1988; Hammel, 1990), obtaining Lyapunov exponents from experimental data (Eckmann and Ruelle, 1985; Eckmann et al., 1986; Sano and Sawada, 1985; Wolf et al., 1985; Bryant et al., 1990; cf. Chapter 4), finding unstable periodic orbits embedded in a chaotic attractor (Gunaratne et al., 1989; Lathrop and Kostlich, 1989; Sommerer et al., 1991a), and controlling chaotic dynamical systems by application of small controls (Ott et al., 1990a,b; Shinbrot et al., 1990; Ditto et al., 1990a; Dressler and Nitsche, 1992; cf. Chapter 4).

3.9   Fat fractals

In this chapter we have primarily been discussing the fractal dimension of sets of zero Lebesgue measure in the phase space. There are, however, Cantor sets with nonzero Lebesgue measure, and such sets often appear in nonlinear dynamics, as we shall see. Farmer (1985) has called these kinds of sets *fat fractals*. Grebogi et al. (1985b) define a set lying in an N-dimensional Euclidian space to be a fat fractal if, for every point $\mathbf{x}$ in the set and every $\varepsilon > 0$, a ball of radius ε centered at the point $\mathbf{x}$ contains a nonzero volume (Lebesgue measure) of points in the set, as well as a nonzero volume outside the set.[10] (If $N = 1$ the 'ball of radius ε' is the interval $[x - \varepsilon, x + \varepsilon]$.) We have already seen examples of fat fractals in Chapter 2, namely the set S_* discussed in Section 2.2 and the set of r-values for which the attractor of the logistic map is chaotic. Other examples will be the set of parameter values yielding quasiperiodic motions (see Chapter 6), and the regions of phase space of a nonintegrable

Hamiltonian system on which there is nonchaotic motion on KAM tori (see Chapter 7).

Since fat fractals have positive Lebesgue measure, their box-counting dimension is the same as the dimension of the space in which they lie, $D_0 = N$. Thus the box-counting dimension of these sets says nothing about the infinitely fine-scaled structure that they possess. One would like to have a quantitative way of characterizing this structure analogous to the box-counting dimension of fractals with zero Legesgue measure ('skinny fractals'). One way of doing this is by the *exterior dimension* definition of Grebogi *et al.* (1985b). These authors begin by noting that, given an ordinary skinny fractal set S_0, the box-counting definition Eq. (3.11) is equivalent to

$$D_0 = N - \lim_{\varepsilon \to 0} \ln V[S(\varepsilon)]/\ln \varepsilon, \tag{3.47}$$

where $S(\varepsilon)$ is obtained by fattening the original set S_0 by an amount ε (i.e., the set $S(\varepsilon)$ consists of the original set S_0 plus all points within a distance ε from S_0 and so $S_0 \equiv S(0)$), and $V[S(\varepsilon)]$ is the N-dimensional volume of this set. To see how Eq. (3.47) arises, say we cover S_0 with $\tilde{N}(\varepsilon)$ cubes from an N-dimensional grid. The volume of all these cubes is $\varepsilon^N \tilde{N}(\varepsilon)$, and this volume scales in roughly the same way with ε as $V[S(\varepsilon)]$. That is,

$$V[S(\varepsilon)] \sim \varepsilon^N \tilde{N}(\varepsilon).$$

Putting this estimate in Eq. (3.47), immediately reproduces the box-counting dimension definition Eq. (3.1). We now define the exterior dimension of S_0 as

$$D_x \equiv N - \lim_{\varepsilon \to 0} \ln V[\bar{S}(\varepsilon)]/\ln \varepsilon, \tag{3.48}$$

where $\bar{S}(\varepsilon) = S(\varepsilon) - S_0$ is what remains if the original set is deleted from the fattened set (hence the name *exterior* dimension). (We assume the set S_0 is closed.) For skinny fractals $V[S_0] = 0$, and we thus have $V[\bar{S}(\varepsilon)] = V[S(\varepsilon)]$ so that the exterior dimension reduces to the box-counting dimension, $D_x = D_0$. However, unlike the box-counting dimension, D_x gives nontrivial results for fat fractals. Note from the definition (3.48) that

$$V[\bar{S}(\varepsilon)] \sim \varepsilon^{N - D_x}. \tag{3.49}$$

One appealing way of interpreting the exterior dimension of a fat fractal is in terms of an *uncertainty exponent*. Say we consider some point $\mathbf{z}$ in a bounded region of space R that also contains the fat fractal set S_0. We are asked to determine whether or not $\mathbf{z}$ is in S_0, but we are also told that the values given for the coordinates of $\mathbf{z}$ have an uncertainty ε. Thus we do not know $\mathbf{z}$ precisely; rather we only know that it lies somewhere in the ball of radius ε centered at the coordinate values (call them $\mathbf{y}$) that we have been given. We can evaluate whether the point $\mathbf{y}$ lies in S_0 or does not lie in S_0. If we say that **z** lies in S_0 because we examine **y** and find that it lies in S_0, we may be wrong. Specifically, if **z** lies in $\bar{S}(\varepsilon)$, then we will commit an error. Now say we choose **z** at random in the bounded region R containing S_0. What is the probability that we commit an error by saying **z** lies in S_0 (or does not lie in S_0) when **y** is determined to lie in S_0 (or to not lie in S_0)? This probability is proportional to $V[\bar{S}(\varepsilon)]$ which, according to Eq. (3.49) scales as $\varepsilon^{\bar{\alpha}}$, where

$$\bar{\alpha} = N - D_x. \tag{3.50}$$

We call $\bar{\alpha}$ the uncertainty exponent. If $\bar{\alpha}$ is small, then a large improvement in accuracy (i.e., reduction in ε) leads to only a relatively small improvement in the ability to determine whether **z** lies in S_0. Thus, it becomes difficult to improve accuracy in the determination of whether **z** lies in S_0 by improving the accuracy of the coordinates if D_x is close to the dimension of the space N. (For further discussion of uncertainty exponents see Chapter 5.)

The above discussion provides a means of evaluating $\bar{\alpha}$ and hence D_x in certain cases. As an example, we consider the measure of the parameter A for which the quadratic map $x_{n+1} = A - x_n^2$ is chaotic. (Since the quadratic map and the logistic map may be related by a simple change of variables, D_x for the set of values of the parameter yielding chaos for the two maps is the same.) The procedure used by Grebogi *et al.* (1985b) to estimate D_x for this set is as follows. First they choose a value of A at random in the chaotic range using a random number generator. They then perturb this value to $A - \varepsilon$ and $A + \varepsilon$. If all three choices yield the same kind of attractor (i.e., all three periodic as indicated by all three having negative Lyapunov exponents, or all three chaotic as indicated by all three having positive Lyapunov exponents), then they say that the A-value is 'certain' for this value of ε. If not (i.e., one of the exponents has a different sign from the other two), then they say it is 'uncertain'. They repeat this procedure for a large number of A-values and evaluate $f(\varepsilon)$, the fraction of these random choices that are uncertain at the given value of ε. They then vary ε over a large range, and determine $f(\varepsilon)$ at several chosen values of ε. Figure 3.14 shows results of numerical experiments on the scaling of $f(\varepsilon)$ with ε. The data on the log–log plot are well fit by a straight line of slope 0.413 ± 0.0005 indicating a power law behavior $f \approx K\varepsilon^{\bar{\alpha}}$ with an uncertainty exponent of $\bar{\alpha} = 0.413 \pm 0.0005$ corresponding to $D_x = 0.587 \pm 0.005$.

To within numerical accuracy the same value was obtained using a different procedure by Farmer (1985) for a quantity related to, but somewhat different from, the exterior dimension definition given here. Farmer conjectured that the fat fractal dimension of the set of chaotic

parameter values is a universal number independent of the function form
of the map for a wide class of one-dimensional maps.

Appendix: Hausdorff dimension

In this appendix we shall define and discuss the definition of dimension
originally given by Hausdorff (1918) and called the *Hausdorff dimension*.
This dimension definition is somewhat more involved than the box-
counting dimension definition, Eq.(3.1), but it has some notable advantages. For example, the set whose elements are the infinite sequence of
points on the real line, $1, \frac{1}{2}, \frac{1}{3}, \ldots$, has a positive box-counting dimension
(see Problem 4). It may be viewed as a deficiency of the box-counting
dimension definition that it does not yield $D_0 = 0$ for this example which
is just a discrete set of points. The Hausdorff dimension, however, yields
zero for this set (Problem 16). For typical invariant sets encountered in
practice in chaotic dynamics, the box-counting and Hausdorff dimensions
are commonly thought to be equal.

In order to define the Hausdorff dimension we first introduce the
Hausdorff measure. Let A be a set in an N-dimensional Cartesian space.
We define the *diameter* of A, denoted $|A|$, as the largest distance between

Figure 3.14 Log–log
plot of $f(\varepsilon)$ versus ε
for the quadratic map
(Grebogi *et al.*,
1985b).

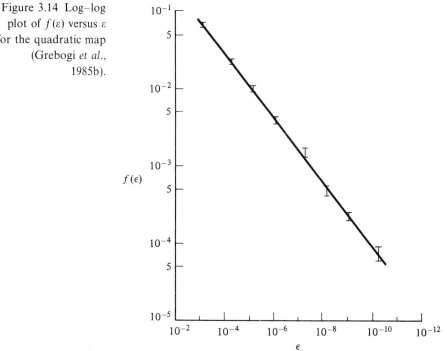

any two points **x** and **y** in A,

$$|A| = \sup_{\mathbf{x},\mathbf{y} \in A} |\mathbf{x} - \mathbf{y}|.$$

Let S_i denote a countable collection of subsets of the Cartesian space such that the diameters ε_i of the S_i are all less than or equal to δ,

$$0 < \varepsilon_i \leq \delta,$$

and such that the S_i are a covering of A, $A \subset \bigcup_i S_i$. Then we define the quality $\Gamma_H^d(\delta)$,

$$\Gamma_H^d(\delta) = \inf_{S_i} \sum_{i=1} \varepsilon_i^d, \tag{3.51a}$$

That is, we look for that collection of covering sets S_i with diameters less than or equal to δ which minimizes the sum in (3.51a) and we denote that minimized sum $\Gamma_H^d(\delta)$. The d-dimensional Hausdorff measure is then defined as

$$\Gamma_H^d = \lim_{\delta \to 0} \Gamma_H^d(\delta). \tag{3.51b}$$

The Hausdorff measure generalizes the usual notions of the total length, area and volume of simple sets. For example, if the set A is a smooth surface of finite area situated in a three-dimensional Cartesian space, then Γ_H^2 is just the area of the set, while Γ_H^d for $d < 2$ is $+\infty$, and Γ_H^d for $d > 2$ is zero.

In general, it can be shown that Γ_H^d is $+\infty$ if d is less than some critical value and is zero if d is greater than that critical value. We denote that critical value D_H and call it the Hausdorff dimension of the set; see Figure 3.15. The value of Γ_H^d at $d = D_H$ can be zero, $+\infty$ or a finite positive number. For instance, in our example above of the smooth surface, $D_H = 2$, and $\Gamma_H^{D_H}$ is a finite positive number, the area of the surface.

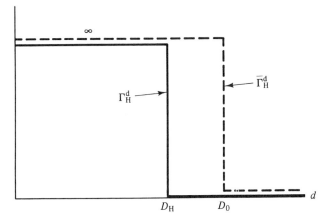

Figure 3.15 The d-dimensional Hausdorff measure Γ_H^d as a function of d (solid line), and the upper bound $\bar{\Gamma}_H^d$ (dashed line) obtained using a covering of cubes from an ε-grid.

Now let us consider the relationship of the box-counting dimension to the Hausdorff dimension. We cover the set A with a particular covering consisting of cubes from a grid of unit size ε in the N-dimensional space. Denoting the cubes, $\bar{S}_i$ we have $|\bar{S}_i| = \varepsilon_i = \varepsilon\sqrt{N}$. Using these cubes in the sum in (3.51a) (and consequently not carrying out the minimization prescribed by the inf in (3.51a)), we have

$$\sum_i \varepsilon_i^d = \tilde{N}(\varepsilon)\varepsilon^d N^{d/2} \equiv \bar{\Gamma}_H^d(\delta)$$

with $\delta = \varepsilon N^{1/2}$. From (3.1) we assume that $\tilde{N}(\varepsilon) \sim \varepsilon^{-D_0}$ for small ε. We then have

$$\bar{\Gamma}_H^d(\delta) \sim \varepsilon^{d-D_0}. \tag{3.52}$$

Thus, $\bar{\Gamma}_H^d \equiv \lim_{\delta \to 0} \varepsilon^{d-D_0}$ is $+\infty$ if $d < D_0$ and is zero if $d < D_0$. Since, in calculating $\bar{\Gamma}_H^d$, we do not carry out the minimization over all possible coverings,

$$\bar{\Gamma}_H^d(\delta) \geq \Gamma_H^d(\delta).$$

Thus, $\bar{\Gamma}_H^d$ must be as shown schematically by the dashed line in Figure 3.15. That is, D_0 is an upper bound on D_H,

$$D_0 \geq D_H. \tag{3.53}$$

As an example, we now calculate the Hausdorff dimension of the chaotic attractor of the generalized baker's map, Figure 3.4. From the definition of the Hausdorff dimension it can be shown that (3.9) also holds for D_H. Thus, we need only calculate the Hausdorff dimension $\hat{D}_H$ of the intersection of the attractor with the x-axis. We denote this intersection $\hat{A}$ and divide it into two disjoint pieces

$$\hat{A} = \hat{A}_a \cup \hat{A}_b,$$

where $\hat{A}_a$ is in the interval $[0, \lambda_a]$, and $\hat{A}_b$ is in the interval $[(1 - \lambda_b), 1]$. Thus, $\Gamma_H^d(\delta)$ for the set $\hat{A}$ can be written

$$\Gamma_H^d(\delta) = \Gamma_{Ha}^d(\delta) + \Gamma_{Hb}^d(\delta), \tag{3.54}$$

where $\Gamma_{Ha}^d(\delta)$ and $\Gamma_{Hb}^d(\delta)$ denote terms in the sum (3.51a) used to cover $\hat{A}_a$ and $\hat{A}_b$ respectively. Noting the similarity property of the attractor we have

$$\Gamma_{Ha}^d(\delta) = \lambda_a^d \Gamma_H^d(\delta/\lambda_a), \tag{3.55}$$

with a similar result holding for $\Gamma_{Hb}^d(\delta)$. On the basis of the limiting behavior, Figure 3.15, we can assume $\Gamma_H^d(\delta)$ to have the following behavior[11] for small δ

$$\Gamma_H^d(\delta) \approx K\delta^{-1(\hat{D}_H - d)}. \tag{3.56}$$

Combining (3.54)–(3.56) we obtain

$$1 = \lambda_a^{\hat{D}_H} + \lambda_b^{\hat{D}_H} \tag{3.57}$$

Comparing (3.57) with (3.12) we see that $\hat{D}_0 = \hat{D}_H$. Hence, the Hausdorff dimension, and the box-counting dimension are identical for the case of the attractor of the generalized baker's map. Thus, Eq. (3.53) holds with the equality applying. It has been widely conjectured that this is the case for the dimensions of typical chaotic attractors. (Although sets for which $D_0 \neq D_H$ can be easily constructed (e.g., see Problems 4 and 16), sets for which $D_0 \neq D_H$ do not seem to arise among invariant sets of typical dynamical systems.)

Problems

1. What is the box-counting dimension of the Cantor set obtained by removing the middle interval of length one half (instead of one third as in Figure 3.2) of the intervals on the previous stage of the construction.

2. Derive Eq. (3.6).

3. What are the box-counting dimensions of the sets, the first few stages of whose constructions are illustrated
   (a) in Figure 3.16,
   (b) in Figure 3.17,
   (c) in Figure 3.18.

4. Consider the set whose elements are the infinite sequence of points $1, \frac{1}{2}, \frac{1}{3}, \frac{1}{4}, \ldots$. What is the box-counting dimension of this set?[12]

5. What is the box-counting dimension of the invariant set in $[0, 1]$ for the one-dimensional map given by
$$x_{n+1} = \begin{cases} 4x_n & \text{for } -\infty < x_n \leq \frac{1}{2}, \\ 2(x_n - \frac{1}{2}) & \text{for } \frac{1}{2} < x_n < +\infty. \end{cases}$$

6. The numbers in the interval $[0, 1]$ can be represented as a ternary decimal
$$x = \sum_{i=1}^{\infty} 3^{-i} a_i = 0.a_1 a_2 a_3 \ldots$$
where $a_i = 0$, 1 or 2. Show that the middle third Cantor set is the subset of numbers in $[0, 1]$ such that $a_i = 1$ never appears in their ternary decimal representation (i.e., only zeros and twos appear).

7. Consider the fractal of Problem 3(a) whose construction is illustrated in Figure 3.16. We put a measure on this fractal as follows. Let $\alpha, \beta, \gamma, \delta$ be positive numbers such that $\alpha + \beta + \gamma + \delta = 1$. At the first stage of construction the box at the upper right has α of the measure, the box at the lower right has β of the measure, the box at the lower left has γ of the measure, and the box at the upper left has δ of the measure. At the next stage of construction each of the four boxes at the first stage splits into four smaller boxes. The sum of the measures in the four smaller boxes is equal to the measure of the larger box containing them at the previous stage. We apportion this measure as before. That is, the fraction of the

measure of the box on the previous stage that is assigned to the upper right smaller box which it contains is α, to the lower right smaller box β, to the lower left smaller box γ, and to the upper left smaller box δ. The same prescription is followed on subsequent stages. What is D_q for this measure?

8. Consider the fractal Problem 3(c) whose construction is illustrated in Figure 3.18. We put a measure on this fractal in a manner similar to that described in Problem 7. Let α and β be positive numbers satisfying $\alpha + 4\beta = 1$. At the first stage of construction we assign a measure α to the middle box of edge length $\frac{1}{2}$ and we assign a measure β to each of the four boxes of edge length $\frac{1}{4}$. On subsequent stages of the construction we apportion the measure of the box on the previous stage amongst the five smaller boxes it contains in a similar way (cf. Problem 7). What is D_q for this measure?

9. Write a computer program to take iterates of the generalized baker's map. Choose $\lambda_a = \lambda_b = \frac{1}{3}$, $\alpha = 0.4$, initial condition $(x_0, y_0) = (1/\sqrt{2}, 1/\sqrt{2})$, iterate the map 20 times, and then plot the next 1000 iterates to get a picture of the attractor.

10. Derive Eq. (3.24) from Eq. (3.23).

Figure 3.16 Construction of the fractal for Problem 3(a).

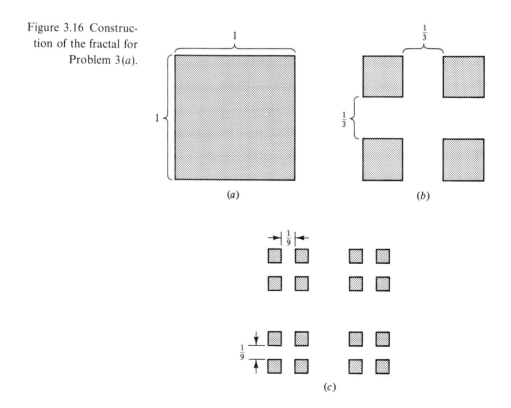

(a)

(b)

(c)

11. Show that the number of strips of width $\lambda_a^m \lambda_b^{n-m}$ for the generalized baker's map is given by the binomial coefficient, Eq. (3.8).

12. Derive Eq. (3.16). (Hint: Show that $d/dq[(1-q)^{-1}\ln \Sigma_i \mu_i^q] \leq 0$. In doing this the following general result may be of use: If $d^2F(x)/dx^2 \geq 0$, then for any set of numbers $p_i \geq 0$ which satisfies $\Sigma_i p_i = 1$ and any other set of numbers x_i, we have $\langle F(x) \rangle \geq F(\langle x \rangle)$ where $\langle F(x) \rangle \equiv \Sigma_i p_i F(x_i)$ and $\langle x \rangle \equiv \Sigma_i p_i x_i$.)

13. Derive Eqs (3.29)–(3.31).

14. Consider an attractor lying in an N-dimensional Cartesian space. Let $C_\varepsilon(\mathbf{x})$ denote an N-dimensional cube of edge length 2ε centered at the point $\mathbf{x}$. Show that

$$\lim_{\varepsilon \to 0} \frac{\ln \mu(B_\varepsilon(\mathbf{x}))}{\ln \varepsilon} = \lim_{\varepsilon \to 0} \frac{\ln \mu(C_\varepsilon(\mathbf{x}))}{\ln \varepsilon}$$

(Hint: Consider the ball of radius ε, $B_\varepsilon(\mathbf{x})$, contained in $C_\varepsilon(\mathbf{x})$ and the ball of radius $(N\varepsilon)^{1/2}$ which contains $C_\varepsilon(\mathbf{x})$.) Thus, $D_p(\mathbf{x})$ can be defined using a cube $C_\varepsilon(\mathbf{x})$ rather than a ball $B_\varepsilon(\mathbf{x})$.

15. Using the result of Problem 14, calculate D_p for the generalized baker's map at the point $(x, y) = (0, \frac{1}{2})$. Is the pointwise dimension at this point the same as the information dimension, and what does this imply about

Figure 3.17 First stages of the construction of the fractal of Problem 3(*b*). This fractal is called a 'Koch curve'.

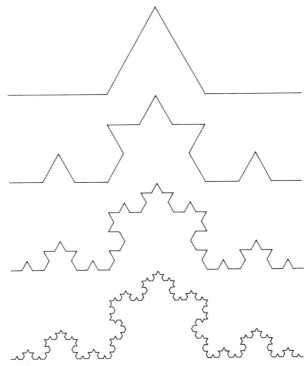

the point $(0, \frac{1}{2})$? (Hint: In taking the limit as ε goes to zero use a subsequence, $\varepsilon_k = \lambda_a^k$, and let the integer k go to infinity.)

16. Show that the Hausdorff dimension of the set in Problem 4 is zero.

17. Find the Hausdorff dimension for the set in Problem 5.

Notes

1. The box-counting dimension may be thought of as a simplified version of the Hausdorff dimension (Hausdorff, 1918), a notion which we will discuss in the appendix to this chapter.

Figure 3.18 First two stages of the construction for the fractal of Problem 3(c).

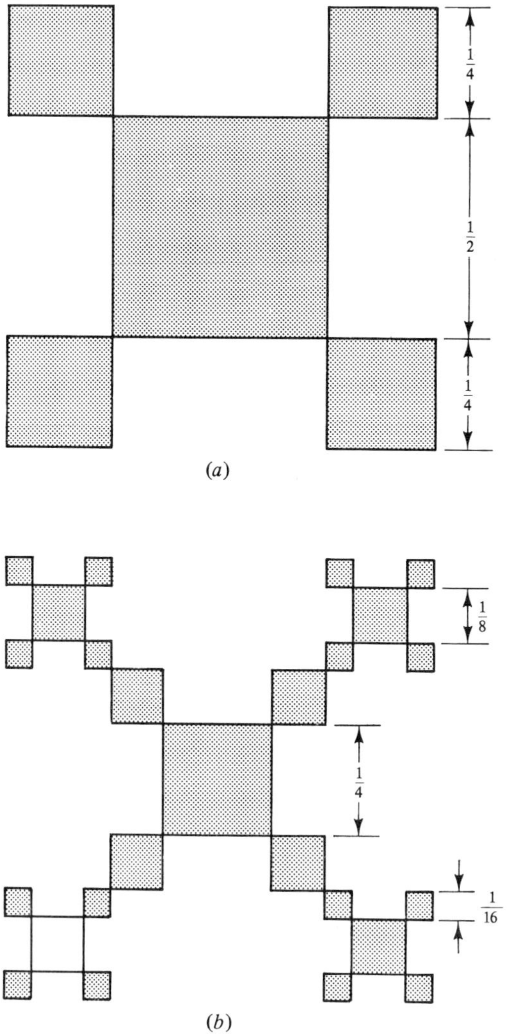

(a)

(b)

2. Henceforth, whenever we write a limit, as in Eq. (3.1), it is to be understood that we are making the assumption that the limit exists.

3. The blow-ups of the Hénon attractor in Figures 1.12(*b*) and (*c*) show apparent self-similarity. We emphasized, however, that this is only because these blow-ups are made about a fixed point of the map that lies on the attractor. Choosing a more typical point on the attractor about which to perform magnification, successive magnification would always reveal a structure looking like many parallel lines as in Figure 1.12, but the picture would not repeat on successive magnifications as in Figures 1.12(*b*) and (*c*).

4. This map was introduced by Farmer *et al.* (1983) as a model for the study of the dimension of strange attractors. The baker's map (as opposed to the generalized baker's map) is an area preserving map of the unit square corresponding to $\alpha = \beta = \lambda_a = \lambda_b = \frac{1}{2}$ in Eqs. (3.7).

5. Sinai's example, Figure 3.6, corresponds to $D_0 = 2$ and $D_1 < 2$. This occurs for the generalized baker's map when $\lambda_a = \lambda_b = \frac{1}{2}$ and $\alpha \neq \frac{1}{2}$ (if $\lambda_a = \lambda_b = \alpha = \frac{1}{2}$, then $D_1 = 2$).

6. See Farmer *et al.* (1983).

7. In calculating $C(\varepsilon)$ for finite k one should restrict the sum in (3.42) by requiring that $|i - j|$ exceed some minimum value dependent on the data set. This is necessary to eliminate dynamical correlations, thus only leaving the geometric correlations that D_2 attempts to characterize.

8. While the dimension obtained for a white noise process is the dimension of the embedding space, this need not be true for other noise processes. In particular, Osborne and Provenzale (1989) have emphasized that 'colored noise' (i.e., noise with a power law frequency power spectrum) can yield fractal correlation dimension spectra under some circumstances.

9. In infinite-dimensional systems or systems with very large dimension it is often the case that one can show that there exists a low-dimensional manifold (the so-called *inertial manifold*) to which the orbit tends and on which the attractor (or attractors) lies. In this case we can regard **y** as specifying points on the inertial manifold and the equation $d\mathbf{x}/dt = \mathbf{F}(\mathbf{x})$ as giving the dynamics on the inertial manifold. For material on inertial manifolds see, for example, Constantin *et al.* (1989) and references therein.

10. For other work on fat fractals see also Umberger and Farmer (1985), Hanson (1987) and Eykholt and Umberger (1988).

11. The use of the scaling ansatz (3.56) is a 'quick and dirty' way of getting the result for D_H. A similar comment applies for our use of (3.22) to obtain D_q. See Farmer *et al.* (1983) for a more rigorous treatment.

12. It may be viewed as a deficiency of the box-counting dimension definition that it does not yield $D_0 = 0$ for this example which is just a discrete set of points. The Hausdorff dimension, defined in the appendix, yields zero for this set. For typical fractal sets encountered in chaotic dynamics, the box-counting and Hausdorff dimensions are commonly thought to be equal.

CHAPTER FOUR

Dynamical properties of chaotic systems

In Chapter 3 we have concentrated on geometric aspects of chaos. In particular, we have discussed the fractal dimension characterization of strange attractors and their natural invariant measures, as well as issues concerning phase space dimensionality and embedding. In this chapter we concentrate on the time evolution dynamics of chaotic orbits. We begin with a discussion of the horseshoe map and symbolic dynamics.

4.1 The horseshoe map and symbolic dynamics

The horseshoe map was introduced by Smale (1967) as a motivating example in his development of *symbolic dynamics* as a basis for understanding a large class of dynamical systems. The horseshoe map M_h is specified geometrically in Figure 4.1. The map takes the square S (Figure 4.1(a)), uniformly stretches it vertically by a factor greater than 2 and uniformly compresses it horizontally by a factor less than $\frac{1}{2}$ (Figure 4.1(b)). Then the long thin strip is bent into a horseshoe shape with all the bending deformations taking place in the cross-hatched regions of Figures 4.1(b) and (c). Then the horseshoe is placed on top of the original square, as shown in Figure 4.1(d). Note that a certain fraction, which we denote $1 - f$, of the original area of the square S is mapped to the region outside the square. If initial conditions are spread over the square with a distribution which is uniform in the vertical direction, then the fraction of initial conditions that generate orbits that do not leave S during n applications of the map is just f^n. This is because a vertically uniform distribution in S remains vertically uniform on application of M_h.

Since $f^n \to 0$ as $n \to \infty$, almost every initial condition with respect to Lebesgue measure eventually leaves the square. (Thus there is no attractor contained in the square.[1]) We are interested in characterizing the invariant set Λ (which is of Lebesgue measure zero) of points which never leave the square. Furthermore, we wish to investigate the orbits followed by points in Λ. In order to do this, we first note that the intersection of the horseshoe with the square represents the regions that points in the square map to if they return to the square on one iterate. These regions are the

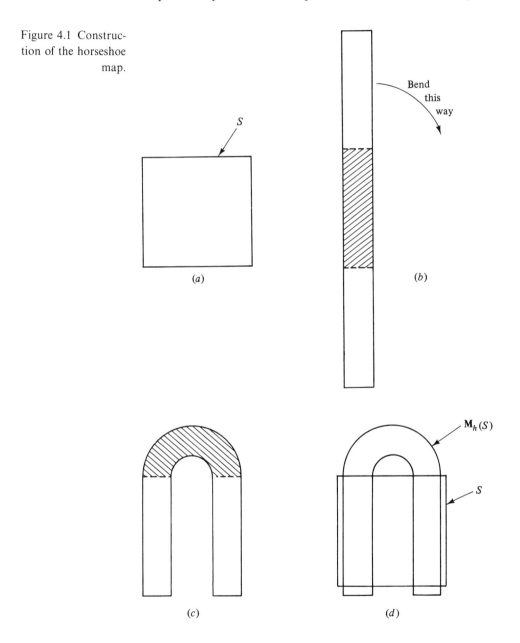

Figure 4.1 Construction of the horseshoe map.

two cross-hatched vertical strips labeled V_0 and V_1 in Figure 4.2(a). We now ask, where did these strips come from? To answer this question we follow the horseshoe construction in Figures 4.1(a)–(d) backward in time (i.e., from (d) to (c) to (b) to (a) in Figure 4.1). Thus, we find that the two vertical strips V_0 and V_1 are the images of two horizontal strips $H_0 = \mathbf{M}_h^{-1}(V_0)$ and $H_1 = \mathbf{M}_h^{-1}(V_1)$, as shown in Figure 4.2(b). Figure 4.2(c) shows what happens if we apply the horseshoe map to the vertical strips V_0 and V_1. Thus, taking the intersection of $\mathbf{M}_h(V_0)$ and $\mathbf{M}_h(V_1)$ with S (Figure 4.2(d)), we see that points originating in the square which remain in the square for two iterates of $\mathbf{M}_h$ are mapped to the four vertical strips labeled $V_{00}, V_{01}, V_{10}, V_{11}$ in Figure 4.2(d). The subscripts on these strips V_{ij} are such that V_{ij} is contained in V_j and $\mathbf{M}_h^{-1}(V_{ij})$ is contained in V_i. Figure 4.2(e) shows the four horizontal strips that the vertical strips V_{ij} came from two iterates previous, $H_{ij} = \mathbf{M}_h^{-2}(V_{ij})$.

Now consider the invariant set Λ and the horizontal and vertical strips, H_0, H_1, V_0, V_1. Since points in Λ never leave S, the forward iterate of Λ must be in the square. Hence Λ is contained in $H_0 \cup H_1$ and is also

Figue 4.2 Vertical and horizontal strips V_i, V_{ij}, H_i and H_{ij} for the horseshoe map.

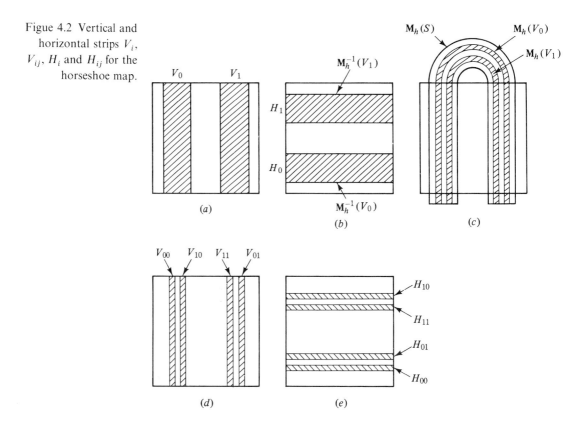

contained in $V_0 \cup V_1$. Thus, Λ is contained in the intersection,

$$(H_0 \cup H_1) \cap (V_0 \cup V_1).$$

This intersection consists of four squares as shown in Figure 4.3(a). Similarly Λ must also lie in the intersection

$$(H_{00} \cup H_{01} \cup H_{11} \cup H_{10}) \cap (V_{00} \cup V_{01} \cup V_{11} \cup V_{10})$$

shown in Figure 4.3(b). This intersection consists of 16 squares, four of which are contained in each of the four squares of Figure 4.3(a). Proceeding in stages of this type, at each successive stage, each square is replaced by four smaller squares that it contains. Taking the limit of repeating this construction an infinite number of times we obtain the invariant set Λ. This set is the intersection of a Cantor set of vertical lines (the Vs in the limit of an infinite number of iterations) with a Cantor set of horizontal lines (the Hs in the limit of an infinite number of iterations).

Figure 4.3(a)
$(H_0 \cup H_1) \cap (V_0 \cup V_1)$
and (b) $(H_{00} \cup H_{01} \cup H_{11} \cup H_{10}) \cap (V_{00} \cup V_{01} \cup V_{10} \cup V_{11})$.

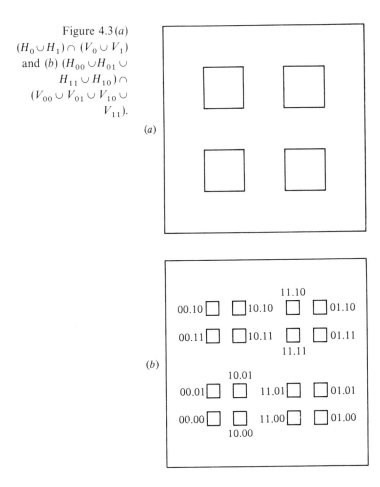

Let $\mathbf{x}$ be a point in the invariant set Λ. Then we claim that we can specify it by a bi-infinite symbol sequence $\mathbf{a}$,

$$\mathbf{a} = \ldots a_{-3}a_{-2}a_{-1} \cdot a_0 a_1 a_2 \ldots \tag{4.1}$$

and each symbol a_i is a function of $\mathbf{x}$ specified by

$$a_i = \begin{cases} 0 & \text{if } \mathbf{M}_h^i(\mathbf{x}) \text{ is in } H_0, \\ 1 & \text{if } \mathbf{M}_h^i(\mathbf{x}) \text{ is in } H_1. \end{cases} \tag{4.2}$$

The above represents a correspondence between bi-infinite symbol sequences $\mathbf{a}$ and points $\mathbf{x}$ in Λ. We denote this correspondence

$$\mathbf{a} = \phi(\mathbf{x}). \tag{4.3}$$

In Figure 4.3(b) we label the 16 rectangles by the symbols $a_{-2}a_{-1} \cdot a_0 a_1$ that correspond to the four middle symbols in (4.1) that all points in Λ that fall in that rectangle must have. The correspondence given by Eqs. (4.1)–(4.3) may be shown to be one to one and continuous (with a suitable definition of a metric on the space of bi-infinite symbol sequences). Define the *shift* operation,

$$\mathbf{a}' = \sigma(\mathbf{a}),$$

where $a_i' = a_{i+1}$. That is $\mathbf{a}'$ is obtained from $\mathbf{a}$ by moving the decimal point in Eq. (4.1) one place to the right. From Eq. (4.2) we have

$$a_i' = \begin{cases} 0 & \text{if } \mathbf{M}_h^{i+1}(\mathbf{x}) = \mathbf{M}_h^i(\mathbf{M}_h(\mathbf{x})) \text{ is in } H_0, \\ 1 & \text{if } \mathbf{M}_h^{i+1}(\mathbf{x}) = \mathbf{M}_h^i(\mathbf{M}_h(\mathbf{x})) \text{ is in } H_1. \end{cases}$$

Hence, $\mathbf{a}'$ is the symbol sequence corresponding to $\mathbf{M}_h(\mathbf{x})$, or $\sigma(\mathbf{a}) = \phi(\mathbf{M}_h(\mathbf{x}))$. We represent the situation schematically in Figure 4.4. Thus, the shift on the bi-infinite symbol space is equivalent to the horseshoe map applied to the invariant set Λ,

$$\mathbf{M}_{h|\Lambda} = \phi^{-1} \cdot \sigma \cdot \phi, \tag{4.4}$$

where $\mathbf{M}_{h|\Lambda}$ symbolizes the restriction of $\mathbf{M}_h$ to the invariant set Λ. Thus, to obtain $\mathbf{x}_{n+1}$ from $\mathbf{x}_n$ we can either apply $\mathbf{M}_h$ to $\mathbf{x}_n$, or else we can obtain $\mathbf{a}_n = \phi(\mathbf{x}_n)$, shift the decimal point to the right to get $\mathbf{a}_{n+1}$ and then obtain $\mathbf{x}_{n+1}$ from $\mathbf{x}_{n+1} = \phi^{-1}(\mathbf{a}_{n+1})$. Furthermore, to obtain $\mathbf{x}_{n+m}$ from $\mathbf{x}_n$ we

Figure 4.4 Equivalence of the shift operation σ and the horseshoe map.

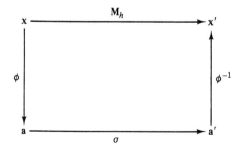

can first get $\mathbf{a}_n = \phi(\mathbf{x}_n)$, then shift the decimal point m places to the right to get $\mathbf{a}_{n+m}$, and then obtain $\mathbf{x}_{n+m}$ from $\mathbf{x}_{n+m} = \phi^{-1}(\mathbf{a}_{n+m})$. For example, fixed points of σ^n are mapped by ϕ^{-1} to fixed points of $\mathbf{M}_h^n$. Since the former are just sequences that repeat after n shifts and since there are 2^n ways of choosing a sequence of n zeros and ones, we see that there are 2^n fixed points of $\mathbf{M}_h^n$ in Λ. This can be shown to imply that the set of points on periodic orbits is dense in the invariant set Λ. In addition, there is an uncountable set of nonperiodic orbits in Λ, and it may be shown that there are orbits which are dense in Λ. (If there is an orbit that is dense in an invariant set then we say that the set is *transitive*.) In Sections 2.1 and 3.1 we have established symbolic dynamics representations for one-dimensional noninvertible maps (Eqs. (2.3) and (3.3)) in which the system state was represented as an infinite sequence $(\cdot a_0 a_1 a_2 \ldots)$, as opposed to the bi-infinite sequence representation, Eq. (4.1). This difference comes about as a result of the noninvertibility of the maps of Sections 2.1 and 3.1. (This is reflected by the fact that the shift operating on $\cdot a_0 a_1 a_2 \ldots$ produces $\cdot a_1 a_2 a_3 \ldots$, and thus there is no information in the new symbol sequence of what a_0 is. Hence, we cannot go backward in time.) The correspondence we have established above between the horseshoe map and the shift of a bi-infinite symbol sequence of zeros and ones is an example of a reduction to symbolic dynamics that can be established for a large class of smooth invertible dynamical systems (in particular, systems having the property of hyperbolicity which we define later in this chapter).

As another example of symbolic dynamics, consider the map $\tilde{\mathbf{M}}$ geometrically shown in Figure 4.5(a). The map is similar to the horseshoe map, but contains an additional intersection with the square S, namely the region $\tilde{\mathbf{M}}(H_2)$. The three regions H_0, H_1 and H_2 shown in Figure 4.5(b) represent initial conditions that return to the square. We can again represent points $\mathbf{x}$ in the invariant set $\tilde{\Lambda}$ of the map $\tilde{\mathbf{M}}$ as points in a bi-infinite symbol sequence, Eq. (4.1), but now we need three symbols;

$$a_i = \begin{cases} 0 & \text{if } \tilde{\mathbf{M}}^i(\mathbf{x}) \text{ is in } H_0, \\ 1 & \text{if } \tilde{\mathbf{M}}^i(\mathbf{x}) \text{ is in } H_1, \\ 2 & \text{if } \tilde{\mathbf{M}}^i(\mathbf{x}) \text{ is in } H_2. \end{cases}$$

Again the operation of the map $\tilde{\mathbf{M}}$ corresponds to a shift σ operating on $\mathbf{a}$. There is, however, an important difference with the horseshoe. This is that all possible sequences of zeros, ones and twos are not allowed. In particular, we see from Figures 4.5(a) and (b) that points of $\tilde{\Lambda}$ that are in H_2 are always mapped by $\tilde{\mathbf{M}}$ to H_0 and not to H_1 or H_2 ($\tilde{\mathbf{M}}(H_2)$ intersects H_0 but does not intersect either H_1 or H_2). Thus, the possible allowable transitions are as shown in Figure 4.5(c). This means that whenever a 2 appears in our symbol sequence it is immediately followed by a zero (i.e.,

$a_i = 2$ implies $a_{i+1} = 0$). We call the symbolic dynamics corresponding to $\tilde{\mathbf{M}}$ a *shift of finite type* on three symbols (the phrase shift of finite type signifies a restriction on the allowed sequences), while we call the symbolic dynamics corresponding to the horseshoe map a *full shift* on two symbols (the word full signifying that there is no restriction on the allowed sequences).

As an application of symbolic dynamics, we mention the work of Levi (1981), who has analyzed a model periodically forced van der Pol equation (i.e., Eq. (1.13) with a periodic function of time on the right hand side). (Levi modifies the equation to facilitate his analysis.) Levi shows that the map obtained from the stroboscopic surface of section obtained by sampling at the forcing period (cf. Chapter 1) possesses an invariant set on which the dynamics is described by a shift of finite type on four symbols. Figure 4.5(d) shows the allowed transitions for Levi's problem.

Figure 4.5(a) The map $\tilde{M}$. (b) $\tilde{M}(S) \cap S$ corresponds to the three strips H_0, H_1, H_2. (c) Allowed transitions for $\tilde{M}$; if x is in H_2 it is always in H_0 after one iterate. (d) Allowed transitions for Levi's problem.

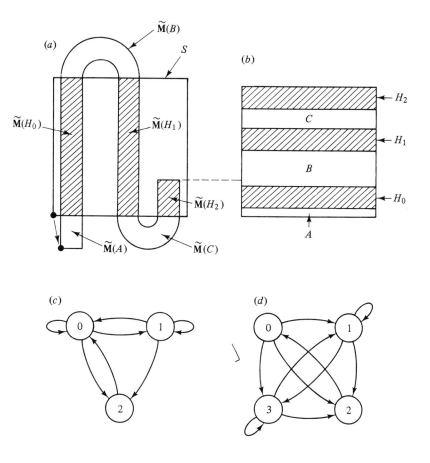

4.2 Linear stability of steady states and periodic orbits

Consider a system of real first-order differential equations $d\mathbf{x}/dt = \mathbf{F}(\mathbf{x})$. A steady state for this system is a point $\mathbf{x} = \mathbf{x}_*$ at which

$$\mathbf{F}(\mathbf{x}_*) = 0.$$

We wish to examine the behaviour of orbits near $\mathbf{x}_*$. Thus we set

$$\mathbf{x}(t) = \mathbf{x}_* + \boldsymbol{\eta}(t),$$

where we assume $\boldsymbol{\eta}(t)$ is small. Substituting this into $d\mathbf{x}/dt = \mathbf{F}(\mathbf{x})$, we expand $\mathbf{F}(\mathbf{x})$ to first order $\boldsymbol{\eta}(t)$,

$$\mathbf{F}(\mathbf{x}_* + \boldsymbol{\eta}) = \mathbf{F}(\mathbf{x}_*) + \mathbf{DF}(\mathbf{x}_*) \cdot \boldsymbol{\eta} + O(\boldsymbol{\eta}^2),$$

where, since $\mathbf{x}_*$ is a steady state, $\mathbf{F}(\mathbf{x}_*) = 0$, and $\mathbf{DF}$ denotes the Jacobian matrix of partial derivatives of $\mathbf{F}$. That is, if we write

$$\mathbf{x} = \begin{bmatrix} x^{(1)} \\ x^{(2)} \\ \vdots \\ x^{(N)} \end{bmatrix}, \quad \mathbf{F}(\mathbf{x}) = \begin{bmatrix} F^{(1)}(x^{(1)}, x^{(2)}, \dots, x^{(N)}) \\ F^{(2)}(x^{(1)}, x^{(2)}, \dots, x^{(N)}) \\ \vdots \\ F^N(x^{(1)}, x^{(2)}, \dots, x^{(N)}) \end{bmatrix},$$

then

$$\mathbf{DF}(\mathbf{x}) = \begin{bmatrix} \partial F^{(1)}/\partial x^{(1)} & \partial F^{(1)}/\partial x^{(2)} & \dots & \partial F^{(1)}/\partial x^{(N)} \\ \partial F^{(2)}/\partial x^{(1)} & \partial F^{(2)}/\partial x^{(2)} & \dots & \partial F^{(2)}/\partial x^{(N)} \\ \vdots & \vdots & & \vdots \\ \partial F^{(N)}/\partial x^{(1)} & \partial F^{(N)}/\partial x^{(2)} & \dots & \partial F^{(N)}/\partial x^{(N)} \end{bmatrix}$$

We obtain the following equation for the time dependence of the perturbation of $\mathbf{x}$ from the steady state

$$d\boldsymbol{\eta}/dt = \mathbf{DF}(\mathbf{x}_*) \cdot \boldsymbol{\eta} + O(\boldsymbol{\eta}^2). \tag{4.5}$$

The linearized stability problem is obtained by neglecting terms of order $\boldsymbol{\eta}^2$ in (4.5) and is of the general form

$$d\mathbf{y}/dt = \mathbf{A} \cdot \mathbf{y}, \tag{4.6}$$

where $\mathbf{y}$ is a real N-dimensional vector and $\mathbf{A}$ is a real time-independent $N \times N$ matrix. If we seek solutions of Eq. (4.6) of the form $\mathbf{y}(t) = \mathbf{e}\exp(st)$, then (4.6) becomes the eigenvalue equation

$$\mathbf{A} \cdot \mathbf{e} = s\mathbf{e}, \tag{4.7}$$

which has nontrivial solutions for values of s satisfying the Nth order polynomial equation

$$D(s) = \det[\mathbf{A} - s\mathbf{I}] = 0, \tag{4.8}$$

where $\mathbf{I}$ denotes the $N \times N$ identity matrix. For our purposes it suffices to consider only the case where $D(s) = 0$ has N *distinct* roots $s = s_k$ for

$k = 1, 2, \ldots, N$, (i.e., $s_k \neq s_j$ if $k \neq j$). For each such root there is an eigenvector $\mathbf{e}_k$, and any time evolution can be represented as

$$\mathbf{y}(t) = \sum_{k=1}^{N} A_k \mathbf{e}_k \exp(s_k t), \qquad (4.9)$$

where the A_k are constant coefficients (that may be complex) determined from the initial condition $\mathbf{y}(0) = \sum_{i=1}^{N} A_k \mathbf{e}_k$. Since the coefficients of the polynomial $D(s)$ are real, the eigenvalues s_k are either real or else occur in complex conjugate pairs.

In the case of complex conjugate pairs of eigenvalues $s_j = s_{j+1}^* = \sigma_j - i\omega_j$, we can also take $\mathbf{e}_j = \mathbf{e}_{j+1}^* = \mathbf{e}_j^R + i\mathbf{e}_j^I$, where the $*$ denotes complex conjugate, and σ_j, ω_j, $\mathbf{e}_j^R$ and $\mathbf{e}_j^I$ are all real. Combining the two solutions, j and $j + 1$, we obtain two linearly independent *real* solutions,

$$\mathbf{g}_j(t) = \tfrac{1}{2}[\mathbf{e}_j \exp(s_j t) + \mathbf{e}_{j+1} \exp(s_{j+1} t)]$$
$$= \mathbf{e}_j^R \exp(\sigma_j t) \cos(\omega_j t) + \mathbf{e}_j^I \exp(\sigma_j t) \sin(\omega_j t), \qquad (4.10a)$$

$$\mathbf{g}_{j+1}(t) = \frac{1}{2i}[\mathbf{e}_j \exp(s_j t) - \mathbf{e}_{j+1} \exp(s_{j+1} t)]$$
$$= \mathbf{e}_j^I \exp(\sigma_j t) \sin(\omega_j t) - \mathbf{e}_j^R \exp(\sigma_j t) \cos(\omega_j t). \qquad (4.10b)$$

If s_j is real ($s_j = \sigma_j$), then we write $\mathbf{g}_j(t) = \mathbf{e}_j \exp(\sigma_j t)$ (where $\mathbf{e}_j$ is real). Equation (4.9) thus becomes

$$\mathbf{y}(t) = \sum_{j=1}^{N} B_j \mathbf{g}_j(t), \qquad (4.11)$$

where (in contrast with the coefficients A_k of Eq. (4.9)) all the B_j are *real*. By use of a similarity transformation

$$\mathbf{z}(t) = \mathbf{T} \cdot \mathbf{y}(t), \qquad (4.12)$$

where $\mathbf{T}$ is a real $N \times N$ matrix, we can recast Eq. (4.6) as

$$d\mathbf{z}/dt = \mathbf{C} \cdot \mathbf{z}, \quad \mathbf{C} = \mathbf{T} \cdot \mathbf{A} \cdot \mathbf{T}^{-1}. \qquad (4.13)$$

where, if there are K real eigenvalues, $\sigma_1, \sigma_2, \ldots, \sigma_k$, and $N - K$ complex conjugate eigenvalues, then the real $N \times N$ matrix $\mathbf{C}$ has the following *canonical form*,

$$\mathbf{C} = \begin{bmatrix} \Sigma & \mathbf{O} \\ \mathbf{O} & \Lambda \end{bmatrix}, \qquad (4.14)$$

where Σ is a $K \times K$ diagonal matrix

$$\Sigma = \begin{bmatrix} \sigma_1 & 0 & 0 & \ldots & 0 \\ 0 & \sigma_2 & 0 & \ldots & 0 \\ 0 & 0 & \sigma_3 & \ldots & 0 \\ \vdots & \vdots & \vdots & & \vdots \\ 0 & 0 & 0 & \ldots & \sigma_K \end{bmatrix}, \qquad (4.15)$$

and Λ is a real matrix of 2×2 blocks along its diagonal

$$\Lambda = \begin{bmatrix} \Lambda_1 & \mathbf{O} & \mathbf{O} & \cdots \\ \mathbf{O} & \Lambda_2 & \mathbf{O} & \cdots \\ \mathbf{O} & \mathbf{O} & \Lambda_3 & \cdots \\ \vdots & \vdots & \vdots & \end{bmatrix}$$ (4.16)

with the 2×2 matrix block Λ_m having the form

$$\Lambda_m = \begin{bmatrix} \sigma_m & \omega_m \\ -\omega_m & \sigma_m \end{bmatrix},$$ (4.17)

and $\mathbf{O}$ being 2×2 matrices of zeros.

For $\mathrm{Re}(s_j) = \sigma_j < 0$, the corresponding solution $\mathbf{g}_j(t)$ approaches the origin asymptotically in time, either spiraling in the plane spanned by $\mathbf{e}_j^R$ and $\mathbf{e}_j^I$ if $\mathrm{Im}(s_j) \neq 0$, or else by moving along the line through the origin in the direction of the real eigenvector $\mathbf{e}_j$ if $\mathrm{Im}(s_j) = 0$. For $\mathrm{Re}(s_j) = \sigma_j > 0$, the corresponding solution $\mathbf{g}_j(t)$ diverges from the origin exponentially in time, either (as for $\sigma_j < 0$) by spiraling or by moving linearly. We call solutions which move away from the origin exponentially with time *unstable* and those that move exponentially toward the origin *stable*. The situation is as illustrated in Figure 4.6.

Note that for the case $\omega_j \neq 0$ any initial condition purely in the subspace spanned by the vectors $\mathbf{e}_j^R$ and $\mathbf{e}_j^I$ remains in that subspace for all time. Hence, that subspace is invariant under the flow $d\mathbf{y}/dt = \mathbf{A} \cdot \mathbf{y}$. Similarly, if the eigenvalue is real ($s_j = \sigma_j$), then an initial condition on the ray from the origin along $\mathbf{e}_j$ remains on that ray, and hence the set of

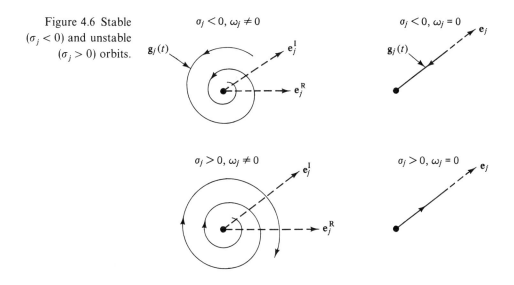

Figure 4.6 Stable ($\sigma_j < 0$) and unstable ($\sigma_j > 0$) orbits.

vectors that are scalar multiples of $\mathbf{e}_j$ is an invariant subspace. We collect all the independent vectors spanning invariant subspaces corresponding to unstable solutions ($\sigma_j > 0$) and denote them

$$\mathbf{u}_1, \mathbf{u}_2, \ldots, \mathbf{u}_{n_u}.$$

Similarly, we collect all the independent vectors spanning invariant subspaces corresponding to stable solutions ($\sigma_j < 0$) and denote them

$$\mathbf{v}_1, \mathbf{v}_2, \ldots, \mathbf{v}_{n_s}$$

If there are eigenvalues whose real parts are zero ($\sigma_j = 0$), we denote the corresponding set of independent vectors spanning this subspace

$$\mathbf{w}_1, \mathbf{w}_2, \ldots, \mathbf{w}_{n_c}$$

All of the **u**s, **v**s and **w**s, taken together, span the whole phase space. Thus

$$n_u + n_s + n_c = N.$$

We define the *unstable subspace* as

$$E^u = \mathrm{span}[\mathbf{u}_1, \mathbf{u}_2, \ldots, \mathbf{u}_{n_u}],$$

(i.e., the space spanned by the vector $\mathbf{u}_1, \mathbf{u}_2, \mathbf{u}_3, \ldots, \mathbf{u}_{n_u}$) the *stable subspace* as

$$E^s = \mathrm{span}[\mathbf{v}_1, \mathbf{v}_2, \ldots, \mathbf{v}_{n_s}],$$

and the *center subspace* as

$$E^c = \mathrm{span}[\mathbf{w}_1, \mathbf{w}_2, \ldots, \mathbf{w}_{n_c}].$$

Figure 4.7 illustrates some cases of stable and unstable subspaces and the corresponding orbits ($n_c = 0$ in Figure 4.7: (*a*) $N = 2$, $n_u = n_s = 1$; (*b*) $N = 3$, $n_u = 1$ and $n_s = 2$, where the stable space corresponds to two real eigenvalues; (*c*) $N = 3$, $n_u = 1$ and $n_s = 2$, where the stable subspace corresponds to a pair of complex conjugate eigenvalues; and (*d*) $N = 3$, $n_s = 1$ and $n_u = 2$, where the unstable subspace corresponds to a pair of complex conjugate eigenvalues).

We now turn from the study of the linear stability of a steady state $\mathbf{x} = \mathbf{x}_*$, to the study of the stability of a periodic orbit,

$$\mathbf{x}(t) = \mathbf{X}_*(t) = \mathbf{X}_*(t + T),$$

where T denotes the period. As for the case of the steady state, we write

$$\mathbf{x}(t) = \mathbf{X}_*(t) + \boldsymbol{\eta}(t)$$

and expand for small $\boldsymbol{\eta}(t)$. We obtain

$$d\boldsymbol{\eta}/dt = \mathbf{DF}(\mathbf{X}_*(t)) \cdot \boldsymbol{\eta} + O(\boldsymbol{\eta}^2), \tag{4.18}$$

which is similar to (4.5) except that now the matrix $\mathbf{DF}(\mathbf{X}_*(t))$ varies periodically in time, whereas $\mathbf{DF}(\mathbf{x}_*)$ in (4.5) is independent of time. The linearized stability problem is of the form

$$d\mathbf{y}/dt = \mathbf{A}(t) \cdot \mathbf{y}, \tag{4.19}$$

where **y** is a real N-dimensional vector and $\mathbf{A}(t)$ is a real time periodic $N \times N$ matrix,

$$\mathbf{A}(t) = \mathbf{A}(t + T).$$

Solutions of (4.19) can be sought in the Floquet form,

$$\mathbf{e}(t) \exp(st),$$

where $\mathbf{e}(t)$ is periodic in time $\mathbf{e}(t) = \mathbf{e}(t + T)$. This defines an eigenvalue problem for eigenvalues s_j and vector eigenfunctions $\mathbf{e}_j(t)$. A development parallel to that for Eq. (4.6) goes through, and stable, unstable, and center subspaces can be analogously defined, although the solution of the Floquet problem is much more difficult. One result for the system (4.18) is immediate, however. Namely, Eq. (4.18) has a solution corresponding to a zero eigenvalue ($s = 0$). To see this, differentiate the equation $d\mathbf{X}_*(t)/dt = \mathbf{F}(\mathbf{X}_*(t))$ with respect to time. This gives an equation of the form of Eq. (4.19), $d\mathbf{e}_0(t)/dt = \mathbf{DF}(\mathbf{X}_*(t)) \cdot \mathbf{e}_0(t)$, where $\mathbf{e}_0(t) \equiv d\mathbf{X}_*(t)/dt$. This zero eigenvalue solution can be interpreted as saying that, if the perturbation $\boldsymbol{\eta}(t)$ puts the perturbed orbit on the closed phase space curve

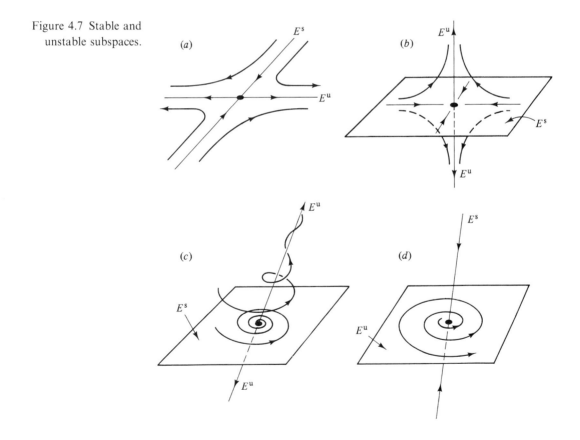

Figure 4.7 Stable and unstable subspaces.

followed by $\mathbf{X}_*(t)$ but slightly displaced from $\mathbf{X}_*(t)$, then $\boldsymbol{\eta}(t)$ varies periodically in time ($s = 0$). This is illustrated in Figure 4.8(a).

Instead of pursuing the Floquet solutions further, we employ a surface of section to reduce the problem $d\mathbf{x}/dt = \mathbf{F}(\mathbf{x})$ to a map $\hat{\mathbf{x}}_{n+1} = \mathbf{M}(\hat{\mathbf{x}}_n)$, where $\hat{\mathbf{x}}_n$ has lower dimensionality than $\mathbf{x}$ by one. As shown in Figure 4.8(b), we assume that the periodic solution $\mathbf{X}_*(t)$ results in a fixed point $\hat{\mathbf{x}}_*$ of the map. Linearizing the map around $\hat{\mathbf{x}}_*$ by writing $\hat{\mathbf{x}}_n = \hat{\mathbf{x}}_* + \hat{\boldsymbol{\eta}}_n$ with $\hat{\boldsymbol{\eta}}_n$ small, we obtain

$$\hat{\boldsymbol{\eta}}_n = \mathbf{DM}(\hat{\mathbf{x}}_*) \cdot \hat{\boldsymbol{\eta}}_n + O(\hat{\boldsymbol{\eta}}_n^2), \tag{4.20}$$

which yields a linearized problem of the form

$$\hat{\mathbf{y}}_{n+1} = \hat{\mathbf{A}} \cdot \mathbf{y}_n. \tag{4.21}$$

Seeking solutions $\hat{\mathbf{y}} = \lambda^n \hat{\mathbf{e}}$, we obtain the eigenvalue equation

$$\hat{\mathbf{A}} \cdot \hat{\mathbf{e}} = \lambda \hat{\mathbf{e}}. \tag{4.22}$$

Again we assume eigenvalue solutions λ_j of the determinantal equation,

$$\hat{D}(\lambda) = \det[\hat{\mathbf{A}} - \lambda \mathbf{I}] = 0, \tag{4.23}$$

and denote the corresponding eigenvectors $\hat{\mathbf{e}}_j$. Directions corresponding to $|\lambda_j| > 1$ are unstable; directions corresponding to $|\lambda_j| < 1$ are stable. Again, we can identify unstable, stable and center subspaces, E^u, E^s and E^c, for the map (e.g., the stable subspace is spanned by the real and imaginary parts of all those vectors $\hat{\mathbf{e}}_j$ for which $|\lambda_j| < 1$).

The map eigenvalues and the Floquet eigenvalues are related by

Figure 4.8(a) The zero eigenvalue solution of (4.18). (b) Surface of section for a periodic orbit.

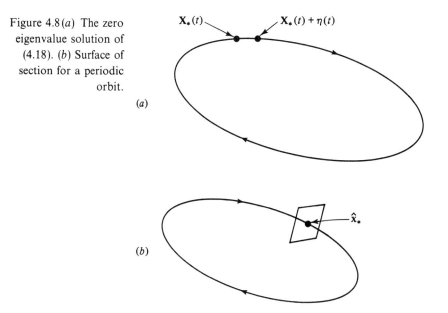

$$\lambda_j = \exp(s_j T), \tag{4.24}$$

and all the s_j of the Floquet problem are included except for the zero eigenvalue illustrated in Figure 4.8(a). The zero eigenvalue is not included because a perturbation $\boldsymbol{\eta}$ that displaces the orbit along the closed curve path followed by $\mathbf{X}_*(t)$ results in no perturbation of the orbit's surface of section piercing at $\hat{\mathbf{x}} = \hat{\mathbf{x}}_*$ (cf. Figure 4.8).

We remark that we have assumed in the above that the periodic orbit results in a fixed point of the surface of section map. If instead it results in a period p orbit (as shown in Figure 4.9 for $p = 3$), $\hat{\mathbf{x}}_0^* \to \hat{\mathbf{x}}_1^* \to \cdots \to \hat{\mathbf{x}}_p^* = \hat{\mathbf{x}}_0^*$, then we select one of the p points $\hat{\mathbf{x}}_j^*$ on the map orbit and examine the pth iterate of the map $\mathbf{M}^p$. For the map $\mathbf{M}^p$ the point $\hat{\mathbf{x}}_p^*$ is a fixed point,

$$\hat{\mathbf{x}}_j^* = \mathbf{M}^p(\hat{\mathbf{x}}_j^*). \tag{4.25}$$

Linearizing about this point, we again have a problem of the form of (4.21), but now $\mathbf{A}$ is identified with $\mathbf{DM}^p(\hat{\mathbf{x}}_j^*)$. We note that the chain rule for differentiation yields

$$\mathbf{DM}^p(\hat{\mathbf{x}}_j^*) = \mathbf{DM}(\hat{\mathbf{x}}_{j-1}^*)\mathbf{DM}(\hat{\mathbf{x}}_{j-1}^*)\ldots\mathbf{DM}(\hat{\mathbf{x}}_0^*)\mathbf{DM}(\hat{\mathbf{x}}_{p-1}^*)\ldots\mathbf{DM}(\hat{\mathbf{x}}_j^*).$$
$$\tag{4.26}$$

This is a matrix version of the one-dimensional map result Eq. (2.7). Finally, we wish to stress that, although we have been discussing the map $\mathbf{M}$ as arising from a surface of section of a continuous time system, our discussion connected with Eqs. (4.20)–(4.23) and (4.25)–(4.26) applies to maps in general, whether they arise via a surface of section or not. (As an example of the latter, Problem 5 deals with the stability of a periodic orbit of the Hénon map.)

Figue 4.9 Period three orbit of the surface of section map $\mathbf{M}$.

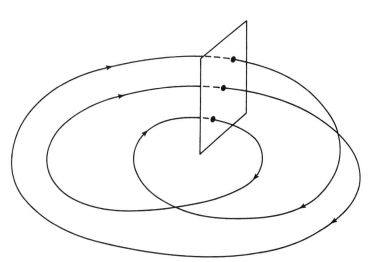

4.3 Stable and unstable manifolds

We define stable and unstable manifolds of steady states and periodic orbits of smooth dynamical systems as follows. The *stable manifold* of a steady state or periodic orbit is the set of points **x** such that the forward orbit starting from **x** approaches the steady state or the closed curve traced out by the periodic orbit. Similarly, the *unstable manifold* of a steady state or periodic orbit is the set of points **x** such that the orbit going backward in time starting from **x** approaches the steady state or the closed curve traced out by the periodic orbit (this assumes invertibility if we are dealing with a map). The existence and smoothness of these manifolds can be proven under very general conditions. Furthermore, stable and unstable manifolds, W^s and W^u, of a steady state or periodic orbit, have the same dimensionality as the linear subspace E^s and E^u and are tangent to them,

$$\dim(W^s) = n_s,$$
$$\dim(W^u) = n_u.$$

Figure 4.10 (*a*) illustrates the situation for a two-dimensional map with a fixed point γ that has one stable and one unstable direction. Figure 4.10(*b*) applies for a situation where $n_s = 2$ and $n_u = 1$ for a fixed point γ of a flow. Also, in Figure 4.10(*b*), we show an orbit in $W^s(\gamma)$ spiraling into the fixed point γ for the case where the two stable eigenvalues are complex conjugates.

For specificity, in what follows we will only be considering the case of periodic orbits of an invertible map where the orbit period is 1 (i.e., fixed points of the map). We now show that stable manifolds cannot intersect stable manifolds, and unstable manifolds cannot intersect unstable manifolds. For the case of self-intersections of an unstable manifold, this follows from the following considerations. Very near the fixed point γ, say within a distance ε, the unstable manifold is a small section of an n_u-dimensional surface tangent to E^u. Call this small piece of the unstable manifold $W^u_\varepsilon(\gamma)$. Since $W^u_\varepsilon(\gamma)$ lies close to the n_u-dimensional plane E^u, it does not intersect itself. Now, continually mapping the small surface $W^u_\varepsilon(\gamma)$ forward in time, it expands in all its n_u unstable directions filling out the whole unstable manifold of γ, $W^u(\gamma)$. Since we assume the map is invertible, two distinct points cannot be mapped to the same point. Thus, $W^u(\gamma)$ cannot intersect itself. Now consider two distinct fixed points γ_1 and γ_2 with unstable manifolds $W^u(\gamma_1)$ and $W^u(\gamma_2)$. These cannot intersect each other because, if they did, then a backward orbit starting at an intersection point would have to approach both γ_1 and γ_2 in the limit of an infinite number of backwards iterates. However, $\gamma_1 \neq \gamma_2$; so this is impossible. Hence there can be no intersections of unstable manifolds, and, applying a similar argument, with the direction of time reversed, there can be no intersection of stable manifolds.

We note, however, that stable and unstable manifolds can intersect each other. In Figure 4.10(c) we show an intersection of stable and unstable manifolds of a fixed point γ of a two-dimensional map. This is

Figure 4.10 Stable and unstable manifolds.

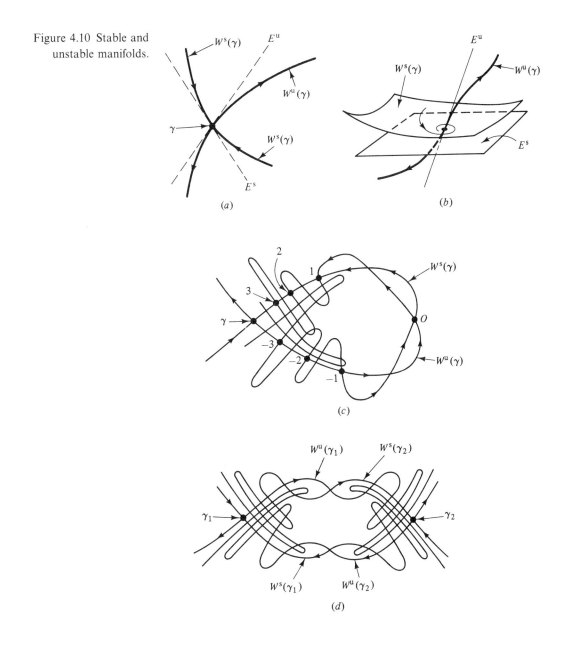

called a *homoclinic intersection*. In Figure 4.10(*d*) we show intersections of the stable and unstable manifolds of one fixed point γ_1 with those of another fixed point γ_2. This is called a *heteroclinic intersection*. The complexity of these diagrams stems from the fact that, if a stable and unstable manifold intersect once, then they must intersect an infinite number of times. To see this, we have labeled one of the intersections O in Figure 4.10(*c*). Since O is on $W^s(\gamma)$ and $W^u(\gamma)$ its subsequent iterates, both forward and backward in time, must also be on $W^s(\gamma)$ and $W^u(\gamma)$, because $W^s(\gamma)$ and $W^u(\gamma)$ are invariant sets by their construction. Thus intersection points map into intersection points. Iterating the point O forward in time, it approaches γ along the stable manifold, successively mapping to the points labeled 1, 2 and 3 in Figure 4.10(*c*). Iterating the point O backward in time, it approaches γ along the unstable manifold, successively mapping to the points labeled -1, -2 and -3. The complicated nature of Figures 4.10(*c*) and (*d*) suggests complicated dynamics when homoclinic or heteroclinic intersections are present. Indeed this is so. Smale (1967) shows that a homoclinic intersection implies horseshoe type dynamics for some sufficiently high iterate of the map. To see this consider the homoclinic intersection for the fixed point γ shown in Figure 4.11(*a*). The manifolds $W^s(\gamma)$ and $W^u(\gamma)$ intersect at the point ξ. Choosing a small rectangle J about the point γ and mapping it forward in time a sufficient number of iterates q_+ we obtain $\mathbf{M}^{q_+}(J)$. Similarly mapping J backward q_- iterates, we obtain the region $\mathbf{M}^{-q_-}(J) \equiv S$. (See Figure 4.11(*b*)). Thus we have the picture shown in Figure 4.11(*c*) which shows that $\mathbf{M}^q$, where $q = q_+ + q_-$, maps S to a horseshoe as in Figure 4.1. Hence, $\mathbf{M}^q$ is a horseshoe map on the long thin rectangle S and has an invariant set in S on which the dynamics is equivalent to a full shift on two symbols. We note that, although we have drawn the shapes of $W^s(\gamma)$ and $W^u(\gamma)$ in Figure 4.11(*a*) to make the horseshoe shape obvious (Figure 4.11(*c*)), the result stated above and proved by Smale depends only on the existence of a homoclinic intersection.[2] Furthermore, a similar result applies in the heteroclinic case, Figure 4.10(*d*).

In Section 4.2 we have discussed the linearized map $\mathbf{y}_{n+1} = \mathbf{A} \cdot \mathbf{y}_n$ about a fixed point $\mathbf{x}_*$, and the splitting of the vector space in which $\mathbf{y}$ lies into subspaces E^s, E^u and E^c that are invariant under the matrix $\mathbf{A}$. We call the vectors $\mathbf{y}$ *tangent vectors*. We call the space in which they lie the *tangent space* of the map at $\mathbf{x} = \mathbf{x}_*$, and we denote this space $T_{\mathbf{x}_*}$.

We say the fixed point $\mathbf{x}_*$ is *hyperbolic* if there is no center subspace E^c. That is, if all the magnitudes of the eigenvalues λ_j are either greater than 1 or less than 1, and $n_u + n_s$ is the dimension of $\mathbf{y}$. In this case, we say that the tangent space $T_{\mathbf{x}_*}$ has a direct sum decomposition into E^s and E^u, $T_{\mathbf{x}_*} = E^s \oplus E^u$. That is, vectors in the space $T_{\mathbf{x}_*}$ can be uniquely specified as the sum of two component vectors, one in the subspace E^s and one in the subspace E^u.

There is a notion of hyperbolicity not only for fixed points, but also for more general invariant sets of a map. Such an invariant set might be, for example, a strange attractor (the strange attractor of the generalized baker's map is hyperbolic), or it may not be an attractor (like the invariant set of the horseshoe map).

We say that an invariant set Σ is *hyperbolic* if there is a direct sum decomposition of T_x into stable and unstable subspace $T_x = E_x^s \oplus E_x^u$ for

Figure 4.11 Construction of a horseshoe from a homoclinic intersection.

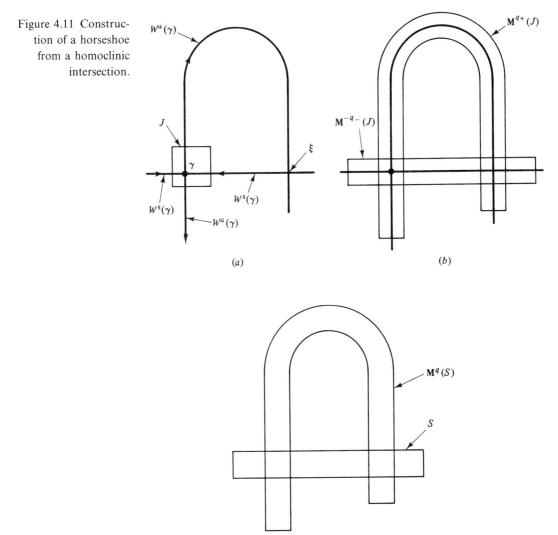

all $\mathbf{x}$ in Σ, such that the splitting into $E_{\mathbf{x}}^{s}$ and $E_{\mathbf{x}}^{u}$ varies continuously with $\mathbf{x}$ in Σ and is invariant in the sense that $\mathbf{DM}(E_{\mathbf{x}}^{s,u}) = E_{M(\mathbf{x})}^{s,u}$, and there are some numbers $K > 0$ and $0 < \rho < 1$ such that the following hold.

(*a*) If $\mathbf{y}$ is in $E_{\mathbf{x}}^{s}$ then

$$|\mathbf{DM}^{n}(\mathbf{x}) \cdot \mathbf{y}| < K\rho^{n}|\mathbf{y}|. \tag{4.27a}$$

(*b*) If $\mathbf{y}$ is in $E_{\mathbf{x}}^{u}$ then

$$|\mathbf{DM}^{-n}(\mathbf{x}) \cdot \mathbf{y}| < K\rho^{n}|\mathbf{y}|. \tag{4.27b}$$

Consider $\mathbf{x}$ as an initial condition. Conditions (*a*) and (*b*) basically say that the orbit originating from another initial condition infinitesimally displaced from $\mathbf{x}$ exponentially approaches the orbit, $\mathbf{M}^{n}(\mathbf{x})$ or exponentially diverges from it if the infinitesimal displacement is in $E_{\mathbf{x}}^{s}$ or $E_{\mathbf{x}}^{u}$, respectively.

Hyperbolic invariant sets are mainly of interest because the property of hyperbolicity allows many interesting mathematically rigorous results to be obtained. Much of what is rigorously known about the structure and dynamics of chaos is only known for cases which satisfy the hyperbolicity conditions. (See, for example, the text by Guckenheimer and Holmes (1983).) Some of these results are the following (some restrictions in addition to hyperbolicity are also required for some of these statements).

(1) Stable and unstable manifolds at $\mathbf{x}$ in Σ, denoted $W^{s}(\mathbf{x})$ and $W^{u}(\mathbf{x})$, can be locally defined. Two points on the same stable manifold, for example, approach each other exponentially in time as illustrated in Figure 4.12. Note that $W^{s,u}(\mathbf{x})$ is tangent to $E_{\mathbf{x}}^{s,u}$ at $\mathbf{x}$.

(2) If small noise is added to a hyperbolic system with a chaotic attractor, then a resulting noisy orbit perturbed from the chaotic attractor can be 'shadowed' by a 'true' orbit of the noiseless system such that the

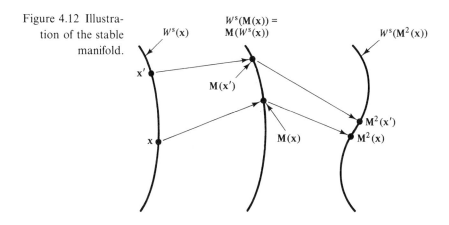

Figure 4.12 Illustration of the stable manifold.

true orbit closely follows the noisy orbit (see Section 1.5 and Problem 2 of Chapter 2 for other discussions of shadowing).

(3) The dynamics on the invariant set can be represented via symbolic dynamics as a full shift or a shift of finite type on a bi-infinite symbol sequence (as illustrated in Figure 4.4).

(4) If the invariant hyperbolic set is an attractor, then a natural measure (as defined in Chapter 2) exists.

(5) The invariant set and its dynamics are *structurally stable* in the sense that small smooth perturbations of the map preserve the dynamics. In particular, if $\mathbf{m}(\mathbf{x})$ is a smooth function of $\mathbf{x}$, then there exists some positive number ε_0 such that the perturbed map, $\mathbf{M}(\mathbf{x}) + \varepsilon\mathbf{m}(\mathbf{x})$, can be transformed to the original map $\mathbf{M}$ by a one to one change of variables for all ε satisfying $|\varepsilon| < \varepsilon_0$. (This change of variables is continuous but may not be differentiable.) In particular, in the range $|\varepsilon| < \varepsilon_0$, the perturbed map and the original map have the same number of periodic orbits for any period, and have the same symbolic dynamics.

An example of chaotic attractors that are apparently not structurally stable are those occurring for the logistic map $x_{n+1} = rx_n(1 - x_n)$. In this case, we saw in Section 2.2 that r-values yielding attracting periodic orbits are thought to be dense in r. Thus, for the case where r is such that there is a chaotic attractor, an arbitrarily small change of r (which can also be said to produce an arbitrarily small change in the map) can completely change the character of the attractor[3] (i.e., from chaotic to periodic).

As mentioned previously, the generalized baker's map and the horseshoe map yield examples of hyperbolic sets. For the generalized baker's map, Eq. (3.7), the Jacobian matrix is

$$\mathbf{DM}(\mathbf{x}) = \begin{bmatrix} \lambda_x(y) & 0 \\ 0 & \lambda_y(y) \end{bmatrix}, \tag{4.28}$$

where $\lambda_x(y) = \lambda_a$ or λ_b for $y < \alpha$ and $y > \alpha$, respectively, and $\lambda_y(y) = \alpha^{-1}$ or β^{-1} for $y < \alpha$ and $y > \alpha$, respectively. Since $\lambda_x(y) < 1$ and $\lambda_y(y) > 1$, the unstable manifolds are vertical lines and the stable manifolds are horizontal lines. Similar considerations apply for the horseshoe, where the stable and unstable manifolds are also horizontal and vertical lines. (In fact they are Cantor sets of horizontal and vertical lines whose intersection is the invariant set.) Another example is the Anosov map,

$$\begin{bmatrix} x_{n+1} \\ y_{n+1} \end{bmatrix} = \begin{bmatrix} 1 & 1 \\ 1 & 2 \end{bmatrix} \begin{bmatrix} x_n \\ y_n \end{bmatrix} \text{ modulo } 1. \tag{4.29}$$

Since x and y are taken modulo 1, they may be viewed as angle variables, and this map is a map acting on the two-dimensional surface of a torus.

The coordinates specifying points on this surface are the two angles x and y, one giving the location the long way around the torus, the other giving the location the short way around the torus, as shown in Figure 4.13 (a). (Here one circuit around is signified by increasing the corresponding angle variable by one, rather than by 2π.) The map is continuous (i.e., two points near each other on the toroidal surface are mapped to two other points that are near each other) by virtue of the fact that the entries of the matrix are integers (note the modulo 1 in (4.29)). This map is hyperbolic and structurally stable. To see this, we note that, by virtue of the linearity of (4.29), the Jacobian matrix $\mathbf{DM}(\mathbf{x})$ is the same as the matrix in (4.29) specifying the map. The eigenvalues of the matrix (4.29) are $\lambda_1 = (3 + \sqrt{5})/2 > 1$ and $\lambda_2 = (3 - \sqrt{5})/2 < 1$. Thus, there are one-dimensional stable and unstable directions that are just the directions parallel to the eigenvectors of the matrix which are $(1, \lambda_{1,2} - 1)$.

For typical initial conditions the map (4.29) generates orbits which eventually come arbitrarily close to any point on the toroidal surface. Furthermore, the typical orbit visits equal areas with equal frequency and hence the natural invariant measure is uniform on the toroidal surface. Note that this map is area preserving,

$$\det \begin{bmatrix} 1 & 1 \\ 1 & 2 \end{bmatrix} = 1$$

The book by Arnold and Avez (1968) contains the illustration of the action of the map (4.29) which we reproduce in Figure 4.13 (b). A picture

Figure 4.13 The cat map. Note the mixing action of the map.

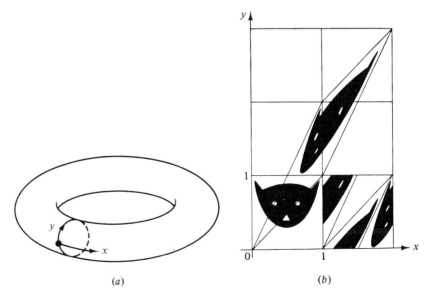

(a) (b)

of the face of a cat is shown on the surface before the map is applied. Neglecting the modulo 1 operations, the square is mapped to a stretched out parallelogram which is returned to the square when the modulo 1 is taken. Because of this picture, (4.29) has been called the 'cat map.'

The map Eq. (3.17) considered by Sinai is a perturbation of Eq. (3.29) (the perturbation is the term $\Delta \cos(2\pi y_n)$). Thus, by the structural stability of (4.29), if Δ is not too large, the attractor for (3.17) is also hyperbolic and structurally stable (we do not know whether this is so for the value $\Delta = 0.1$ used for the plot in Figure 3.6).

While hyperbolic sets are very convenient mathematically, it is unfortunately the case that much of the chaotic phenomena seen in systems occurring in practice is nonhyperbolic and apparently not structurally stable. This seems to be the case for almost all practically interesting chaotic attractors examined to date. On the other hand, in cases of *nonattracting* chaotic sets, such as those arising in problems of chaotic scattering and fractal basin boundaries (see Chapter 5) hyperbolicity seems to be fairly common. As an example of a nonhyperbolic chaotic attractor we mention the Hénon attractor (Figure 1.12). The reason why the Hénon attractor fails to be hyperbolic is that there are points $\mathbf{x}$ on the attractor at which the stable and unstable manifolds $W^s(\mathbf{x})$ and $W^u(\mathbf{r})$ are tangent. We can regard the attractor itself as being the closure of the unstable manifold of points on the attractor.[4] Numerical calculations of stable manifolds of the attractor reveal the structure shown in Figure 4.14, which, according to the discussion in the caption, shows that there are tangencies of stable and unstable manifolds. We require for hyperbolicity that $E_\mathbf{x}^s \oplus E_\mathbf{x}^u$ span the tangent space at every point $\mathbf{x}$ on the attractor. Since the tangents to $W^s(\mathbf{x})$ and $W^u(\mathbf{x})$ coincide for $\mathbf{x}$ at such points, $E_\mathbf{x}^s$ and $E_\mathbf{x}^u$ cannot be defined at tangency points, and the Hénon attractor is thus not hyperbolic.

4.4 Lyapunov exponents

Lyapunov exponents give a means of characterizing the stretching and contracting characteristics of attractors and other invariant sets. First consider the case of a map $\mathbf{M}$. Let $\mathbf{x}_0$ be an initial condition and $\mathbf{x}_n$ ($n = 0, 1, 2, \ldots$) the corresponding orbit. If we consider an infinitesimal displacement from $\mathbf{x}_0$ in the direction of a tangent vector $\mathbf{y}_0$, then the evolution of the tangent vector, given by

$$\mathbf{y}_{n+1} = \mathbf{DM}(\mathbf{x}_n) \cdot \mathbf{y}_n, \tag{4.30}$$

determines the evolution of the infinitesimal displacement of the orbit from the unperturbed orbit $\mathbf{x}_n$. In particular, $\mathbf{y}_n/|\mathbf{y}_n|$ gives the direction of the infinitesimal displacement of the orbit from $\mathbf{x}_n$ and $|\mathbf{y}_n|/|\mathbf{y}_0|$ is the factor by which the infinitesimal displacement grows ($|\mathbf{y}_n| > |\mathbf{y}_0|$) or shrinks ($|\mathbf{y}_n| < |\mathbf{y}_0|$). From (4.30), we have $\mathbf{y}_n = \mathbf{DM}^n(\mathbf{x}_0) \cdot \mathbf{y}_0$, where

$$\mathbf{DM}^n(\mathbf{x}_0) = \mathbf{DM}(\mathbf{x}_{n-1}) \cdot \mathbf{DM}(\mathbf{x}_{n-2}) \cdot \ldots \cdot \mathbf{DM}(\mathbf{x}_0).$$

We define the Lyapunov exponent for initial condition $\mathbf{x}_0$ and initial orientation of the infinitesimal displacement given by $\mathbf{u}_0 = \mathbf{y}_0/|\mathbf{y}_0|$ as

$$h(\mathbf{x}_0, \mathbf{u}_0) = \lim_{n \to \infty} \frac{1}{n} \ln(|\mathbf{y}_n|/|\mathbf{y}_0|)$$

$$= \lim_{n \to \infty} \frac{1}{n} \ln|\mathbf{DM}^n(\mathbf{x}_0) \cdot \mathbf{u}_0|. \qquad (4.31)$$

If the dimension of the map is N, then there will be N or less distinct Lyapunov exponents for a given $\mathbf{x}_0$, and which one of these exponent values applies depends on the initial orientation $\mathbf{u}_0$. (In Chapter 2 we have already discussed Lyapunov exponents for one-dimensional maps ($N = 1$) in which case there is only one exponent.)

To see why there are several possible values of the Lyapunov exponent

Figure 4.14 The Hénon attractor is shown together with a numerically calculated finite length segment of the stable manifold of a point on the attractor (You, 1991). Since almost every point in the basin of attraction goes to the attractor, the stable manifold segment will come arbitrarily close to any point in the basin areas as the segment length is increased. Also, as the segment length is increased, more and more near tangencies with the unstable manifold are produced. Since other stable manifold segments are locally parallel to the calculated segment (they cannot cross), a near tangency for the calculated segment generally indicates an exact tangency for some other segment.

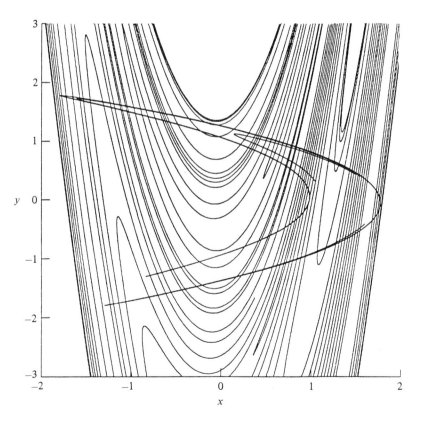

depending on the orientation of $\mathbf{u}_0$, say n is large and approximate $h(\mathbf{x}_0, \mathbf{u}_0)$ as

$$h(\mathbf{x}_0, \mathbf{u}_0) \simeq \bar{h}_n(\mathbf{x}_0, \mathbf{u}_0) \equiv \frac{1}{n} \ln |\mathbf{DM}^n(\mathbf{x}_0) \cdot \mathbf{u}_0|$$

$$= \frac{1}{2n} \ln [\mathbf{u}_0^\dagger \cdot \mathbf{H}_n(\mathbf{x}_0) \cdot \mathbf{u}_0], \tag{4.32}$$

where $\mathbf{H}_n(\mathbf{x}_0) = [\mathbf{DM}^n(\mathbf{x}_0)]^\dagger \mathbf{DM}^n(\mathbf{x}_0)$, and $\dagger$ denotes the transpose. Since $\mathbf{H}_n(\mathbf{x}_0)$ is a real nonnegative hermitian matrix, its eigenvalues are real and nonnegative, and its eigenvectors may be taken to be real. Choosing $\mathbf{u}_0$ to lie in the direction of an eigenvector of $\mathbf{H}_n(\mathbf{x}_0)$, we obtain values of the approximate Lyapunov exponent corresponding to each eigenvector. We denote these values $\bar{h}_{jn}(\mathbf{x}_0) = (2n)^{-1} \ln H_{jn}$, where H_{jn} denotes an eigenvalue of $\mathbf{H}_n(\mathbf{x}_0)$, and we order the subscript labeling of the $\bar{h}_{jn}(\mathbf{x}_0)$ such that $\bar{h}_{1n}(\mathbf{x}_0) \geq \bar{h}_{2n}(\mathbf{x}_0) \geq \cdots \geq \bar{h}_{Nn}(\mathbf{x}_0)$. Thus $\bar{h}_{1n}$ is the largest exponent and $\bar{h}_{Nn}$ is the smallest (if $\bar{h}_{Nn} < 0$, then $\bar{h}_{Nn}$ is the most negative exponent). Letting n approach infinity the approximations $\bar{h}_{jn}(\mathbf{x}_0)$ approach the Lyapunov exponents, which we denote[5]

$$h_1(\mathbf{x}_0) \geq h_2(\mathbf{x}_0) \geq \cdots \geq h_N(\mathbf{x}_0). \tag{4.33}$$

(Often, in the literature, one encounters reference to Lyapunov *numbers*, $\lambda_j(\mathbf{x}_0)$. These are given in terms of the Lyapunov exponents by $\lambda_j(\mathbf{x}_0) = \exp[h_j(\mathbf{x}_0)]$.)

Note that if we choose $\mathbf{u}_0$ in some arbitrary way, it will typically have a mixture of all eigenvector components

$$\mathbf{u}_0 = \sum_{j=1}^N a_j \mathbf{e}_j, \tag{4.34}$$

where $\mathbf{e}_j$ are the eigenvectors of $\mathbf{H}_n(\mathbf{x}_0)$ which we take to be orthonomal. Then we have

$$\mathbf{u}_0^\dagger \cdot \mathbf{H}_n(\mathbf{x}_0) \cdot \mathbf{u}_0 = \sum_{j=1}^N a_j^2 \exp[2n\bar{h}_{jn}(\mathbf{x}_0)].$$

For sufficiently large n, the dominant term in the sum is $j = 1$ since that corresponds to the largest exponent (for specificity we assume that the $\bar{h}_{jn}(\mathbf{x}_0)$ are distinct),

$$\mathbf{u}_0^\dagger \cdot \mathbf{H}_n(\mathbf{x}_0 \cdot \mathbf{u}_0) \simeq a_1 \exp[2n\bar{h}_{1n}(\mathbf{x}_0)],$$

which when placed in (4.32) yields $\bar{h}_{1n}(\mathbf{x}_0)$. Thus, an arbitrary choice of $\mathbf{u}_0$ in (4.31) invariably produces $h_1(\mathbf{x}_0)$. To obtain $h_2(\mathbf{x}_0)$, we can restrict $\mathbf{u}_0$ to lie in the subspace orthogonal to $\mathbf{e}_1$; that is $a_1 = 0$ in (4.34). This yields

$$\mathbf{u}_0^\dagger \cdot \mathbf{H}_n(\mathbf{x}_0) \cdot \mathbf{u}_0 \simeq a_2 \exp[2n\bar{h}_{2n}(\mathbf{x}_0)]$$

(provided n is sufficiently large and $a_2 \neq 0$), which when placed in (4.32) yields $\bar{h}_{2n}(\mathbf{x}_0)$. Proceeding in this way one can imagine, at least in principle, obtaining all the Lyapunov exponents. In practice, when attempting to calculate Lyapunov exponents numerically, special techniques are called for (Benettin *et al.*, 1980). We shall discuss these at the end of this section.

Say we sprinkle initial conditions in a small ball around x_0, and then evolve each initial condition under the map $\mathbf{M}$ for n iterates. Considering the initial ball radius to be infinitesimal, the initial ball evolves into an ellipsoid. This is illustrated in Figure 4.15 for the case $N = 2$ and $h_1(x_0) > 0 > h_2(x_0)$. In the limit of large time the Lyapunov exponents give the time rate of exponential growth or shrinking of the principal axes of the evolving ellipsoid.

Oseledec's multiplicative ergodic theorem (1968) guarantees the existence of the limits used in defining the Lyapunov exponents under very general circumstances. In particular, if μ is an ergodic measure (Section 3.3.3), the Lyapunov exponent values $h_i(x_0)$ obtained from (4.30) and (4.31) are the same set of values for almost every x_0 with respect to the measure μ (see, for example, Ruelle (1989)). For the case of the natural measure on an attractor, this implies that the Lyapunov exponents with respect to the measure are also the same set of values for all x_0 in the basin of attraction of the attractor, except for a set of Lebesgue measure zero. Henceforth, we will often drop the x_0 dependence of $h_i(x_0)$ and write h_i, with the understanding that the h_i are the Lyapunov exponents that apply for almost every x_0 with respect to Lebesgue measure in the basin of attraction of the attractor. Thus, we can speak of the Lyapunov exponents of an attractor without reference to a specific initial condition. We define the attractor to be *chaotic* if it has a positive Lyapunov exponent (i.e., $h_1 > 0$). In this case typical, infinitesimally displaced, initial conditions separate from each other exponentially in time, with the infinitesimal distance between them on average growing as $\exp(nh_1)$. Hence, we also refer to the condition $h_1 > 0$ as implying *exponential sensitivity to initial conditions* for the attractor. (Two initial conditions separated by a small

Figure 4.15 Evolution of an initial infinitesimal ball after n iterations of the map.

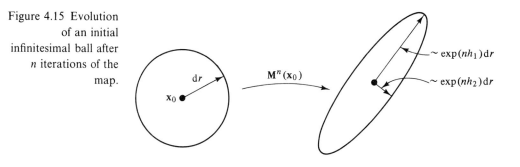

distance $\varepsilon > 0$ (not infinitesimal) will initially separate exponentially, but (assuming the initial conditions lie on a bounded attractor) exponential separation only holds for distances small compared to the attractor size (as in Figure 1.15).)

Consider a periodic orbit of a map $\mathbf{M}$, $\mathbf{x}_0^* \to \mathbf{x}_1^* \to \cdots \to \mathbf{x}_p^* = \mathbf{x}_0^*$. The Lyapunov exponents for the periodic orbit are

$$h_i = (1/p) \ln |\lambda_i|,$$

where λ_i are the eigenvalues of the matrix $\mathbf{DM}^p$ evaluated at one of the points $\mathbf{x} = \mathbf{x}_j^*$. For a chaotic attractor we have seen that there can be infinitely many unstable periodic orbits embedded within the attractor (Section 2.1). Each of these unstable periodic orbits typically yields a set of Lyapunov exponents, $h_i = (1/p) \ln |\lambda_i|$, that are different from those which apply for almost every initial condition with respect to Lebesgue measure in the basin. Thus, they (together with their stable manifolds) are part of the 'exceptional' zero Lebesgue measure set which does not yield the Lyapunov exponents of the chaotic attractor (assuming the chaotic attractor has a natural measure).

As an example, let us calculate the Lyapunov exponents for the generalized baker's map, Eqs. (3.7). The Jacobian matrix of the map is diagonal (Eq. (4.28)) and hence so is $\mathbf{H}_n(\mathbf{x})$,

$$\mathbf{H}_n(\mathbf{x}) = \begin{bmatrix} H_x(\mathbf{x}) & 0 \\ 0 & H_y(\mathbf{x}) \end{bmatrix},$$

$$H_x(\mathbf{x}) = (\lambda_a^{n_1} \lambda_b^{n_2})^2 < 1,$$

$$H_y(\mathbf{x}) = (\alpha^{-n_1} \beta^{-n_2})^2 > 1,$$

where n_1 and n_2 are the number of times the orbit of length $n = n_1 + n_2$ which starts at $\mathbf{x}$ falls below and above the horizontal line $y = \alpha$. From (4.31) choosing $\mathbf{u}_0$ purely in the y-direction and purely in the x-direction, we have

$$h_1 = \lim_{n \to \infty} \left(\frac{n_1}{n} \ln \frac{1}{\alpha} + \frac{n_2}{n} \ln \frac{1}{\beta} \right),$$

$$h_2 = \lim_{n \to \infty} \left(\frac{n_1}{n} \ln \lambda_a + \frac{n_2}{n} \ln \lambda_b \right)$$

For a typical $\mathbf{x}$ the quantity $\lim_{n \to \infty} (n_1/n)$ is just the natural measure of the attractor in $y < \alpha$. Thus, $\lim_{n \to \infty} (n_1/n) = \alpha$. Similarly $\lim_{n \to \infty} (n_2/n)$ is the natural measure in $y > \alpha$, and thus $\lim_{n \to \infty} (n_2/n) = \beta$. The Lyapunov exponents for the generalized baker's map are thus

$$h_1 = \alpha \ln \frac{1}{\alpha} + \beta \ln \frac{1}{\beta} > 0, \tag{4.35a}$$

$$h_2 = \alpha \ln \lambda_a + \beta \ln \lambda_b < 0. \tag{4.35b}$$

On the other hand, the point $x = y = 0$ is a fixed point on the attractor, and if we use this for the initial condition of our orbit, we obtain $n_1/n = 1$, $n_2/n = 0$. Thus, this 'exceptional' initial condition yields values of the exponents that are different from those in (4.35); namely, we obtain $h_1 = \ln \alpha^{-1}$ and $h_2 = \ln \lambda_a$.

It has been conjectured by Kaplan and Yorke (1979b) that there is a relationship giving the fractal dimension of a typical chaotic attractor in terms of Lyapunov exponents (see also Farmer *et al.* (1983)). Let K be the largest integer such that (recall our ordering Eq. (4.33))

$$\sum_{j=1}^{K} h_j \geq 0.$$

Define the quantity D_L called the *Lyapunov dimension*,

$$D_L = K + \frac{1}{|h_{K+1}|} \sum_{j=1}^{K} h_j. \tag{4.36}$$

The conjecture is that the Lyapunov dimension is the same as the information dimension of the attractor for 'typical attractors,'

$$D_1 = D_L. \tag{4.37}$$

In the case of a two-dimensional map with $h_1 > 0 > h_2$ and $h_1 + h_2 < 0$ (e.g., the Hénon map),

$$D_L = 1 + h_1/|h_2|. \tag{4.38}$$

(The condition $h_1 + h_2 < 0$ says that on average areas are contracted by the map.) We can motivate (4.37) for this case as follows. Say we cover some fraction $\theta < 1$ of the attractor natural measure with $N(\varepsilon, \theta)$ small boxes of edge length ε, where we choose the boxes so as to minimize $N(\varepsilon, \theta)$. Now consider one of these boxes and iterate it forward in time by n iterates. In the linear approximation, typically valid for $\varepsilon \exp(nh_1)$ much less than the size of the attractor, the box will be stretched in length by an amount of order $\exp(nh_1)$, and will be contracted in width by an amount of order $\exp(nh_2)$, becoming a long thin parallelogram as shown in Figure 4.16. Now consider covering the parallelogram with smaller boxes of edge $\varepsilon \exp(nh_2)$. This requires of the order of $\exp[n(h_1 - h_2)] = \exp[n(h_1 + |h_2|)]$ such boxes. Since the natural measure is invariant, the measures in the original square and in the parallelogram are the same. Thus, we obtain an estimate for the number of boxes of edge length $\varepsilon \exp(nh_2)$ needed to cover the fraction θ of the natural measure,

$$N(\varepsilon \exp(nh_2), \theta) \sim \exp[n(h_1 + |h_2|)]N(\varepsilon, \theta).$$

Assuming $D_0(\theta) = D_1$ and $N(\varepsilon, \theta) \sim \varepsilon^{-D_1}$, the above gives

$$[\varepsilon \exp(nh_2)]^{-D_1} \sim \exp[n(h_1 + |h_2|)]\varepsilon^{-D_1}$$

from which, by taking logarithms, we immediately obtain (4.37). Note that (4.37) also holds for the generalized baker's map as can be checked by substituting h_1 and h_2 from (4.35) in (4.38) and comparing with the result Eq. (3.24) for the information dimension of the generalized baker's map. No rigorous proof of the Kaplan–Yorke conjecture exists, but some rigorous results related to it have been obtained by Young (1982) and by Ladrappier (1981). Numerical evidence in the case of two-dimensional maps (Russell *et al.*, 1980) also supports it.

One question that naturally arises in the above heuristic derivation of the conjecture $D_L = D_1$ is why not take $\theta = 1$, in which case we would have $D_L = D_0$. The point is that the Lyapunov exponents h_1 and h_2 of the attractor apply for almost every point with respect to the natural measure on the attractor. On the other hand, there is typically a natural measure zero set on a chaotic attractor which yields larger stretching than h_1. For small ε this natural measure zero set makes $N(\varepsilon, 1)$ much larger than $N(\varepsilon, \theta)$, $\theta < 1$. Thus points on the attractor with nontypical stretching dominate in determining D_0. In Figure 4.16 we have assumed that the stretching experienced by a box is given by the 'typical' values h_1 and h_2, and this is appropriate for D_1 but not D_0. Indeed $D_L = D_0$ is contradicted by the example of the generalized baker's map.

In all of the above we have been discussing the Lyapunov exponents of maps. For the case of a continuous time system in the form of an autonomous system of first-order ordinary differential equations, $d\mathbf{x}/dt = \mathbf{F}(\mathbf{x})$, all of the above considerations carry through with Eq. (4.31) replaced by

$$h(\mathbf{x}_0, \mathbf{u}_0) = \lim_{t \to \infty} \frac{1}{t} \ln(|\mathbf{y}(t)|/|\mathbf{y}_0|),$$

$$= \lim_{t \to \infty} \frac{1}{t} \ln|\hat{\mathbf{O}}(\mathbf{x}_0, t) \cdot \mathbf{u}_0|, \tag{4.39}$$

where $d\mathbf{y}(t)/dt = \mathbf{DF}(\mathbf{x}(t)) \cdot \mathbf{y}(t)$, $\mathbf{x}_0 = \mathbf{x}(0)$, $\mathbf{y}_0 = \mathbf{y}(0)$, $\mathbf{u}_0 = \mathbf{y}_0/|\mathbf{y}_0|$, and $\hat{\mathbf{O}}(\mathbf{x}_0, t)$ is the matrix solution of the equation

$$d\hat{\mathbf{O}}/dt = \mathbf{DF}(\mathbf{x}(t)) \cdot \hat{\mathbf{O}}$$

Figure 4.16 Box stretched into a long thin parallelogram.

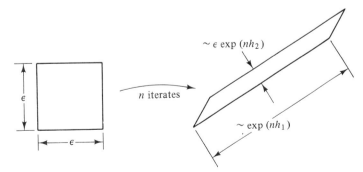

subject to the initial condition

$$\hat{\mathbf{O}}(\mathbf{x}_0, 0) = \mathbf{I}.$$

From the above we have $\mathbf{y}(t) = \hat{\mathbf{O}}(\mathbf{x}_0, t) \cdot \mathbf{y}_0$, and $\hat{\mathbf{O}}$ plays the roles of $\mathbf{DM}^n$ in our treatment of maps. For a chaotic attractor of a flow there is one Lyapunov exponent which is zero, corresponding to an infinitesimal displacement along the flow (Section 4.3). Thus, for example, the Lorenz attractor (Eqs. (2.30)) has $h_1 > 0$, $h_2 = 0$, $h_3 < 0$ and $h_1 + h_2 + h_3 < 0$ yielding

$$D_L = 2 + h_1/|h_3|.$$

A useful fact to keep in mind is that, in cases where there is uniform phase space contraction, there is a relationship among the Lyapunov exponents. For example, for the Hénon map Eq. (1.14), the determinant of the Jacobian is the constant $-B$ independent of the orbit position. Thus $\mathbf{H}_n(\mathbf{x})$ has determinant B^{2n}. Since the determinant of $\mathbf{H}_n(\mathbf{x})$ is the product of its eigenvalues, Eq. (4.32) yields

$$h_1 + h_2 = \ln|B|$$

for the Hénon map. Thus a numerical calculation of h_1 immediately yields h_2 by the above formula without need for an additional numerical calculation of h_2. A similar situation arises for the Lorenz attractor Eqs. (2.30) for which phase space volumes contract at the exponential rate $-(1 + \tilde{\sigma} + \tilde{b})$ (cf. Section 2.4.1). In this case we therefore have

$$h_1 + h_3 = -(1 + \tilde{\sigma} + \tilde{b})$$

with $h_2 = 0$.

A technique for numerically calculating Lyapunov exponents of chaotic orbits has been given by Benettin *et al.* (1980). First consider calculating the largest exponent h_1. Choosing $\mathbf{y}_0$ arbitrarily (so that it has a component in the direction of maximum exponential growth), and iterating (4.30) for a long time, $|\mathbf{y}_n|$ typically becomes so large that one encounters computer overflow if $h_1 > 0$. This problem can be overcome by renormalizing $|\mathbf{y}|$ to 1 periodically. That is, at every time $\tau_j = j\tau$ ($j = 0, 1, 2, 3, \ldots$), where τ is some arbitrarily chosen time interval (not too large), we divide the tangent vector by its magnitude α_j to renormalize it to a vector of magnitude 1. Storing the α_j, we obtain the largest Lyapunov exponent as (cf. Eq. (4.31))

$$h_1 = \lim_{l \to \infty} \frac{1}{l\tau} \sum_{j=1}^{l} \ln \alpha_j, \tag{4.40}$$

and we approximate h_1 by

$$h_1 \simeq \frac{1}{l\tau} \sum_{j=1}^{l} \ln \alpha_j$$

for some sufficiently large l such that numerically the result appears to have converged to within some acceptable tolerance. To calculate the second Lyapunov exponent we choose *two* independent arbitrary starting vectors $\mathbf{y}_0^{(1)}$ and $\mathbf{y}_0^{(2)}$. Two such vectors define the two-dimensional area A_0 of a parallelogram lying in the N-dimensional phase space. Iterating these two vectors n times, we obtain $\mathbf{y}_n^{(1)}$ and $\mathbf{y}_n^{(2)}$ which define a parallelogram of area A_n. Since $\mathbf{y}_0^{(1)}$ and $\mathbf{y}_0^{(2)}$ are assumed to have components in the directions $\mathbf{e}_1$ and $\mathbf{e}_2$, the original parallelogram will be distorted as in Figure 4.17 so that $A_n \sim \exp[n(h_1 + h_2)]A_0$. Thus we have

$$h_1 + h_2 = \lim_{n \to \infty} \frac{1}{n} \ln(A_n/A_0). \tag{4.41}$$

Hence, if we have calculated an estimate of h_1, then numerical calculation of an estimate of the right-hand side of Eq. (4.41) yields an estimate of h_2. There are two difficulties: (1) as before, $\mathbf{y}_n^{(1)}$ and $\mathbf{y}_n^{(2)}$ tend to become very large as n is increased, and (2) their orientations become more and more coincident (since, for large n, their predominant growth is in the direction corresponding to h_1). Problem (2) implies that computer roundoff errors will eventually obliterate the difference between the directions $\mathbf{y}_n^{(1)}/|\mathbf{y}_n^{(1)}|$ and $\mathbf{y}_n^{(2)}/|\mathbf{y}_n^{(2)}|$ (cf. Figure 4.17). To circumvent these problems, the previous technique for calculating h_1 can be naturally extended by generalization of the normalization procedure. At each time τ_j we replace the evolving pair of vectors by two *orthonormal* vectors in the two-dimensional linear space spanned by the evolving vectors. We then obtain

$$h_1 + h_2 \cong \frac{1}{l\tau} \sum_{j=1}^{l} \ln \alpha_j^{(2)},$$

Figure 4.17(a) Ortho-normal vectors at time $n = \tau_j$. (b) Vectors at time $n = \tau_{j+1}$ before normalization procedure is applied.

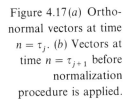

$n = \tau_j$

Area = 1

(a)

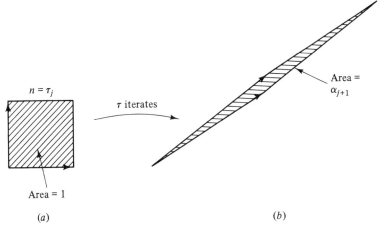

τ iterates

Area = α_{j+1}

(b)

where $\alpha_j^{(2)}$ is the parallelogram area (before normalization) at time τ_j. For h_k ($k = 3, 4, \ldots$) the procedure is basically the same. We evolve k vectors, keeping track of the k-dimensional parallelepiped volume which they define, and normalizing the set of k evolved vectors at time τ_j. In this case at every time τ_j we replace the evolved vectors by the corresponding orthonormal set defined through the Gram–Schmidt orthonormalization procedure, which preserves the linear subspace spanned by the evolved vectors before normalization. Thus, we obtain

$$\sum_{k'=1}^{k} h_{k'} \simeq \frac{1}{l\tau} \sum_{j=1}^{l} \ln \alpha_j^{(k)},$$

where $\alpha_j^{(k)}$ is the volume of the k-dimensional parallelepiped just before normalization. Subtracting the result for $k - 1$ from the result for k we have for $k \geq 2$,

$$h_k \cong \frac{1}{l\tau} \sum_{j=1}^{l} \ln (\alpha_j^{(k)}/\alpha_j^{(k-1)}).$$

The Gram–Schmidt orthogonalization procedure is summarized in the appendix. Thus, we can, in principle, calculate as many of the Lyapunov exponents as we desire. See Benettin *et al.* (1980) for further discussion and numerical examples of this procedure. An alternate numerical method for calculating Lyapunov exponents is given by Greene and Kim (1987) and by Eckmann and Ruelle (1985).

4.5 Entropies

In this section we discuss two other quantities that, like the Lyapunov exponents, serve as a way of quantifying chaos. These are the metric entropy and the topological entropy. In particular, the metric and topological entropies are positive for chaotic systems and are zero for nonchaotic systems. Both the metric entropy and the topological entropy have played a fundamental role in the mathematical theory of chaos. So far, however, these entropies have been less useful than the Lyapunov exponents for examining situations occurring in practice. This is because their values are generally much harder to determine. (The numerical determination of the topological entropy, while comparatively difficult, has been carried out for systems solved on computers in several cases (Newhouse, 1986; Chen *et al.*, 1990a; Biham and Wenzel, 1989; Kovács and Tél, 1990; Chen *et al.*, 1991).) In the following discussion we only consider discrete time dynamical systems (i.e., maps). We begin by discussing the metric entropy.

The metric entropy (Kolmogorov, 1958; Sinai, 1976, 1959) was introduced by Kolmogorov and is also sometimes called the K–S entropy after Kolmogorov and Sinai. The metric entropy can be thought of as a number measuring the time rate of creation of information as a chaotic orbit evolves. This statement is to be understood in the following sense. Due to the sensitivity to initial conditions in chaotic systems, nearby orbits diverge. If we can only distinguish orbit locations in phase space to within some given accuracy, then the initial conditions for two orbits may appear to be the same. As their orbits evolve forward in time, however, they will eventually move far enough apart that they may be distinguished as different. Alternatively, as an orbit is iterated, by observing its location with the given accuracy that we have, initially insignificant digits in the specification of the initial condition will eventually make themselves felt. Thus, assuming that we can calculate exactly and that we know the dynamical equations giving an orbit, if we view that orbit with limited precision, we can, in principle, use our observations to obtain more and more information about the initial unresolved digits specifying the initial condition. It is in this sense that we say that a chaotic orbit creates information, and we shall illustrate this with a concrete example subsequently.

The definition of the metric entropy is based on Shannon's formulation of the degree of uncertainty in being able to predict the outcome of a probabilistic event. Say an experiment has r possible outcomes, and let $p_1, p_2, \ldots, p_r$, be the probabilities of each outcome. (Think of the experiment as spinning a roulette wheel with numbers $1, 2, \ldots, r$ assigned to r segments (possibly of unequal length) composing the periphery of the wheel and with the p_is proportional to the length of their respective segments on the wheel.) The Shannon entropy gives a number which characterizes the amount of uncertainty that we have concerning which outcome will result. The number is

$$H_S = \sum_{i=1}^{r} p_i \ln(1/p_i) \qquad (4.42)$$

(where we define $p \ln(1/p) \equiv 0$ if $p = 0$). For example, if $p_1 = 1$ and $p_2 = p_3 = \cdots = p_r = 0$, then there is no uncertainty, since we know that event 1 always occurs. Thus, we can predict the outcome with complete confidence. In this case, Eq. (4.42) gives $H_S = 0$. The most uncertain case is when all r events are equally probable, $p_i = 1/r$ for $i = 1, 2, \ldots, r$. In this case, (4.42) gives $H_S = \ln r$. In general, by virtue of $p_1 + \cdots + p_r = 1$, the function of the p_is given by H_S defined in (4.42) will lie between 0 and $\ln r$, and we say the outcome is more uncertain (harder to predict) if H_S is larger (i.e., closer to $\ln r$).

Say we have an invariant probability measure μ for some dynamical system. The metric entropy for that measure is denoted $h(\mu)$ and is defined as follows.

Let W be a bounded region containing the probability measure which is invariant under a map $\mathbf{M}$. Let W be partitioned into r disjoint components, $W = W_1 \cup W_2 \cup \ldots \cup W_r$. We can then form an entropy function for the partition $\{W_i\}$,

$$H(\{W_i\}) = \sum_{i=1}^{r} \mu(W_i) \ln[\mu(W_i)]^{-1}. \tag{4.43a}$$

Now we construct a succession of partitions $\{W_i^{(n)}\}$ of smaller and smaller size by the following procedure. We take our original partition and form the sets $\mathbf{M}^{-1}(W_k)$. Then, for each pair of integers j and k $(j, k = 1, 2, \ldots, r)$, we form the r^2 intersections

$$W_j \cap M^{-1}(W_k).$$

Collecting all the *nonempty* intersections thus formed, we obtain the partition $\{W_i^{(2)}\}$. The next stage of partition $\{W_i^{(3)}\}$ is obtained by forming the r^3 intersections $(j, k, l = 1, 2, \ldots, r)$

$$W_j \cap \mathbf{M}^{-1}(W_k) \cap \mathbf{M}^{-2}(W_l),$$

and so on, so that for $\{W_i^{(n)}\}$ we form the intersections,

$$W_{i_1} \cap \mathbf{M}^{-1}(W_{i_2}) \cap \mathbf{M}^{-2}(W_{i_3}) \cap \ldots \cap \mathbf{M}^{-(n-1)}(W_{i_n}),$$

$i_1, i_2, \ldots, i_n = 1, 2, \ldots, r$. Next we write

$$h(\mu, \{W_i\}) = \lim_{n \to \infty} \frac{1}{n} H(\{W_i^{(n)}\}). \tag{4.43b}$$

The quantity $h(\mu, \{W_i\})$ depends on the original partition $\{W_i\}$. To obtain the metric entropy we maximize $h(\mu, \{W_i\})$ over all possible initial partitions $\{W_i\}$,

$$h(\mu) = \sup_{\{W_i\}} h(\mu, \{W_i\}). \tag{4.43c}$$

We now give a specific example of the construction of the successive stages of a partition. We use for this example the natural measure for the chaotic attractor of the generalizaed baker's map (see Figure 3.4). Figure 4.18(a) shows the unit square on which the generalized baker's map operates. For our initial partition we choose the two horizontal strips $0 \le y \le \alpha$ and $\alpha \le y \le 1$ shown as W_1 and W_2 in the figure, where the vertical width of W_1 is α and the vertical width of W_2 is $(1 - \alpha) = \beta$. The set $\mathbf{M}^{-1}(W_1)$, which is the set mapping to W_1 on one iterate of the generalized baker's map, is composed of the two strips (one of width α^2, the other of width $\alpha\beta$) shown crosshatched in Figure 4.18(b). Similarly, Figure 4.18(c) shows $\mathbf{M}^{-1}(W_2)$. Forming the intersections $W_j \cap \mathbf{M}^{-1}(W_k)$ $(i, k = 1, 2)$ we obtain the new partition $W_1^{(2)}, W_2^{(2)}, W_3^{(2)}, W_4^{(2)}$, shown in Figure 4.18(d). Continuing in this manner, one finds that $\{W_i^{(n)}\}$ consists of 2^n horizontal strips, $Z(n, m)$ of which we have widths

$\alpha^m \beta^{n-m}; m = 0, 1, 2, \ldots, n$ (see Eq. (3.8)). Since the natural measure for the generalized baker's map is uniform in y, the measure of each strip is simply equal to its vertical width. Thus, $\mu(W_1) = \alpha$ and $\mu(W_2) = \beta$. Hence, we obtain from (4.43a)

$$H(\{W_i\}) = \alpha \ln(1/\alpha) + \beta \ln(1/\beta).$$

Doing the same for the partition $\{W_i^{(2)}\}$ shown in Figure 4.18(d), we have $H(\{W_i^{(2)}\}) = \alpha^2 \ln \alpha^{-2} + 2\alpha\beta \ln(\alpha\beta)^{-1} + \beta^2 \ln \beta^{-2} = 2H(\{W_i\})$. In fact for $\{W_i^{(n)}\}$ we obtain (Problem 6)

$$H(\{W_i^{(n)}\}) = n[\alpha \ln(1/\alpha) + \beta \ln(1/\beta)] = nH(\{W_i\})$$

for all n. (While in general $H(\{W_i^{(n)}\}) \sim n$ for large n, the result $H(\{W_i^{(n)}\}) = nH(\{W_i\})$ only applies in very special cases.) From (4.43b) this yields $h(\mu, \{W_i\}) = H(\{W_i\})$. It may be shown that the partition we have chosen is, in fact, optimal in the sense that it yields the maximum value prescribed by Eq. (4.43c). Thus, for example, we have for the metric entropy

$$h(\mu) = \alpha \ln(1/\alpha) + \beta \ln(1/\beta). \tag{4.44}$$

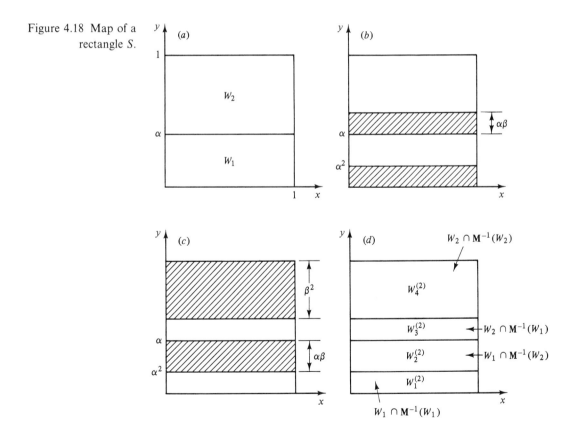

Figure 4.18 Map of a rectangle S.

Say that we can observe whether the orbit of the generalized baker's map lies in W_1 or W_2 on any given iterate. Then the information associated with such a specification of the initial condition is $H(\{W_i\}) = \alpha \ln(1/\alpha) + \beta \ln(1/\beta)$. Now say we observe the orbit and its first iterate and find which of the components, W_1 or W_2 they lie in. These observations of the initial condition and its first iterate determine a more precise knowledge of the location of the initial condition; namely, they determine which of the four narrower strips $W_i^{(2)}$ the initial condition lies in. (For example, if the initial condition is in W_1 and the first iterate is in W_2, then knowledge of the map as specified in Figure 3.4 determines that the initial condition is in $W_2^{(2)}$). The information concerning the initial condition obtainable in such specifications of the initial condition and its first iterate is $H(\{W_i^{(2)}\}) = 2[\alpha \ln(1/\alpha) + \beta \ln(1/\beta)]$, and the information *gained* by observing the first iterate is $H(\{W_i^{(2)}\}) - H(\{W_i\}) = \alpha \ln(1/\alpha) + \beta \ln(1/\beta)$. In fact, for the generalized baker's map, we obtain for all n the result $H(\{W_i^{(n+1)}\}) - H(\{W_i^{(n)}\}) = \alpha \ln(1/\alpha) + \beta \ln(1/\beta)$, which is the metric entropy. Thus, the metric entropy gives the gain of information concerning the location of the initial condition per iterate, assuming that we can only observe with limited accuracy (i.e., we can only observe whether the orbit lies in W_1 or W_2). This interpretation holds in general, independent of the map considered, as we now show. Consider an initial condition and its first $n - 1$ iterates and assume that we observe which component of the initial partition $\{W_i\}$ is visited on each iterate. Let W_{i_m} denote the component visited on iterate m, $\mathbf{x}_m = \mathbf{M}^m(\mathbf{x}_0) \subset W_{i_m}$. Having made such a set of observations, we can specify the region in which the initial condition must lie if it visits the W_{i_m} ($m = 0, 1, \ldots, n - 1$) in the observed order: it must lie in the component of $\{W_i^{(n)}\}$ given by

$$W_{i_0} \cap \mathbf{M}^{-1}(W_{i_1}) \cap \ldots \cap \mathbf{M}^{-(n-1)}(W_{i_{n-1}}).$$

Thus, from (4.42), the information about the initial condition associated with such an observatioin of the first $n - 1$ iterates is $H(\{W_i^{(n)}\})$. Using the following general result for limits of the form (4.43b),

$$\lim_{n \to \infty} \left[\frac{1}{n} H(\{W_i^{(n)}\}) \right] = \lim_{n \to \infty} [H(\{W_i^{(n)}\}) - H(\{W_i^{(n-1)}\})],$$

and we see that we can interpret $h(\mu)$ as the information gained per iterate for large iterate number.

Comparing (4.44) with the expression for the positive Lyapunov exponent of the generalized baker's map, Eq. (4.35a), we see that they are the same, $h(\mu) = h_1$. In general, it has been proven that the metric entropy is at most the sum of the positive Lyapunov exponents (e.g., Ruelle (1989)),

$$h(\mu) \leq \sum_{h_i > 0} h_i. \tag{4.45a}$$

For the generalized baker's map we have found that (4.45a) holds with the equality applying, $h(\mu) = h_1$. Pesin (1976) has shown that

$$h(\mu) = \sum_{h_i > 0} h_i \tag{4.45b}$$

applies for typical Hamiltonian systems (where the measure of interest is just the volume fraction of the relevant chaotic ergodic region; see Chapter 7). Subsequently, it was shown (e.g., Ruelle (1989)) that (4.45b) applies for *Axiom A* attractors of dissipative dynamical systems. (An attractor is said to satisfy *Axiom A* if it is hyperbolic and if the periodic orbits are dense in the attractor. The generalized baker's map satisfies these conditions.) We note, however, that it is unknown whether or not (4.45b) holds even for nonhyperbolic cases as simple as the attractor of the Hénon map (Figure 1.12).

Young (1982) has obtained an interesting rigorous result which relates the metric entropy to the information dimension of an ergodic invariant measure μ of a smooth two-dimensional invertible map with Lyapunov exponents $h_1 > 0 > h_2$. Namely, Young proves that

$$D_1 = h(\mu)\left(\frac{1}{h_1} + \frac{1}{|h_2|}\right). \tag{4.46}$$

(Here, h_1 and h_2 are the Lyapunov exponents obtained for almost every $\mathbf{x}$ with respect to the measure μ.) Specializing to the case of the natural measure of a chaotic attractor, and comparing (4.46) with the Ka-plan–Yorke conjecture (4.38), we see that for the case of the natural measure of a two-dimensional smooth invertible map with $h_1 > 0 > h_2$, the Kaplan–Yorke conjecture reduces to the conjecture that $h(\mu) = h_1$. Numerical experiments (e.g., on the Hénon attractor) which tend to confirm the Kaplan–Yorke conjecture thus may be taken as support of the equality of $h(\mu)$ and h_1.

We now discuss the topological entropy (originally introduced by Adler, Konheim and McAndrew (1965)). The definition of the topological entropy for a map $\mathbf{M}$ is based on the same construction of a succession of finer and finer partitions as was used in the definition of $h(\mu)$. In particular, we start with some partition $\{W_i\}$ and construct the succession of partitions $\{W_i^{(n)}\}$ as before. Let $N^{(n)}(\{W_i\})$ be the number of (nonempty) components of the partition $\{W_i^{(n)}\}$ derived from $\{W_i\}$, and let

$$h_{\mathrm{T}}(\mathbf{M}, \{W_i\}) = \lim_{n \to \infty} \left[\frac{1}{n} \ln N^{(n)}(\{W_i\})\right] \tag{4.47a}$$

Now maximizing over all possible beginning partitions, we obtain the topological entropy of the map $\mathbf{M}$,

$$h_T(\mathbf{M}) = \sup_{\{W_i\}} h(\mathbf{M}, \{W_i\}). \tag{4.47b}$$

In particular, for the generalized baker's map example (Figure 4.18) we have $N^{(n)}(\{W_i\}) = 2^n$ and $h_T = \ln 2$ ($h_T = \ln 2$ also holds for the invariant set of the horseshoe map). We note that the value $h_T = \ln 2$ is greater than or equal to $h(\mu) = \alpha \ln(1/\alpha) + \beta \ln(1/\beta)$ (maximizing the expression $\alpha \ln \alpha^{-1} + (1 - \alpha)\ln(1 - \alpha)^{-1}$ over α yields $\ln 2$ at $\alpha = \frac{1}{2}$). In fact, it is generally true that

$$h_T \geq h(\mu).$$

Also, it can be shown that the topological entropies of $\mathbf{M}$ and $\mathbf{M}^{-1}$ are the same,

$$h_T(\mathbf{M}) = h_T(\mathbf{M}^{-1}). \tag{4.48}$$

One of the key aspects of the topological entropy and the metric entropy is their invariance under certain classes of transformations of the map. Indeed, it was these invariance properties which originally motivated the introduction of these quantities. In particular, h_T is the same for the map $\mathbf{M}$ and for any map derived from $\mathbf{M}$ by a continuous, invertible (but not necessarily differentiable) change of the phase space variables. In such a case, we say that the derived map is *topologically conjugate* to $\mathbf{M}$. As an example, the function ϕ in Eq. (4.3) gives a topological conjugacy between the horseshoe map and the shift map (operating on the space of bi-infinite symbol sequences with two symbols). Hence, the two maps have the same topological entropy (namely, $\ln 2$). Thus, if, as was done for the horseshoe, we can reduce a system to a shift on an appropriate bi-infinite symbol space, then the topological entropy of the original map is fully determined by its symbolic dynamics. Hence, we see that the topological entropy of a map is useful from the point of view that it restricts the symbolic dynamics representation the map might have. The metric entropy, on the other hand, is invariant under isomorphisms; that is, one to one changes of variables (not necessarily continuous). The introduction of the metric entropy solved an important open mathematical question of the time. Namely, it was unknown whether two dynamical systems like the full shift on two symbols with equal probability measures ($\frac{1}{2}$ and $\frac{1}{2}$) on each of the two symbols and the full shift on three symbols with equal probabilities on each of the three symbols were isomorphic. With Kolmogorov's introduction of the metric entropy the answer became immediate: since their metric entropies are $\ln 2$ and $\ln 3$, the equal probability full two shift and the equal probability full three shift cannot be isomorphic.

4.6 Controlling chaos

In a practical situation involving a real physical apparatus one can
imagine that it will often be the case that it is desired that chaos be avoided
and/or that the system dynamics be changed in some way so that
improved performance is obtained. Given a chaotic attractor, one might
consider making some large, possibly costly, change in the system to
achieve the desired objective. Here we assume this is not an option. Rather
we consider the case where we have accessible to us some parameter of the
system which we can vary only relatively slightly.[6] Can we achieve what
we want in this type of circumstance? As shown by Ott *et al.* (1990a, b)
and Romeiras *et al.* (1991), the answer is often yes. The key point is that
chaotic attractors typically have embedded within them a dense set of
unstable periodic orbits. Because we can make only small changes, we
cannot create new orbits with very different properties from the existing
ones. Thus, the approach is to determine first some of the low-period
unstable periodic orbits embedded in the chaotic attractor. Then we
examine these orbits and choose one which yields improved system
performance. Finally, we attempt to program our small parameter
adjustments so as to stabilize this unstable periodic orbit.

As an illustration, consider the case of a two-dimensional map $\mathbf{M}(\mathbf{x}, p)$
where p is a system parameter. At $p = \bar{p}$ the map $\mathbf{M}$ has a chaotic attractor
which has embedded within it an unstable fixed point $\mathbf{x}_F(\bar{p})$. (The
extension to higher period is straightforward.) Say we want to convert the
chaotic motion to a stable orbit $\mathbf{x} = \mathbf{x}_F(\bar{p})$ by small variations of the
parameter p. Since we vary p at each step we replace p by $p_n = \bar{p} + q_n$,
where we restrict the perturbation q_n to satisfy $|q_n| < q_*$. That is, the
maximum allowed perturbation of p is q_*, which we regard as small.
Linearizing the map $\mathbf{x}_{n+1} = \mathbf{M}(\mathbf{x}_n, p)$ about $\mathbf{x} = \mathbf{x}_F(\bar{p})$ and $p = \bar{p}$, we have

$$\mathbf{x}_{n+1} \simeq q_n \mathbf{g} + [\lambda_u \mathbf{e}_u \mathbf{f}_u + \lambda_s \mathbf{e}_s \mathbf{f}_s] \cdot [\mathbf{x}_n - q_n \mathbf{g}], \qquad (4.49)$$

where we have chosen coordinates so that $\mathbf{x}_F(\bar{p}) = 0$, and the quantities
appearing in (4.49) are as follows: $\mathbf{g} = \partial \mathbf{x}_F / \partial p|_{p=\bar{p}}$, λ_s and λ_u are the stable
and unstable eigenvalues, $\mathbf{e}_s$ and $\mathbf{e}_u$ are the stable and unstable eigen-
vectors, and $\mathbf{f}_s$ and $\mathbf{f}_u$ are contravariant basis vectors (defined by
$\mathbf{f}_s \cdot \mathbf{e}_s = \mathbf{f}_u \cdot \mathbf{e}_u = 1$, $\mathbf{f}_s \cdot \mathbf{e}_u = \mathbf{f}_u \cdot \mathbf{e}_s = 0$). Assume that $\mathbf{x}_n$ falls near the fixed
point so that (4.49) applies. We then attempt to pick q_n so that $\mathbf{x}_{n+1}$ falls
approximately on the stable manifold of $\mathbf{x}_F(\bar{p}) = 0$. That is, we choose q_n
so that $\mathbf{f}_u \cdot \mathbf{x}_{n+1} = 0$. If $\mathbf{x}_{n+1}$ falls on the stable manifold of $\mathbf{x} = 0$, we can
then set $q_n = 0$, and the orbit will approach the fixed point at the
geometrical rate λ_s. Dotting (4.49) with $\mathbf{f}_u$ we obtain the following
equation for q_n,

$$q_n = q(\mathbf{x}_n) \equiv \lambda_u (\lambda_u - 1)^{-1} (\mathbf{x}_n \cdot \mathbf{f}_u)/(\mathbf{g} \cdot \mathbf{f}_u), \qquad (4.50)$$

which we use for q_n when $|q(\mathbf{x}_n)| < q_*$. When $|q(\mathbf{x}_n)| > q_*$, we set $q_n = 0$. Thus, for small q_*, a typical initial condition will generate a chaotic orbit which is the same as for the uncontrolled case until $\mathbf{x}_n$ falls within a narrow slab $|x_n^{\mathrm{u}}| < x_*$, where $x_n^{\mathrm{u}} = \mathbf{f}_{\mathrm{u}} \cdot \mathbf{x}_n$ and $x_* = q_* |(1 - \lambda_{\mathrm{u}}^{-1}) \mathbf{g} \cdot \mathbf{f}_{\mathrm{u}}|$. At this time the control (4.50) will be activated. Even then, however, the orbit may not be brought to the fixed point because of nonlinearities not included in (4.50). In this event the orbit will leave the slab and continue to move chaotically as if there was no control. Eventually (due to ergodicity of the uncontrolled attractor) the orbit will fall near enough to the desired fixed point that attraction to it is obtained.

Thus, we create a stable orbit, but it is preceded by a chaotic transient. The length of such a chaotic transient depends sensitively on initial conditions, and, for randomly chosen initial conditions, it has an average $\langle \tau \rangle$ which scales as

$$\langle \tau \rangle \sim q_*^{-\gamma},$$

where the exponent γ is given by

$$\gamma = 1 + \tfrac{1}{2}(\ln |\lambda_{\mathrm{u}}| / \ln |\lambda_{\mathrm{s}}|^{-1}) \tag{4.51}$$

(see Ott *et al.* (1990a)).

The above procedure specifying the control q_n is a special case of the general technique known as 'pole placement' in the theory of control systems (see Ogata (1990)). For an N-dimensional map $\mathbf{M}(\mathbf{x}, p)$, linearization around a fixed point $\mathbf{x}_{\mathrm{F}}(\bar{p})$ and the nominal parameter value $\bar{p}$ yields

$$\boldsymbol{\delta}_{n+1} = \mathbf{A}\boldsymbol{\delta}_n + \mathbf{B}q_n,$$

where $\boldsymbol{\delta} = \mathbf{x} - \mathbf{x}_{\mathrm{F}}(\bar{p})$, $\mathbf{A}$ is the $N \times N$ matrix of partial derivatives of $\mathbf{M}$ with respect to $\mathbf{x}$, $\mathbf{D}_{\mathbf{x}}\mathbf{M}(\mathbf{x}_{\mathrm{F}}(\bar{p}), \bar{p})$, and $\mathbf{B}$ is the N vector derivative of $\mathbf{M}$ with respect to p, $D_p \mathbf{M}(\mathbf{x}_{\mathrm{F}}(\bar{p}), \bar{p})$. For a linear control we have in general

$$q_n = \mathbf{K} \, \boldsymbol{\delta}_n,$$

where $\mathbf{K}$ is an N-dimensional row vector. Thus, we have $\boldsymbol{\delta}_{n+1} = (\mathbf{A} + \mathbf{B}\mathbf{K})\boldsymbol{\delta}_n$, and we desire to choose $\mathbf{K}$ so that the matrix $\mathbf{A}' = (\mathbf{A} + \mathbf{B}\mathbf{K})$ is stable (i.e., has eigenvalues of magnitude less than 1). If $\mathbf{A}$ and $\mathbf{B}$ satisfy a certain condition called 'controllability' (this condition is typically satisfied), then the pole placement technique allows one to determine a control vector $\mathbf{K}$ which yields *any* set of eigenvalues that we may choose for the matrix $\mathbf{A}'$. Equation (4.50) corresponds to the choice wherein the unstable eigenvalue of $\mathbf{A}$ is made zero, while the stable eigenvalue is unaltered by the control. See Romeiras *et al.* (1991) for implementation of the pole placement technique for controlling a chaotic system governed by a four-dimensional map (this system, called the kicked double rotor, is described in Chapter 5). For this map and the chosen parameter values there are 36 unstable fixed points embedded within the chaotic attractor. Choosing one of the unstable fixed points for control, q_n was set equal to $\mathbf{K} \cdot \boldsymbol{\delta}_n$ whenever $|\mathbf{K} \cdot \boldsymbol{\delta}_n| < q_*$ and $q_n = 0$ otherwise. Figure 4.19 shows results for one component of $\mathbf{x}$ versus iterate number for the case where we first control one fixed point (labeled (1)) for iterates 0–1000 and then successively switch the control, at iterate numbers 2000, 3000 and 4000, to stabilize three other fixed points (labeled (2), (3) and (4)). We note that after switching the control, the orbit goes into transient chaotic motion, but then eventually approaches close enough to the desired fixed point orbit, after which time it locks into it. Thus, not only can a small control stabilize a chaotic system in a desired motion, but it provides the flexibility of being able to switch the system behavior from one type of periodic motion to another.

Recently, this method has been implemented in an experiment on a periodically forced magnetoelastic ribbon by Ditto *et al.* (1990a). It is especially noteworthy that no reliable mathematical model is currently available for their system, but that the method was nevertheless applied by making use of the experimental delay coordinate embedding technique. In particular, this technique yielded a surface of section map as well as estimates of the locations of periodic orbits, their eigenvalues, and the vectors $\mathbf{e}_{u,s}$ and $\mathbf{f}_{u,s}$. Furthermore, the control achieved is quite insensitive to errors in the estimates of these quantities and is also not greatly effected

Figure 4.19 A component of $\mathbf{x}$ versus iterate number for successive control of four different fixed points in the time intervals labeled (1), (2), (3), and (4) (Romeiras *et al.*, 1991).

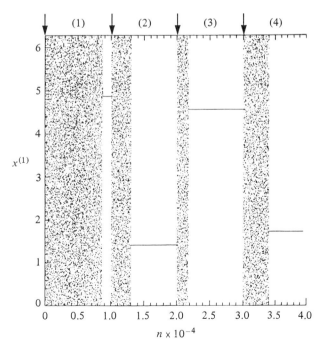

by noise if q_* is not too small. For additional work on controlling chaotic dynamics by stabilization of orbits embedded in the attractor see Dressler and Nitsche (1992), Fowler (1989) and Singer *et al.* (1991).

Appendix: Gram–Schmidt orthogonalization

In this appendix we briefly state the Gram–Schmidt orthogonalization procedure used in calculating Lyapunov exponents. Say we are given k linearly independent vectors $\mathbf{v}_1, \mathbf{v}_2, \ldots, \mathbf{v}_k$ which lie in a vector space of dimension $N \geq k$. We wish to find a set of k orthonormal basis vectors $\mathbf{e}_1, \mathbf{e}_2, \ldots, \mathbf{e}_k$ for the subspace spanned by the vectors $\mathbf{v}_1, \ldots, \mathbf{v}_k$. That is, we wish to determine the $\mathbf{e}_i$ such that $\mathbf{e}_i^\dagger \mathbf{e}_j = \delta_{ij}$ where δ_{ij} is the Kronecker delta and each $\mathbf{e}_i$ is a linear combination of the vs for $i = 1, \ldots, k$. A solution to this problem is

$$\mathbf{e}_i = \left[\mathbf{v}_i - \sum_{j=1}^{i-1} (\mathbf{v}_i \cdot \mathbf{e}_j)\mathbf{e}_j \right] \Big/ \beta_j,$$

$$\beta_j = \left\| \mathbf{v}_i - \sum_{j=1}^{i-1} (\mathbf{v}_i \cdot \mathbf{e}_j)\mathbf{e}_j \right\|,$$

with $\mathbf{e}_1 = \mathbf{v}_1/\|\mathbf{v}_1\|$ and $\|\mathbf{w}\| = (\mathbf{w}^\dagger \mathbf{w})^{1/2}$. Thus, iterating the equation for $\mathbf{e}_i$, we can determine the k orthonormal basis vectors by starting at $i = 1$ and proceeding to successively larger i. The volume of the k-dimensional parallelopiped defined by the vectors $\mathbf{v}_1, \ldots, \mathbf{v}_k$ is

$$\alpha^{(k)} = \beta_1 \beta_2 \ldots \beta_k.$$

Problems

1. For the maps illustrated in Figures 4.20(a) and (b) describe the symbolic dynamics for the invariant set in the original rectangle S. How many distinct periodic orbits of period four are there for the map in Figure 4.20(a)? For the map in Figure 4.20(b), specify the restrictions on the allowed symbol sequences by drawing a figure like those in Figures 4.5(c) and (d).

2. Show that the horseshoe map has an uncountable set of nonperiodic orbits in the invariant set Λ.

3. Consider linear maps $\mathbf{x}_{n+1} = \mathbf{M}(\mathbf{x}_n)$ where $\mathbf{M}(\mathbf{x}) = \mathbf{A} \cdot \mathbf{x}$ where $\mathbf{A}$ is a matrix. For the following matrices $\mathbf{A}$ describe the dynamics and say whether or not the dynamics is hyperbolic.

   (a)
   $$\begin{pmatrix} 1 & -1 \\ 1 & 1 \end{pmatrix},$$

   (b)
   $$\begin{pmatrix} \frac{1}{2} & \frac{1}{2} \\ -1 & 1 \end{pmatrix},$$

   (c)
   $$\begin{pmatrix} 3 & 2 \\ \frac{5}{2} & 2 \end{pmatrix}.$$

4. Consider the two-dimensional map (Kaplan and Yorke, 1979b)

$$x_{n+1} = 3x_n \text{ modulo } 1,$$
$$y_{n+1} = \lambda y_n + 2\cos(2\pi x_n).$$

(a) What are the Lyapunov exponents of this map?
(b) What is the Lyapunov dimension of the map if $\lambda = \frac{1}{4}$? What is it if $\lambda = \frac{1}{2}$?
(c) Show that this map has an attractor which lies in $|y| \leq 2/(1 - |\lambda|)$ provided $|\lambda| < 1$.

5. For the Hénon map there are two fixed points. Find them. Which one is on the attractor (refer to Figure 1.12)? For the one on the attractor

Figure 4.20 Two maps of a rectangle S. As for the horseshoe, we assume that the action of the map is uniformly stretching vertically and uniformly compressing horizontally for the regions of S that map back to S (e.g., H_0, H_1 and H_2 in (b)).

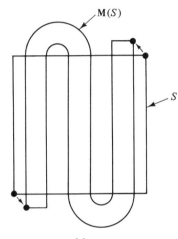

(a)

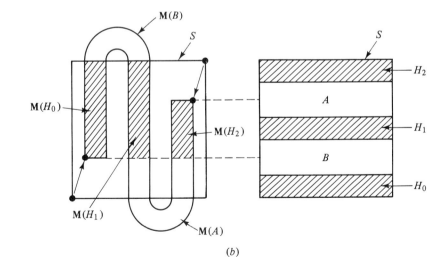

(b)

compute the eigenvalues and eigenvectors of the linearized matrix. What striking fact does one see by comparing the direction of the unstable eigenvector with the direction of the striations of the attractor (Figures 1.12(b) and (c))?

6. Write a computer program to calculate the largest Lyapunov number h_1 of the Hénon map. Using the orbit originating from the initial condition $(x_0, y_0) = (0,0)$ plot a graph of h_1 versus B for $A = 1.4$ and $0 \leq B \leq 0.3$. Using the fact that $h_1 + h_2 = \ln|B|$ also plot the Lyapunov dimension in this range.

7. Consider the forced damped pendulum equation, Eq. (1.6), with parameters $v = 0.22$, $T = 2.7$ and $f = 1/2\pi$ (these are the parameters used for Figure 1.7). Numerically, it is found that $h_1 \simeq 0.135$. What are h_2, h_3 and the Lyapunov dimension?

8. What are the explicit expressions for the quantities **g** and **f** in Section 4.6 for the Hénon attractor of Figure 1.12 and the fixed point on the attractor (see Problem 5)? Identify p with A and take $\bar{p} = 1.3$ such that $A = 1.3 + q$.

Notes

1. The generalized baker's map of Chapter 3 is closely related to the horseshoe map. The latter has the advantage of being smooth, while the generalized baker's map involves discontinuities in the map function at the line $y = \alpha$. Thus, horseshoe type dynamics can occur in typical smooth systems. On the other hand, the discontinuity in the generalized baker's map allows the presence of chaotic attractors which are not present in the horseshoe.

2. In particular, if J is chosen small enough, one can always ensure that the action of the map $\mathbf{M}^q$ on the intersections $S \cap \mathbf{M}^q(S)$ is hyperbolic (defined in Eqs. (4.27)).

3. In particular, if there is a chaotic attractor at r_1 and $r_2 > r_1$, the map at r_2 has more periodic orbits than the map at r_1. This follows because windows imply the occurrence of forward tangent bifurcations and period doublings creating new periodic orbits.

4. For the case of a chaotic attractor, the attractor must contain the unstable manifold of every point on the attractor. For example, for the Hénon attractor there is an unstable saddle fixed point located in the middle of the rectangle in Figure 1.12(c), and it has been suggested and somewhat supported by numerical calculations that the attractor coincides with the closure of the unstable manifold of that fixed point.

5. For the case of a hyperbolic invariant set satisfying Eqs. (4.27) there are no Lyapunov exponents in the range $\ln \rho^{-1} \geq h \geq -\ln \rho^{-1}$.

6. For example, we can think of a chemical reactor (see Section 2.4.3) in which the uncontrolled time dependence is chaotic. We wish to control the dynamics but we only have access to one 'knob' which regulates the inflow rate of one of the chemicals, and we are strongly limited in how much we can turn the knob.

CHAPTER FIVE

Nonattracting chaotic sets

We have already encountered situations where chaotic motion was nonattracting. For example, the map Eq. (3.3) had an invariant Cantor set in $[0, 1]$, but all initial conditions except for a set of Lebesgue measure zero eventually leave the interval $[0, 1]$ and then approach $x = \pm \infty$. Similarly, the horseshoe map has an invariant set in the square S (cf. Figure 4.1), but again all initial conditions except for a set of Lebesgue measure zero eventually leave the square.[1] The invariant sets for these two cases are examples of nonattracting chaotic sets. While it is clear that chaotic attractors have practically important observable consequences, it may not at this point be clear that nonattracting chaotic sets also have practically important observable consequences. Perhaps the three most prominent consequences of nonattracting chaotic sets are the phenomena of *chaotic transients*, *fractal basin boundaries*, and *chaotic scattering*.

The term chaotic transient refers to the fact that an orbit can spend a long time in the vicinity of a nonattracting chaotic set before it leaves, possibly moving off to some nonchaotic attractor which governs its motion ever after. During the initial phase, when the orbit is in the vicinity of the nonattracting chaotic set, its motion can appear to be very irregular and is, for most purposes, indistinguishable from motion on a chaotic attractor.

Say we sprinkle a large number of initial conditions with a uniform distribution in some phase space region W containing the nonattracting chaotic set. Then the length of the chaotic transient that a given one of these orbits experiences depends on its initial condition. The number $N(\tau)$ of orbits still in the chaotic transient phase of their orbit after a time τ typically decays exponentially with τ, $N(\tau) \sim \exp - (\tau/\langle \tau \rangle)$, for large τ. Thus, the fraction of orbits $P(\tau)\,d\tau$ with chaotic transient lengths between τ and $\tau + d\tau$ is

$$P(\tau) = dN(\tau)/d\tau \sim \exp - (\tau/\langle \tau \rangle), \qquad (5.1)$$

where we call $\langle \tau \rangle$ the average lifetime of a chaotic transient. We can also interpret $P(\tau)$ as the probability distribution of τ, given that we choose an initial condition randomly in the region W containing the nonattracting chaotic set. We have already seen examples of the exponential decay law (5.1) for the case of the map Eq. (3.3) and the horseshoe map (cf. Figure 4.1). In particular, referring to Eq. (3.5), we see that, for this example,

$$\langle \tau \rangle = [\ln (1 - \Delta)^{-1}]^{-1},$$

Hence, the average transient lifetime can be long if Δ is small. In such a case observations of an orbit for some appreciable time duration of the order of $\langle \tau \rangle$ or less may not be sufficient to distinguish a chaotic transient from a chaotic attractor.

We shall be discussing chaotic transients in greater detail in Chapter 8. In this chapter we will concentrate on fractal basin boundaries and chaotic scattering. We will also present general results relating the Lyapunov exponents and the average decay time $\langle \tau \rangle$ to the fractal dimensions of nonattracting chaotic sets. (A useful review dealing with some of this material has been written by Tél (1991).)

5.1   Fractal basin boundaries

Dynamical systems can have multiple attractors, and which of these is approached depends on the initial condition of the particular orbit. The closure of the set of initial conditions which approach a given attractor is the basin of attraction for that attractor. From this definition it is clear that the orbit through an initial condition inside a given basin must remain inside that basin. Thus, basins of attraction are invariant sets.

As an example, consider the case of a particle moving in one dimension under the action of friction and the two-well potential $V(x)$ illustrated in Figure 5.1(a). Almost every initial condition comes to rest at one of the

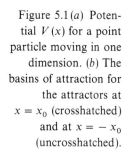

Figure 5.1(a) Potential $V(x)$ for a point particle moving in one dimension. (b) The basins of attraction for the attractors at $x = x_0$ (crosshatched) and at $x = -x_0$ (uncrosshatched).

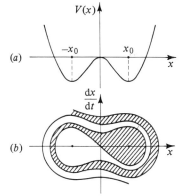

two stable equilibrium points $x = x_0$ or $x = -x_0$. Figure 5.1(b) schematically shows the basins of attraction for these two attractors in the position–velocity phase space of the system. Initial conditions starting in the crosshatched region are attracted to the attractor at $x = +x_0$, $dx/dt = 0$, while initial conditions starting in the uncrosshatched region are attracted to the attractor at $x = -x_0$, $dx/dt = 0$. The boundary separating these two regions (the 'basin boundary') is, in this case, a simple curve. This curve goes through the unstable fixed point $x = 0$. Initial conditions on the basin boundary generate orbits that eventually approach the unstable fixed point $x = 0$, $dx/dt = 0$. Thus, *the basin boundary is the stable manifold of an unstable invariant set*. In this case the unstable invariant set is particularly simple (it is the point $x = 0$, $dx/dt = 0$). We shall see, however, that the above statement also holds when the unstable invariant set is chaotic.

For the example of Figure 5.1 the basin boundary was a simple curve. We now give several pictorial examples showing that basin boundaries can be much more complicated than is the case for Figure 5.1.

Figure 5.2(a) shows the basins of attraction for the map (Grebogi *et al.*, 1983a; McDonald *et al.*, 1985),

$$\theta_{n+1} = \theta_n + a \sin 2\theta_n - b \sin 4\theta_n - x_n \sin \theta_n, \tag{5.2a}$$

$$x_{n+1} = -J \cos \theta_n, \tag{5.2b}$$

where $J_0 = 0.3$, $a = 1.32$ and $b = 0.90$. This map has two fixed points, $(\theta, x) = (0, -J_0)$ and (π, J_0), which are attracting. Figure 5.2 was constructed using a 256×256 grid of initial conditions. For each initial condition the map was iterated a large number of times. It was found that all the initial conditions generate orbits which go to one of the two fixed point attractors. Thus, we conclude that these are the only attractors for this map. If an initial condition yields an orbit which goes to $(0, -J_0)$, then a black dot is plotted at the location of the initial condition. If the orbit goes to the other attractor, then no dot is plotted. (The size of the plotted points on the grid is such that, if all points were plotted, the entire region would be black.) Thus, the black and blank regions are essentially pictures of the two basins of attraction to within the accuracy of the grid used. The graininess in this figure is due to the finite resolution used. At any rate it is apparent that very fine-scale structure in the basins of attraction is present. Furthermore, this fine-scale structure is evidently present on all scales, as revealed by examining magnifications of successively smaller and smaller regions of the phase space which contain fine scale structure. Figure 5.2(b) shows such a magnification. We see that on a small scale the basins evidently consist of many narrow black and blank parallel strips of varying widths. In fact, as we shall see, the basin boundary on this scale may be regarded as a Cantor set of parallel lines (separating the black and blank regions), and the fractal dimension of this basin boundary has been numerically computed to be approximately 1.8 (Grebogi *et al.*, 1983a).

Figure 5.3 shows the basin structure for the forced damped pendulum equation

$$\mathrm{d}^2\theta/\mathrm{d}t^2 + 0.1\,\mathrm{d}\theta/\mathrm{d}t + \sin\theta = f\cos t$$

for two cases, $f = 1.2$ (Figure 5.3(*a*)) and $f = 2.0$ (Figure 5.3(*b*)) (Grebogi, Ott and Yorke, 1987c). In both cases there are two periodic attractors that have the same period as the forcing (namely 2π). The orbit for one of these two attractors has average clockwise motion (negative average value of $\dot\theta$), while the orbit for the other attractor has average counterclockwise motion. In Figure 5.3 the black region represents initial ($t = 0$) values of θ and $\dot\theta$ that asymptote to the attractor whose orbit has average counterclockwise motion. Again, we see that there is small scaled structure on which the black and blank regions appear to be finely interwoven. This is again a manifestation of the fractal nature of the basin boundaries. Numerical experiments on the forced damped pendulum

Figure 5.2(*a*) Basins of attraction for Eqs. (5.2). A_+ and A_- denote the two fixed point attractors. (*b*) Magnification by a factor of 10^5 of the region in (*a*) given by $1.92200 \le \theta \le 1.92201$ and $-0.50000 \le x \le -0.49999$ (McDonald *et al.*, 1985).

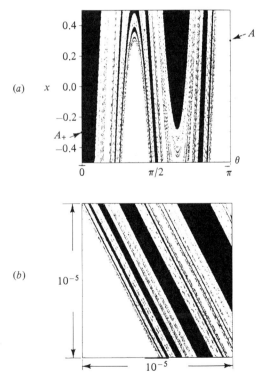

equation show that fractal basin boundaries are extremely common for this system.

As a further illustrative example, consider the 'kicked double rotor' mechanical system illustrated in Figure 5.4. A fixed pivot is attached to a bar with moment of inertia I_1. The free end of this bar is attached by a pivot to the middle of a second bar of moment of inertia I_2. An impulsive upward vertical force, $F = f \sum_n \delta(t - nT)$, is periodically applied to one end of the second bar at time instants $t = 0, T, 2T, 3T, \dots$. There is friction at the two pivots with coefficients v_1 and v_2. Examining the positions (θ, ϕ) and the angular velocities $(d\theta/dt, d\phi/dt)$ just after an impulsive kick, we can analytically derive a four-dimensional map giving the positions and angular velocities just after the $(n + 1)$th kick in terms of their values just after the nth kick (Grebogi et al., 1987a). Figure 5.5 shows the basin structure for this map for a particular set of the parameters $f, I_1,$

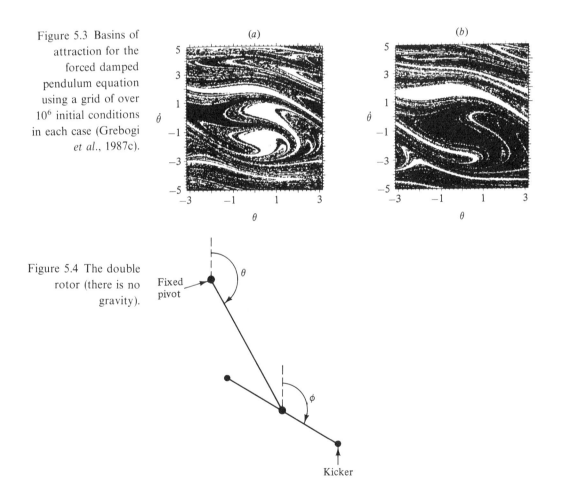

Figure 5.3 Basins of attraction for the forced damped pendulum equation using a grid of over 10^6 initial conditions in each case (Grebogi et al., 1987c).

Figure 5.4 The double rotor (there is no gravity).

I_2, v_1 and v_2. For this choice of parameters there are two attractors; one is the stable fixed point $\theta = \phi = 0$, $d\theta/dt = d\phi/dt = 0$ (both arms are oriented straight up), while the other attractor is chaotic. The plot in Figure 5.5(a) and the magnification in Figure 5.5(b) show initial conditions on a two-dimensional surface in the four-dimensional phase space (namely, the surface $d\theta/dt = d\phi/dt = 0$), with the black region corresponding to the basin of the fixed point attractor and the blank region corresponding to the basin of the chaotic attractor. Thus, we can regard Figure 5.5(a) as a 'slice' by a two-dimensional plane cutting across the four-dimensional phase space. Numerically, it is found that the boundary between the black and blank regions in Figure 5.5 has dimension 1.9, corresponding to a dimension of the basin boundary in the full four-dimensional phase space[2] of 3.9.

Fractal basin boundaries also occur for one-dimensional maps. Consider the map shown in Figure 5.6(a), where the map function consists of straight lines in $[0,\frac{1}{5}]$, $[\frac{2}{5},\frac{3}{5}]$, $[\frac{4}{5},1]$. This map has two attracting fixed points, labeled A_+ and A_-. The region $x \geq 1$ is part of the basin of attraction for A_+ and that the region $x \leq 0$ is part of the basin of attraction for A_-. We now focus on the structure of the basins in $[0,1]$. Since the interval $[\frac{1}{5},\frac{2}{5}]$ maps to $x \geq 1$, it is in the basin of A_+. Since the interval $[\frac{3}{5},\frac{4}{5}]$ maps to $x \leq 1$, it is in the basin of A_-. This is indicated in Figure 5.6(b). We now ask, which intervals map to $[\frac{1}{5},\frac{2}{5}]$ and which to $[\frac{3}{5},\frac{4}{5}]$? These are the six intervals of length $\frac{1}{25}$ shown in Figure 5.6(b). We see that, at this stage of the construction, the intervals assigned to the two basins alternate between the basin of A_+ and the basin of A_- as we move from $x = 0$ to $x = 1$. In fact, this is true at every stage of the construction. Thus, we build up a very complicated, finely interwoven basin structure, and the boundary between the two basins is the nonattracting invariant

Figure 5.5 Basins of attraction for the kicked double rotor; (b) shows a manifestation of a small subregion of (a) (Grebogi et al., 1987a).

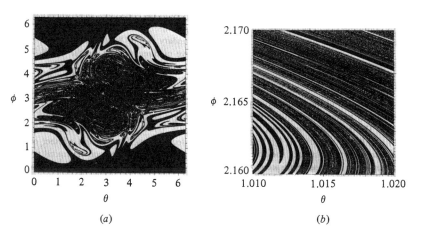

(a)   (b)

Cantor set of points which never leave the interval $[0, 1]$. (The dimension of this Cantor set is $(\ln 3)/(\ln 5)$.)

As a final example of a fractal basin boundary, consider the logistic map, $x_{n+1} = M(x_n) \equiv rx_n(1 - x_n)$, in the range of r-values for which there is an attracting period three orbit. Although there is only one attractor in this case (the period three orbit), we can create a situation where there are three attractors by considering the map $M^3(x)$ rather than $M(x)$ (see Figure 2.13). In this case there are three fixed point attractors, which are just the three components of the attracting period three orbit of $M(x)$, and the boundary separating their basins is fractal (McDonald *et al.*, 1985; Park *et al.*, 1989; Napiórkowski, 1986). Figure 5.7 shows the basin of the middle fixed point attractor of $M^3(x)$ (blank regions) as a function of r.

For further discussion and examples of fractal basin boundaries see McDonald *et al.* (1985) and the book by Gumowski and Mira (1980).

Figure 5.6(*a*) One-dimensional map with a fractal basin boundary. (*b*) The basin in $[0, 1]$.

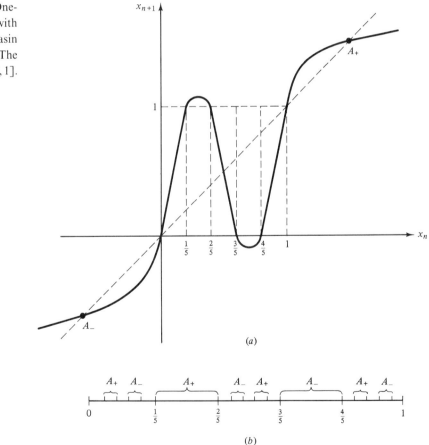

(a)

(b)

5.2   Final state sensitivity

The small scale alternation between different basins that we have seen in the above examples can present a problem when one attempts to predict the future state of a dynamical system. In particular, in the presence of fractal basin boundaries, a small uncertainty in initial conditions can cause anomalously large degradation in one's ability to determine which attractor is approached. In order to make this quantitative, first consider the case of a simple nonfractal basin boundary Σ for two fixed point attractors A_+ and A_-, as shown schematically in Figure 5.8. Say our initial conditions are uncertain by an amount ε in the sense that, when we say that the initial condition is $\mathbf{x} = \mathbf{x}_0$, what we really know is only that the initial condition lies somewhere in $|\mathbf{x} - \mathbf{x}_0| \leq \varepsilon$. For the situation in Figure 5.8, under uncertainty ε, we know for sure that initial condition 1 goes to attractor A_+. On the other hand, the point labeled 2 in the figure lies in the basin of attractor A_-, but because of the ε uncertainty, the actual orbit may go to either attractor A_+ or attractor A_-. We call initial condition 1 ε-certain and initial condition 2 ε-uncertain. Clearly, initial conditions that are ε-uncertain are those which lie within a distance ε of the basin boundary Σ. If we were to pick an initial condition at random in the rectangle shown in Figure 5.8, the probability of obtaining an ε-uncertain initial condition is the fraction of the area (or, in higher dimensionality, volume) of the phase space which lies within ε of the boundary Σ. Denote this fraction $f(\varepsilon)$. For a simple nonfractal boundary, as in Figure 5.8, $f(\varepsilon)$ scales linearly with ε, $f(\varepsilon) \sim \varepsilon$. Thus improvement of the initial condition accuracy say by a factor of 10 (i.e., reduction of ε by 10), reduces $f(\varepsilon)$ and hence our probability of potential error by a factor of 10. However, as we show in the appendix, when the boundary is fractal, $f(\varepsilon)$ has a different scaling with ε;

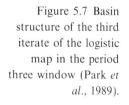

Figure 5.7 Basin structure of the third iterate of the logistic map in the period three window (Park *et al.*, 1989).

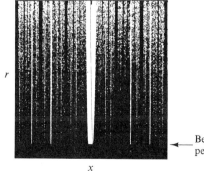

 Beginning of the period three window

x

$$f(\varepsilon) \sim \varepsilon^\alpha, \tag{5.3a}$$

$$\alpha = N - D_0, \tag{5.3b}$$

where N is the phase space dimensionality and D_0 is the box-counting dimension of the basin boundary. For a nonfractal boundary $D_0 = N - 1$ and $\alpha = 1$. For fractal basin boundaries, such as those in Figures 5.2–5.4, $D_0 > N - 1$ and hence $\alpha < 1$. For example, for the situation in Figure 5.5 we have $N = 4, D_0 \approx 3.9$, and hence $\alpha = 0.1$. Thus $f(\varepsilon) \sim \varepsilon^{0.1}$. In this case there is relatively little one can do to reduce $f(\varepsilon)$ by improving the accuracy of initial conditions. In the case of Figure 5.5 ($\alpha \approx 0.1$), to reduce $f(\varepsilon)$ by a factor of 10 requires a reduction of ε by a factor of 10^{10}! If $\alpha < 1$ (i.e., the boundary is fractal), then we say there is *final state sensitivity*, and, as the example above makes clear, the situation with respect to potential improvement in prediction by increasing initial condition accuracy is less favorable the smaller α is. (Note in Eq. (5.3b) that the dimension D_0 satisfies $D_0 \geq N - 1$, since the basin boundary must divide the phase space; hence α cannot exceed 1.) We call α the 'uncertainty exponent'.

The dimension of a fractal basin boundary can be numerically calculated on the basis of the above discussion (McDonald *et al.*, 1983). For example, for the case of the basin boundary shown in Figure 5.2 we proceed as follows. Consider an initial condition (θ_0, x_0), and perturb its x coordinate by an amount ε producing two new initial conditions, $(\theta_0, x_0 - \varepsilon)$ and $(\theta_0, x_0 + \varepsilon)$. Now iterate the map and determine which attractor (A_+ or A_-) each of the three initial conditions goes to. If they do not all go to the same attractor, then we count the original initial condition as uncertain. Now, we randomly choose a large number of initial conditions in the rectangle of Figure 5.3. We then determine the fraction $\bar{f}(\varepsilon)$ of these that are uncertain, and we repeat this for several values of ε. From the definitions of $f(\varepsilon)$ and $\bar{f}(\varepsilon)$, we expect that f is approximately proportional to $\bar{f}$ (for further discussion see Grebogi *et al.* (1988a)) so that α can be extracted from the scaling of $\bar{f}$ with ε. Figure 5.9

Figure 5.8 Region of phase space divided by a nonfractal basin boundary Σ.

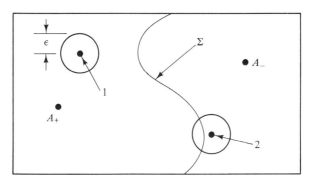

shows results from a set of numerical experiments plotted on log–log axes. The straight line fit indicates that $\bar{f}$ scales as a power of ε, and the slope of the line gives the power α. The result is $\alpha \approx 0.2$, from which Eq. (5.3) yields $D_0 \approx 1.8$.

Even when error in initial conditions is essentially absent, errors in the specification of parameter values specifying the system may be present (e.g., the parameter f in the pendulum equation used in Figure 5.3; $d^2\theta/dt^2 + 0.1\, d\theta/dt + \sin\theta = f\cos t$). A small error in a system parameter might alter the location of the basin boundary so that a fixed initial condition shifts from one basin to another. In a finite region of parameter space, the fraction of randomly chosen parameter values which produces such a change when perturbed by a parameter error δ is some uncertain fraction which we denote $f_p(\delta)$. If the basin boundary dimension is approximately constant in the region of parameter space examined, then the scaling of $f_p(\delta)$ is the same as that for $f(\varepsilon)$; $f_p(\delta) \sim \delta^\alpha$ with

Figure 5.9 $\bar{f}$ versus ε (McDonald *et al.*, 1985).

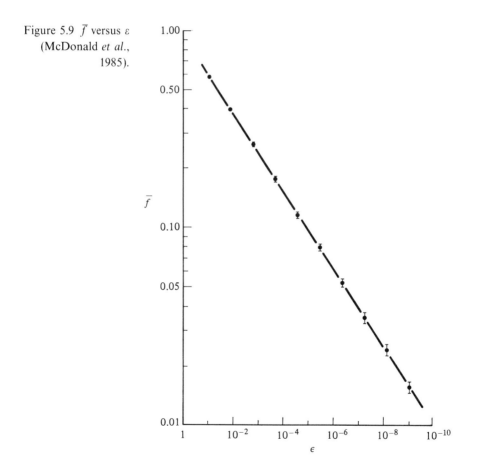

$\alpha = N - D_0$. Moon (1984) has experimentally examined the parameter dependence of the system in Figure 1.1 to see which attractor a fixed initial condition goes to, and he has concluded, on this basis, that the basin boundary is fractal.

In addition to final state and parameter sensitivity, another practical consequence of fractal basin boundaries and nonattracting chaotic sets has been investigated in the Josephson junction experiments of Iansiti *et al.* (1985). These authors find that, when periodic attractors are near a fractal basin boundary, noise can cause frequent kicks of the orbit into the region of finely interwoven basin structure. This leads to an orbit which resembles a chaotic orbit on a strange attractor even when the noise is relatively small.

5.3 Structure of fractal basin boundaries

We now give a description of how the dynamics of the map Eq. (5.2) leads to the fractal basin boundary structure in Figure 5.2. Figure 5.10(*a*) schematically shows a region of the phase space in $0 \leq \theta \leq \pi$ (and narrower in *x* than the region shown in Figure 5.2). In addition to the two fixed point attractors A_+ and A_-, there are also three other fixed points which are not attracting. These three, labeled S_+, S_- and S_0, are saddles; that is, they have a one-dimensional stable manifold and a one-dimensional unstable manifold. We are particularly interested in the saddles S_+ and S_- segments of whose stable manifolds *ab* and *cd* are indicated in the figure. The entire region to the left of *ab* (right of *cd*) can be shown to be part of the basin of attraction of the fixed point attractor A_+ (A_-). The question now becomes, what is the basin structure in the region $Q = abcd$ which lies between the two stable manifold segments *ab* and *cd*? (*Q* is shown crosshatched in Figure 5.10(*a*).) For the purpose of addressing this question, we show an expanded schematic view of the region *Q* in Figure 5.10(*b*). The action of the map on *Q* is to take it into the S-shaped crosshatched region shown in Figure 5.10(*b*), where the map takes *a* to *a'*, *b* to *b'*, *c* to *c'* and *d* to *d'*. The stable manifold segments *ab* and *cd* divide the S-shaped region $\mathbf{M}(Q)$ into five subregions, labeled I', II', III', IV' and V'. The region II' lies to the right of the stable manifold of S_- and so is in the basin of attraction of A_-. Similarly, region IV' is in the basin of A_+. We now ask, what are the preimages of these regions? In particular, the preimage of II' (which we denote II) will be in the basin of A_-, and the preimage of the region IV' (denoted IV) will be in the basin of A_+. These preimages are shown in Figure 5.11. Since the region $\mathbf{M}(Q) \cap$ II is in the basin of A_- its preimage, $\mathbf{M}^{-1}[\mathbf{M}(Q) \cap$ II] is also in the basin of A_-. This preimage is also shown in Figure 5.11 as the three narrow crosshatched vertical strips. Similarly, $\mathbf{M}^{-1}[\mathbf{M}(Q) \cap \text{IV}]$ is the three narrow shaded vertical strips and is part of the basin of A_+. Proceeding iteratively in this way we build up successively finer and finer scale resolution of the basin structure. Note that the shaded and crosshatched vertical strips alternate as we move horizontally across Q, and that this is true at all stages of the construction.[3] Note the similarity of the action of the map on the region Q in Figure 5.10 with the horseshoe map (imagine turning Figure 4.1(d) on its side). The main difference is that $\mathbf{M}(Q) \cap Q$ consists of *three* strips for the case in Figure 5.10, while the action of the horseshoe map on the square produces a region (the horseshoe) which intersects the original square in *two* strips. A symbolic dynamics of the chaotic invariant set for Figure 5.10 can be worked out in analogy to the horseshoe analysis (cf. Problem 1 of Chapter 4), and is a full shift on three symbols (in contrast with the two symbols of the horseshoe map). As in the horseshoe, we may

Figure 5.10 Schematic illustration of the construction of the basin structure for the map Eq. (5.2).

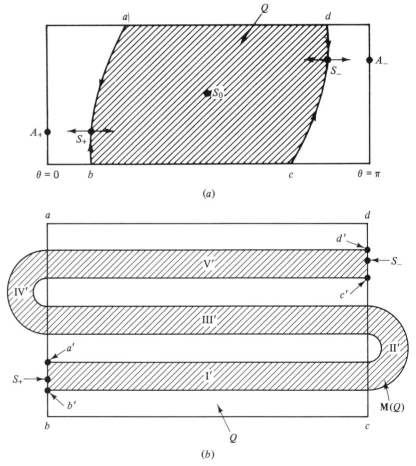

(a)

(b)

think of the chaotic invariant set as the intersection of a Cantor set of lines running vertically with a Cantor set of lines running horizontally. Furthermore, the Cantor set of vertically oriented lines constitutes the basin boundary in the region Q, and is also the stable manifold of the chaotic invariant set. (The horizontal lines are the unstable manifold.) Thus, we see that, in both this example and the example of the two well potential (Figure 5.1), the basin boundary is the stable manifold for a nonattracting invariant set (i.e., the point $x = dx/dt = 0$ for Figure 5.1 and a nonattracting chaotic invariant set for the case of Figure 5.10).

We emphasize that the type of basin structure we have found here, locally consisting of a Cantor set of smooth curves, is very common, but it is not the only type of structure that fractal basin boundaries for typical dynamical systems can have. In particular, McDonald *et al.* (1985) and Grebogi *et al.* (1983b, 1985a) give an example where the basin boundary can be analytically calculated and is a continuous, but nowhere differentiable, curve. The example they consider is the following map,

$$x_{n+1} = \lambda_x x_n \;(\text{mod }1), \tag{5.4a}$$

$$y_{n+1} = \lambda_y y_n + \cos(2\pi x_n), \tag{5.4b}$$

with λ_y and λ_x greater than 1 and λ_x an integer greater than λ_y.

This map has no attractors with finited y (cf. Problem 4 of Chapter 4). Almost every initial condition generates an orbit which approaches either $y = +\infty$ or $y = -\infty$. We regard $y = +\infty$ and $y = -\infty$ as the two attractors for this system. The basins of attraction are shown in Figure 5.12 (the $y = -\infty$ attractor is shown black). The analysis shows that the basin boundary is the continuous curve given by

$$y = -\sum_{j=1}^{\infty} \lambda_y^{-1} \cos(2\pi \lambda_x^{j-1} x). \tag{5.5}$$

Figure 5.11 The basin of A_- is shown cross-hatched, and the basin of A_+ is shown shaded.

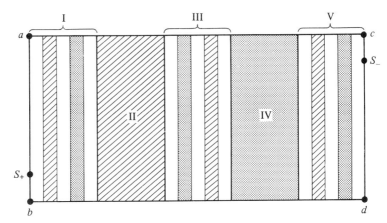

The sum converges since $\lambda_y > 1$. The derivative dy/dx does not exist, however; differentiating inside the sum produces the sum

$$\frac{2\pi}{\lambda_x} \sum_{j=1}^{\infty} \left(\frac{\lambda_x}{\lambda_y}\right)^j \sin\left(2\pi\lambda_x^{j-1}x\right),$$

which does not converge since we have assumed $(\lambda_x/\lambda_y) > 1$. Equation (5.5) is called a Weierstrass curve and has fractal dimension

$$D_0 = 2 - [(\ln\lambda_y)/(\ln\lambda_x)]$$

(which is $D_0 = 1.62\ldots$ for the parameters of Figure 5.12).

Another type of basin structure that is common is the case where the same basin boundary can have different dimensions in different regions. Furthermore, in a certain sense, these regions of different dimension can be intertwined on arbitrarily fine scale (Grebogi *et al.*, 1987a). An example illustrating this phenomenon is the basin boundary of the kicked double rotor shown in the cross section in Figure 5.5. In Figure 5.5(*a*) we see that the boundary between the black and blank areas in the region $0 \le (\theta, \phi) \le 1$ appears to be a simple smooth curve ($D_0 = 1$) sharply dividing the two basins. On the other hand the very mixed appearance in the central region surrounding the point $\theta = \phi = \pi$ suggests that the boundary is fractal there. Indeed application of the numerical final state sensitivity technique to the region $0 \le (\theta, \phi) \le 2\pi$ yields a dimension of the boundary of approximately 1.9 (in the $d\theta/dt = d\phi/dt = 0$ cross section). Note, however, that, when we consider two sets, S_a and S_b, of different dimensions, d_a and d_b, the dimension of the union of the two sets is the larger of the dimensions of the two sets,

$$\dim(S_a \cup S_b) = \max(d_a, d_b).$$

Figure 5.12 Basins for Eqs. (5.4) with $\lambda_z = 3$ and $\lambda_y = 1.5$ (McDonald *et al.*, 1985).

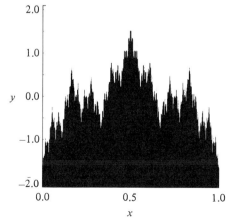

Thus, there is no contradiction with our observation that the dimension in $0 \leq (\theta, \phi) \leq 1$ is 1. (Indeed applying the final state sensitivity technique to the region $0 \leq (\theta, \phi) \leq 1$ yields $D_0 = 1$.) Now, consider the magnification shown in Figure 5.5(b). The dimension of the boundary in this small region is again $D_0 \approx 1.9$. Note, however, that there are areas within this small region where the basin boundary is apparently one-dimensional (e.g. $1.010 \leq \theta \leq 1.012$, $2.160 \leq \phi \leq 2.162$). Moreover, this situation is general for the double rotor: Given any square subregion within $0 \leq (\theta, \phi) \leq 2\pi$ which contains part of the basin boundary, the boundary in that square is nonfractal ($D_0 = 1$) or fractal, and if it is fractal its dimension is always the same ($D_0 \approx 1.9$). Furthermore, no matter how small the square is, if the dimension of the boundary in the square is fractal, then there is some smaller square within it for which the contained boundary is not fractal ($D_0 = 1$). Thus, regions of the basin boundary with different dimension are interwoven on arbitrarily fine scale. For further discussion of this phenomenon and how it comes about as a result of the dynamics see Grebogi *et al.* (1987a, 1988a).

So far we have been discussing fractal boundaries that separate the basins of different attractors. We wish to point out, however, that fractal boundaries can also occur even in conservative (nondissipative) systems for which attractors do not exist. As a simple example of this type, consider the motion of a point particle without friction moving along straight line orbit segments in a region bounded by hard walls (shown in Figure 5.13(a)) at which the particle experiences specular reflection on each encounter (Bleher *et al.*, 1988). We examine initial conditions on the dashed horizontal line segment shown in Figure 5.13(a). The initial position x_0 is measured from the center of the line and the initial velocity

Figure 5.13(a) Particle moving in a region with reflecting walls and two holes A and B. (b) Initial conditions in the black region exit through hole A ($r = 1$, $L = 4$, $a = 0.1$, $b = 0.2$) (Bleher *et al.*, 1988).

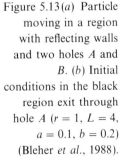

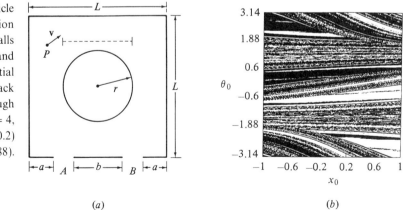

(a)

(b)

vector angle θ_0 is measured clockwise from the vertical. Figure 5.13(b) shows the regions of this initial condition space for which the particle exits through hole A (black) and for which it exits through hole B (blank). The dimension of the boundary separating these two regions is numerically found to be approximately 1.8. Blow-ups, however, reveal that there is the same sort of fine-scaled interweaving of fractal ($D_0 \approx 1.8$) and nonfractal ($D_0 = 1$) boundary regions as for the kicked double rotor example.

5.4 Chaotic scattering

In this section we consider the classical scattering problem for a conservative dynamical system.[4] The simplest example of this problem deals with the motion without friction of a point particle in a potential $V(\mathbf{x})$ for which $V(\mathbf{x})$ is zero, or else very small, outside of some finite region of space which we call the scattering region. Thus, the particle moves along a straight line (or an approximately straight line) sufficiently far outside the scattering region. We envision that a particle moves toward the scattering region from outside it, interacts with the scatterer, and then leaves the scattering region. The question to be addressed is how does the motion far from the scatterer after scattering depend on the motion far from the scatterer before scattering? As an example, consider Figure 5.14 which shows a scattering problem in two dimensions. The incident particle has a velocity parallel to the x-axis at a vertical displacement $y = b$. After interacting with the scatterer, the particle moves off to infinity with its velocity vector making an angle ϕ to the x-axis. We refer to the quantities b and ϕ as the impact parameter and the scattering angle, and we wish to investigate the character of the functional dependence of ϕ on b.

As an example consider the potential (Bleher *et al.*, 1990)

$$V(x, y) = x^2 y^2 \exp[-(x^2 + y^2)] \tag{5.6}$$

shown in Figure 5.15. This potential consists of four potential 'hills' with equal maxima at (x, y)-coordinate locations $(1, 1)$, $(1, -1)$, $(-1, 1)$, and $(-1, -1)$. The maximum value of the potential is $E_m = 1/e^2$. For large distances $r = (x^2 + y^2)^{1/2}$ from the origin, $V(x, y)$ approaches zero rapidly with increasing r. Figure 5.16(a) shows a plot of the *scattering function*, ϕ versus b, for the case where the incident particle energy E is larger than E_m. We observe for this case ($E/E_m = 1.626$) that the scattering function is a smooth curve. Furthermore, it is also found to be a smooth curve for all $E > E_m$. Figure 5.16(b) shows the scattering function for a case where $E < E_m$. We observe that the numerically computed dependence of ϕ on b is poorly resolved in the regions $0.6 \gtrsim \pm b \geq 0.2$. To understand why this might be so, we note that Figure 5.16 is constructed by choosing a large number ($\sim 10^4$) of b-values evenly spaced along the interval of the b-axis shown in the plot. We then integrate the equation of motion for a particle of mass m, $m\,d^2\mathbf{x}/dt^2 = -\nabla V(\mathbf{x})$, for incident particles far from the potential for each b-value, and obtain the corresponding scattering angles ϕ. We then plot these angles to obtain the figure. Thus, the speckling of individually discernible points seen in Figure 5.16(b) in the region $0.6 \geq \pm b \geq 0.2$ might be taken to imply that the curve ϕ versus b varies too rapidly to be resolved on the scale determined by the spacing of b-values used to construct the figure. In this view one might still hope that sufficient resolution would reveal a smooth curve as in Figure 5.16(a). That this is not the case can be seen in Figures 5.17(a)–(c) which show successive magnifications of unresolved regions. Evidently magnification of a portion of unresolved region of Figure

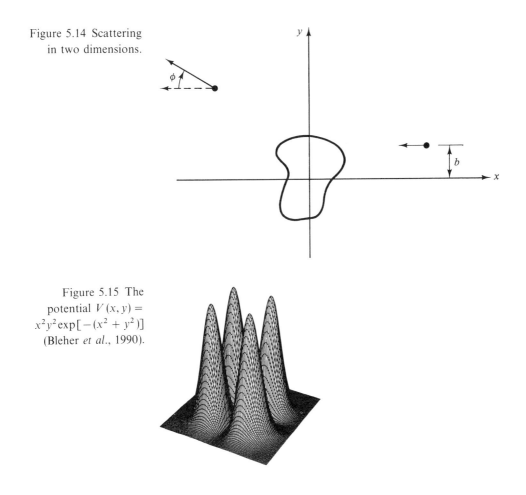

Figure 5.14 Scattering in two dimensions.

Figure 5.15 The potential $V(x, y) = x^2 y^2 \exp[-(x^2 + y^2)]$ (Bleher *et al.*, 1990).

5.16(*b*) by a factor of order 10^3 (Figure 5.17(*c*)) does not reveal a smooth curve. (This persists on still further magnification.) We call a value $b = b_s$ a singularity of the scattering function, if, for an interval $[b_s - (\Delta b/2), b_s + (\Delta b/2)]$, a plot of the scattering function made as in Figures 5.16 and 5.17 always shows unresolved regions for any interval length Δb, and, in particular, for *arbitrarily small* Δb. Another, more precise, way of defining b_s as a singularity of the scattering function is to say that, in any small interval $[b_s - (\Delta b/2), b_s + (\Delta b/2)]$, there is a pair of *b*-values which yields scattering angles whose difference exceeds some value $K > 0$ which is *independent* of Δb. (That is arbitrarily small differences in *b* yield ϕ-values which differ by order 1.) The interesting result concerning the scattering function shown in Figure 5.16(*b*) is that the set of singular *b*-values is a fractal. Bleher *et al.* (1990) calculate a fractal dimension of approximately 0.67 for the singular set. We call the phenomenon seen in Figure 5.16(*b*) *chaotic scattering* as distinguished from the case of *regular scattering* (Figure 5.16(*a*)). (The transition from regular to chaotic scattering as the energy is lowered from the value in Figure 5.16(*a*) to the value in Figure 5.16(*b*) will be discussed in Chapter 8.)

The chaotic scattering phenomenology we have described above is a

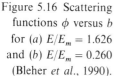

Figure 5.16 Scattering functions ϕ versus b for (*a*) $E/E_m = 1.626$ and (*b*) $E/E_m = 0.260$ (Bleher *et al.*, 1990).

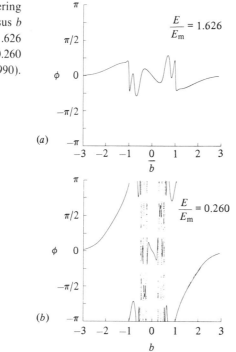

general feature of a large class of problems. Chaotic scattering has appeared in numerous applications including celestial mechanics (Petit and Hénon, 1986), the scattering of vortices in fluids (Eckhardt and Aref, 1988), scattering of microwaves (Doron *et al.*, 1990), the conversion of magnetic field energy to heat in solar plasmas (Lau and Finn, 1991). chemical reactions (Noid *et al.*, 1986), collisions between nuclei (Rapisarda and Baldo, 1991), and conductance fluctuations in very tiny two-dimensional conductor junctions (Jalabert *et al.*, 1990). The latter three examples are cases where it becomes important to consider the

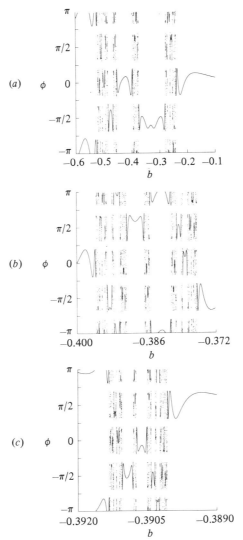

Figure 5.17 Successive magnifications of the scattering function (*a*) for a small *b* interval in Figure 5.16(*b*), (*b*) for a small *b* interval in Figure 5.17(*a*), and (*c*) for a small *b* interval in Figure 5.17(*b*) (Bleher *et al.*, 1990).

quantum mechanical treatment of a problem whose classical counterpart exhibits chaotic scattering. For further material on the quantum aspects of chaotic scattering see Blümel (1991), Cvitanović and Eckhardt (1989), Gaspard and Rice (1989a,b,c), and Blümel and Smilansky (1988).

5.5   The dynamics of chaotic scattering

How does the dynamics of the scattering problem lead to the phenomena we have observed in Figures 5.16(b) and 5.17? In order to gain some insight into this question we plot in Figure 5.18 the 'time delay' (the amount of time that a particle spends in the scattering region bouncing between the hills) as a function of the impact parameter b for the potential (5.6) with the same particle energy as for Figures 5.16(b) and 5.17. We see that the regions of poor resolution of the scattering function (cf. Figures 5.16(b) and 5.17) coincide with b-values for which the time delay is long. Indeed careful examination of magnifications suggests that the singularities of the scattering function coincide with the values of b where the time delay is infinite. Very near a value of b for which the time delay is infinite the time delay will be very long, indicating that the incident particle

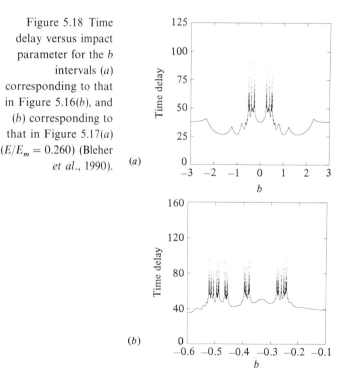

Figure 5.18 Time delay versus impact parameter for the b intervals (a) corresponding to that in Figure 5.16(b), and (b) corresponding to that in Figure 5.17(a) ($E/E_m = 0.260$) (Bleher et al., 1990).

experiences many bounces between potential hills before leaving the scattering region. Say we choose a b-value yielding a long time delay for which the particle experiences say 1000 bounces before exiting the scattering region. Now change b very slightly so as to increase the delay time by a small percentage yielding say 1001 bounces before the particle exits the scattering region. The presence of this one extra bounce means that the scattering angle for the two cases can be completely different. Hence, we expect arbitrarily rapid variations of ϕ with b near a b-value yielding an infinite time delay, and we may thus conclude that these values coincide with the singularities of the scattering function. The effect is illustrated in Figure 5.19 which shows two orbit trajectories whose b values differ by 10^{-8}. The orbit in Figure 5.19(a) ($b = -0.39013269$) experiences about 14 bounces (depending on how you define a bounce). The orbit in Figure 5.19(b) ($b = -0.39013268$) is very close to that in Figure 5.19(a) for the first 13 or so bounces but subsequently experiences about 4 more bounces than the orbit in Figure 5.19(a). The two orbits have completely different scattering angles, one yielding scattering upward (Figure 5.19(a)) and the other yielding scattering downward.

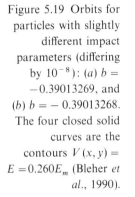

Figure 5.19 Orbits for particles with slightly different impact parameters (differing by 10^{-8}): (a) $b = -0.39013269$, and (b) $b = -0.39013268$. The four closed solid curves are the contours $V(x, y) = E = 0.260E_m$ (Bleher *et al.*, 1990).

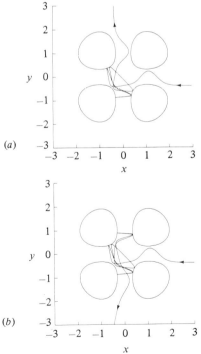

The interpretation of these results is as follows. The equations of motion are four-dimensional,

$$m \, dv/dt = -\nabla V(x), \tag{5.7a}$$

$$dx/dt = v, \tag{5.7b}$$

where $x = (x, y)$ and $v = (v_x, v_y)$, but because the particle energy,

$$E = \tfrac{1}{2} m v^2 + V(x), \tag{5.8}$$

is conserved, we may regard the phase space as being three-dimensional. (For example, we can regard the phase space as consisting of the three variables, x, y, θ, where θ is the angle the vector v makes with the positive x-axis. These three variables uniquely determine the system state, x, y, v_x, v_y, because (5.8) gives $|v|$ in terms of x and $y, |v| = [2(E - V(x))/m]^{1/2}$.) The presence of infinite time delays on a fractal set of b-values is due to the existence of a nonattracting chaotic invariant set that is in a bounded region of phase space. Orbits on this invariant set bounce forever between the hills never leaving the scattering region both for $t \to +\infty$ and for $t \to -\infty$. This chaotic set is essentially the intersection of its stable and unstable manifolds, each of which locally consists of a Cantor set of approximately parallel two-dimensional surfaces in the three-dimensional phase space. Thus, the stable and unstable manifolds are each fractal sets of dimension between 2 and 3.

We have, in numerically obtaining our scattering function plots, taken initial conditions at some large x-value $x = x_0$ and have chosen the initial angle θ_0 between v and the positive x-axis to be $\theta_0 = \pi$ (i.e., $v_{y_0} = 0$ and $v_{x_0} < 0$; see Figure 5.14). This defines a line in the space (x, y, θ) which we regard as the phase space. This line of initial conditions generically intersects the stable manifold of the nonattracting chaotic invariant set in a Cantor set of dimension between zero and one[2] (cf. Eq. (3.46)). It is this intersection set that is the set of singular b-values of the scattering function. Since these b-values correspond to initial conditions on the stable manifold of the chaotic invariant set, the orbits they generate approach the invariant set as $t \to +\infty$; hence they never leave the scattering region.

Figure 5.20(a) shows a numerical plot of the $y = 0$ cross section of the stable manifold of the chaotic invariant set. This plot is created by taking a grid of initial conditions in (x_0, θ_0) and integrating them forward in time. Then only those initial conditions yielding long delay times are plotted. We observe that the stable manifold intersection appears as smooth (and swirling) along one dimension with (poorly resolved) fine-scale (presumably fractal) structure transverse to that direction. Figure 5.20(b) shows a similar plot of the intersection of the unstable manifold with the $y = 0$ plane obtained by integrating initial conditions on the grid backwards in time and again plotting those initial conditions whose orbits remain in the scattering region for a long time. We see that the unstable manifold picture is a mirror image (through the line $x_0 = 0$) of the stable manifold picture. This is a result of the time reversal symmetry[5] of Eqs. (5.7) (they are invariant to the transformation $\mathbf{v} \rightarrow -\mathbf{v}, t \rightarrow -t$) and the symmetry of the potential (5.6). In particular this means that the stable and unstable manifolds have the same fractal dimension. Figure 5.20(c) shows the

Figure 5.20 Intersection with the $y = 0$ surface of section of (a) the stable manifold, (b) the unstable manifold, and (c) the nonattracting chaotic invariant set (Bleher *et al.*, 1990).

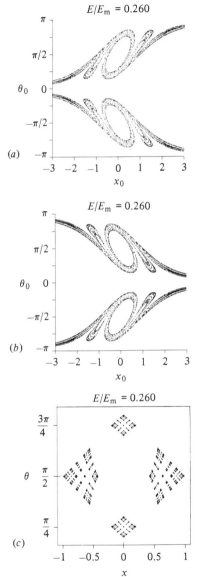

intersection of the chaotic invariant set with the plane $y = 0$. This picture is consistent with the invariant set being the intersection of its computed stable and unstable manifolds (i.e., the set shown in Figure 5.20(c) is the intersection of the sets shown in Figures 5.20(a) and (b)). Apparently, these intersections occur with angles bounded well away from zero. Hence, there appear to be no tangencies between the stable and unstable manifolds, thus supporting the idea that, in this case, the dynamics on the invariant set is hyperbolic. (See Bleher *et al.* (1990) for a description of how Figure 5.20(c) is numerically computed. This computation makes use of a numerical technique for obtaining unstable chaotic sets which is discussed and analyzed by Nusse and Yorke (1989); see also Hsu *et al.* (1988).)

The existence of a Cantor set of singular b-values for the scattering function implies that it will often be very difficult to obtain accurate values of the scattering angle if there are small random experimental errors in the specification of b. This situation is similar to that which exists when there are fractal basin boundaries.[6] Indeed we can employ a modification of the uncertainty exponent technique of Section 5.2 to obtain the fractal dimension of the singular set. We observe that, for our example (Eq. (5.6) with $E/E_m = 0.260$), small perturbations about a singular b-value can lead to either upward scattering ($0 \leq \phi \leq \pi$ as in Figure 5.19(a)) or downward scattering ($-\pi \leq \phi \leq 0$ as in Figure 5.19(b)). Thus, we randomly choose many values of b in an interval containing the Cantor set. We then perturb each value by an amount ε and determine whether the scattering is upward or downward for each of the three impact parameter values, $b - \varepsilon$, b and $b + \varepsilon$. If all three scatter upward or all three scatter downward, we say that the b-value is ε-certain, and, if not, we say it is ε-uncertain. We do this for several ε-values and plot on a log–log scale the fraction of uncertain b-values $\tilde{f}(\varepsilon)$. The result is shown in Figure 5.21 which shows a good straight line fit to the data indicating a power law dependence $\tilde{f}(\varepsilon) \sim \varepsilon^\alpha$. The exponent α is related to the dimension of the set of singular b-values by

$$D_0 = 1 - \alpha \qquad (5.9)$$

(i.e., Eq. (5.3b) with $N = 1$ corresponding to the fact that the initial conditions of the scattering function lie along a line in the three-dimensional phase space). The straight line fit in Figure 5.21 yields a slope of $\alpha = 0.33$ corresponding to a fractal dimension of $D_0 = 0.67$ for the scattering function of Figure 5.16(b) and 5.17. The dimension of the stable and unstable manifolds in the full three-dimensional phase space is $2 + D_0$, and the dimension of their intersection (which is the dimension D_{cs} of the chaotic set) is (cf. Eq. (3.46))

$$D_{cs} = 2D_0 + 1. \tag{5.10}$$

The dimension of the intersection of the chaotic set with the $y = 0$ plane[2] (i.e., the dimension of the set plotted in Figure 5.20(c)) is $2D_0$.

In all of our discussion of chaotic scattering we have been concerned with a particular illustrative example of scattering in two degrees of freedom (i.e., the two spatial dimensions x and y). The phenomena we see are typical for two-degree-of-freedom scattering. Other works on chaotic scattering have also tended to be for examples with two degrees of freedom. The possibility of new chaotic scattering phenomena in systems with more than two degrees of freedom remains largely unexplored. An exception is the paper of Chen *et al.* (1990b) who consider the question of whether the presence of a chaotic invariant set in the phase space implies a fractal set of singularities in a scattering function plot (i.e., a plot giving an after-scattering variable as a function of a single before-scattering variable). They find that, when the number of degrees of freedom is greater than 2, the scattering function typically does not exhibit fractal behavior, even when the invariant set is fractal and chaotic, unless the fractal dimension of the invariant set is large enough,

$$D_{cs} > 2M - 3, \tag{5.11}$$

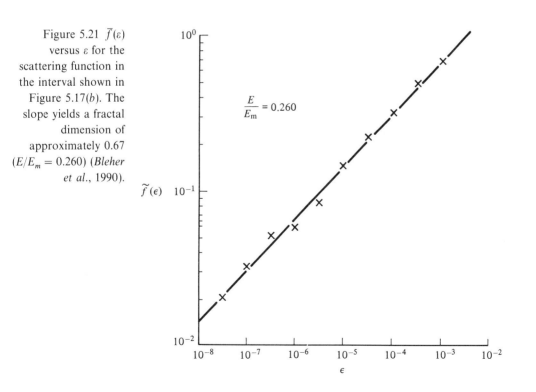

Figure 5.21 $\tilde{f}(\varepsilon)$ versus ε for the scattering function in the interval shown in Figure 5.17(b). The slope yields a fractal dimension of approximately 0.67 ($E/E_m = 0.260$) (*Bleher et al.*, 1990).

where M is the number of degrees of freedom.[7] Since D_{cs} for $M = 2$ is greater than 1 in the fractal case, we see that Eq. (5.11) is always satisfied in chaotic two-degree-of-freedom potential scattering problems. In the case of three-degree-of-freedom systems, however, we require $D_{cs} > 3$. Chen *et al.* illustrate this numerically for the simple three-dimensional scattering system consisting of four hard reflecting spheres of equal radii with centers located on the vertices of a regular tetrahedron, as illustrated in Figure 5.22. They show numerically that as the sphere radius increases, D_{cs} increases from below 3 to above 3, and this is accompanied by the appearance of fractal behavior in typical scattering functions.

5.6 The dimensions of nonattracting chaotic sets and their stable and unstable manifolds

We have seen in Chapter 4 that there is an apparent relationship between the Lyapunov exponents of a chaotic attractor and its information dimension (Eqs. (4.36)–(4.38)). In this section we will show that the same is true for nonattracting chaotic invariant sets of the type that arise in chaotic transients, fractal basin boundaries, and chaotic scattering (Kantz and Grassberger, 1985; Bohr and Rand, 1987; Hsu *et al.*, 1988). In particular, we treat the case of a smooth two-dimensional map $\mathbf{M}(\mathbf{x})$ which has a nonattracting variant chaotic set Λ. In Figure 5.23 we schematically picture the invariant set as being the intersection of stable and unstable manifolds. Let B, also shown in the figure, be a compact set containing the invariant set such that under the action of $\mathbf{M}$ almost all points (with respect to Lebesgue measure) eventually leave B and never return. The only initial conditions that generate forward orbits which remain forever in B are those which lie on the invariant set and its stable manifold. Thus, part of B must map out of B and the Lebesgue measure (area) which remains in B is decaying.

Say, we randomly sprinkle $N(0)$ points uniformly in B. After iterating these points for t iterates, only $N(t) < N(0)$ have not yet left. The average

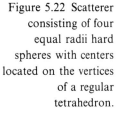

Figure 5.22 Scatterer consisting of four equal radii hard spheres with centers located on the vertices of a regular tetrahedron.

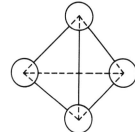

decay time (see Eq. (5.1)) is

$$\frac{1}{\langle \tau \rangle} = \lim_{t \to \infty} \frac{1}{t} \lim_{N(0) \to \infty} \ln[N(0)/N(t)]. \qquad (5.12)$$

Lyapunov exponent approximations can be calculated for any finite time t by taking the $N(t)$ orbits still in the box B and using their initial conditions to calculate exponents over the time interval $0 \le n \le t$. Averaging these over the $N(t)$ initial conditions, letting $N(0)$ approach infinity, and then letting t approach infinity, we obtain the Lyapunov exponents with respect to the natural measure on the nonattracting chaotic invariant set (this measure is defined subsequently). Following the notation of Eqs. (4.32),

$$\bar{h}_t(\mathbf{u}_0) = \frac{1}{t} \frac{1}{N(t)} \sum_{i=1}^{N(t)} \ln |\mathbf{DM}^t(\mathbf{x}_{0i}) \cdot \mathbf{u}_0|,$$

$$h(\mathbf{u}_0) = \lim_{t \to \infty} \lim_{N(0) \to \infty} \bar{h}_t(\mathbf{u}_0),$$

and $h(\mathbf{u}_0)$ has two possible values $h_1 > 0 > h_2$ which depend on $\mathbf{u}_0$. In particular, assuming hyperbolicity, $h(\mathbf{u}_0) = h_2$ if $\mathbf{u}_0$ is tangent to the stable manifold and $h(\mathbf{u}_0) = h_1$ otherwise.

We now define natural measures on the stable manifold, on the unstable manifold, and on the invariant set Λ itself. We denote these measures by μ_s, μ_u and μ, respectively. We then heuristically relate the information dimensions of these measures to the quantities h_1, h_2 and $\langle \tau \rangle$.

We define the stable manifold natural measure of a set A to be

$$\mu_s(A) = \lim_{t \to \infty} \lim_{N(0) \to \infty} N_s(A, t)/N(t), \qquad (5.13)$$

where $N_s(A, t)$ is the number of the remaining $N(t)$ orbit points whose initial conditions lie in the set A. For large but finite time t the $N(t)$ orbit points still in B are arranged in narrow strips of width of order $\exp(h_2 t)$ along the unstable manifold running horizontally the full length across the box B (see Figure 5.23). Iterating these orbits backward t iterates to see

Figure 5.23 Schematic of the nonattracting chaotic invariant set.

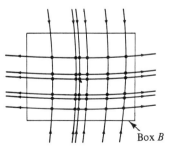

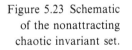

the initial conditions that they came from, we find that these initial conditions are arranged in narrow strips of width of order $\exp(-h_1 t)$ along the stable manifold running vertically the full height across B. (To see this, it is useful to think of the example of the horseshoe map and to associate our box B with the set S in Figure 4.1. In particular, refer to Figures 4.2(d) and (e).) The projection of this set of initial conditions along the stable manifold onto a horizontal line is a fattened Cantor-like set of intervals of size $\varepsilon \sim \exp(-h_1 t)$. Let $1 + d_s$ be the information dimension of the measure μ_s. There will be of the order of ε^{-d_s} intervals. Hence, the total length occupied by the fattened Cantor set is of the order of ε^{1-d_s}. This length is proportional to the fraction of the $N(0)$ initial conditions that have not yet left B, $\varepsilon^{1-d_s} \sim N(t)/N(0) \sim \exp(-t/\langle \tau \rangle)$. Thus,

$$\exp[-th_1(1 - d_s)] \sim \exp(-t/\langle \tau \rangle).$$

Hence, taking logarithms we obtain the following formula for the dimension of the stable manifold,

$$d_s = 1 - \frac{1}{\langle \tau \rangle h_1} \tag{5.14}$$

For the unstable manifold we iterate t times and consider the image of the points remaining in B. This image, as we have said, consists of horizontal strips along the unstable manifold. These strips have vertical widths of the order of $\varepsilon \sim \exp(h_2 t)$. We define the natural measure μ_u on the unstable manifold as

$$\mu_u(A) = \lim_{t \to \infty} \lim_{N(0) \to \infty} N_u(A, t)/N(t), \tag{5.15}$$

where $N_u(A, t)$ is the number of orbit points in $A \cap B$ at time t. The density of points in the horizontal strip is larger than the density of the original sprinkling of the $N(0)$ points at $t = 0$ if the map is area contracting. In particular, areas are typically contracted by an amount $\exp[t(h_1 + h_2)]$ where $(h_1 + h_2) < 0$ for contraction; see Figure 4.16. Thus, letting $(1 + d_u)$ denote the information dimension of the measure μ_u, we have that the fraction of points remaining in B is roughly

$$\frac{\varepsilon^{1-d_u}}{\exp[(h_1 + h_2)t]} \sim \exp(-t/\langle \tau \rangle)$$

which yields

$$d_u = \frac{h_1 - 1/\langle \tau \rangle}{|h_2|} \tag{5.16}$$

To define the natural measure μ on the chaotic invariant set itself, we first pick a number ξ in the range $0 < \xi < 1$ (e.g., we might chose $\xi = \frac{1}{2}$). We then have

$$\mu(A) = \lim_{t \to \infty} \lim_{N(0) \to \infty} N_\xi(A,t)/N(t), \tag{5.17}$$

where $N_\xi(A,t)$ is the number of orbit points in $A \cap B$ at time ξt. For t and $N(0)$ large, trajectories that remain in B would stay near the invariant set for most of the time between zero and t, except at the beginning when they are approaching the invariant set along the stable manifold, and at the end when they are exiting along the stable manifold. Thus, the measure μ, defined in Eq. (5.17) is expected to be independent of ξ, as long as $0 < \xi < 1$ (note that Eq. (5.17) gives μ_s if $\xi = 0$ and μ_u if $\xi = 1$). Since the invariant set is the intersection of its stable and unstable manifold, we conclude from Eqs. (5.14) and (5.16) that the information dimension of μ is

$$d = d_s + d_u = \left(h_1 - \frac{1}{\langle \tau \rangle} \right) \left(\frac{1}{h_1} - \frac{1}{h_2} \right). \tag{5.18}$$

Note that for an attractor $\langle \tau \rangle = \infty$ and Eq. (5.18) reduces to the Kaplan–Yorke result Eq. (4.38). Note also that in an area preserving map (as would result for a surface of section for a conservative system) $h_1 + h_2 = 0$. Thus, Eqs. (5.14) and (5.16) give $d_u = d_s = d/2 = 1 - 1/(\langle \tau \rangle h_1)$. This is the case, for example, for chaotic scattering (Figure 5.20). Comparing Eq. (5.18) with Young's formula Eq. (4.46), we conclude that the metric entropy of the measure μ is

$$h(\mu) = h_1 - \frac{1}{\langle \tau \rangle}. \tag{5.19}$$

The derivations of the dimension formulae Eqs. (5.14), (5.16) and (5.18) given above are heuristic. Thus, it is worthwhile to test them numerically. This has been done for all three formulae using the Hénon Eqs. (1.14), in the paper of Hsu *et al.* (1988) and for (5.18) by Kantz and Grassberger (1985). The case $A = 1.42$ and $B = 0.3$ studied by Hénon (Figure 1.12) has a strange attractor. Increasing A slightly it is observed that there is no bounded attractor, and almost all initial conditions with respect to Lebesgue measure generate orbits which go to infinity. In this case there is a nonattracting chaotic invariant set. Numerical experiments checking the formulae for d_s, d_u and d using $A = 1.6$, $B = 0.3$ and $A = 3.0$, $B = 0.3$ were performed and yielded data which tended to support the formulae. Figures 5.24 from Hsu *et al.* (1988) show the nonattracting chaotic invariant set, its stable manifold, and its unstable manifold for $A = 1.6$ and $B = 0.3$.

Appendix: Derivation of Eqs. (5.3)

Here we derive Eqs. (5.3). Let $B(\varepsilon, \Sigma)$ be the set of points within ε of a closed bounded set Σ whose box-counting dimension is D_0 (cf. Grebogi *et al.*, 1988a). We cover the region of the N-dimensional space in which the set lies by a grid of spacing ε. Each point x in Σ lies in a cube of the grid. Any point y within ε of x must therefore lie in one of the 3^N cubes which are the original cube containing x or a cube touching the original cube. Thus, the volume of $B(\varepsilon, \Sigma)$ satisfies,

$$\text{Vol}\,[B(\varepsilon, \Sigma)] \leq 3^N \varepsilon^N \bar{N}(\varepsilon)$$

where $\bar{N}(\varepsilon)$ is the number of cubes needed to cover Σ. Now, cover Σ with a grid of cubes of edge length $\varepsilon/N^{1/2}$. Any two points within such a cube are separated by a distance of at most ε. Thus, every cube of the grid used in covering Σ lies within $B(\varepsilon, \Sigma)$,

$$\text{Vol}\,[B(\varepsilon, \Sigma)] \geq (\varepsilon/N^{1/2})^N \bar{N}(\varepsilon/N^{1/2}).$$

Hence,

$$N\frac{\ln N^{-1/2}}{\ln \varepsilon} + \frac{\ln \bar{N}(\varepsilon/N^{1/2})}{\ln(\varepsilon/N^{1/2}) + \ln N^{1/2}} + N \geq \frac{\ln\{\text{Vol}\,[B(\varepsilon, \Sigma)]\}}{\ln \varepsilon}$$

$$\geq \frac{\ln 3^N}{\ln \varepsilon} + \frac{\ln \bar{N}(\varepsilon)}{\ln \varepsilon} + N$$

Letting ε approach zero and noting the definition of the box-counting dimension D_0 Eq. (3.1), we have

$$\lim_{\varepsilon \to 0} \frac{\ln \text{Vol}\,[B(\varepsilon, \Sigma)]}{\ln \varepsilon} = N - D_0.$$

Figure 5.24 (a) The invariant set for the Hénon map with $A = 1.6$ and $B = 0.3$. (b) A magnification of the invariant set in the small rectangle shown in (a). (c) The stable manifold. (d) The unstable manifold. (Hsu *et al.*, 1988.)

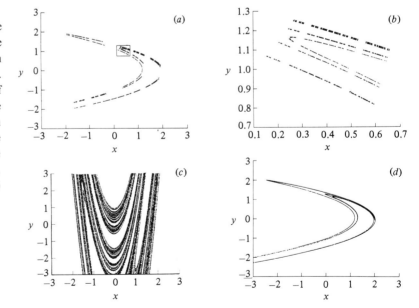

Since $f(\varepsilon)$ is proportional to Vol $[B(\varepsilon, \Sigma)]$, we have

$$\lim_{\varepsilon \to 0} \frac{\ln f(\varepsilon)}{\ln \varepsilon} = N - D_0 = \alpha,$$

which we have abbreviated in Eq. (5.3a) as $f(\varepsilon) \sim \varepsilon^{\alpha}$.

Problems

1. Describe the basin boundary structure for the following systems. (*a*) The map given by Eq. (3.3) and Figure 3.3 regarding $x = +\infty$ and $x = -\infty$ as two attractors. (*b*) The map shown in Figure 5.25, where O, U and W are unstable fixed points; B is an attracting fixed point; points in (O, U) tend to B; points in $(-\infty, 0)$ tend to $-\infty$; and we assume that $-\infty$ and B are the only attractors. In particular show for both (*a*) and (*b*) that regions of the basin boundary which are fractal are interwoven on arbitrarily fine scale with nonfractal regions of the boundary.

2. Consider a sequence of points λ^n $(n = 0, 1, 2, \ldots)$ in $[0, 1]$ where $0 < \lambda < 1$. Define two sets A and B, where $A = \cup_{m=0}^{\infty} [\lambda^{2m+1}, \lambda^{2m}]$ and

Figure 5.25 The map for Problem 1(*b*).

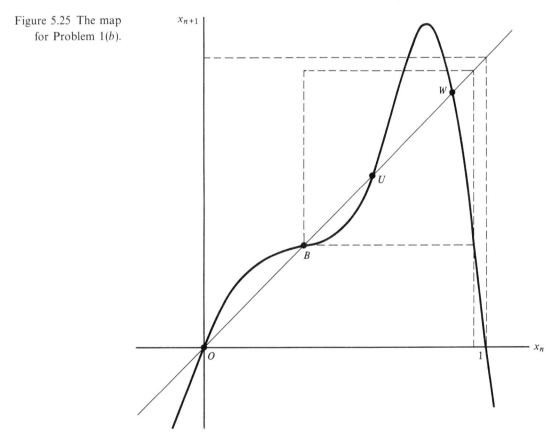

$B = \cup_{m=1}^{\infty}[\lambda^{2m}, \lambda^{2m-1}]$. Say we pick a point at random in $[0,1]$ and we specify that point to have an uncertainty ε. Show that the probability $f(\varepsilon)$ of a possible error in determining whether the randomly chosen point lies in A or in B is roughly given by $f(\varepsilon) \approx K\varepsilon \log(1/\varepsilon)$. (*Note*: We define the symbol $\sim$, so that this scaling is included in the statement $f(\varepsilon) \sim \varepsilon$ by virtue of $\lim_{\varepsilon \to 0}[(\log f(\varepsilon))/(\log \varepsilon)] = 1$ for $f(\varepsilon) = K\varepsilon \log(1/\varepsilon)$.)

3. Consider a particle which experiences two-dimensional free motion without friction between perfectly elastic bounces off three hard cylinders (Figure 5.26). At each bounce the angle of incidence is equal to the angle of reflection. Consider initial conditions on the y-axis in the range $-K > y > K$ ($K > r_0$) directed parallel to the x-axis to the right. Argue that the set of initial conditions that bounce forever between the cylinders is a Cantor set. To do this consider the set of initial conditions that experience at least one bounce, the set that experiences at least two bounces, three bounces, etc. Argue that the set experiencing at least n bounces consists of 2^{n-1} small intervals, and that each of these contain two of the two small intervals of initial conditions which experience $n + 1$ bounces.

Figure 5.26 Scattering from three hard cylinders symmetrically arranged with centers at the corners of an equilateral triangle.

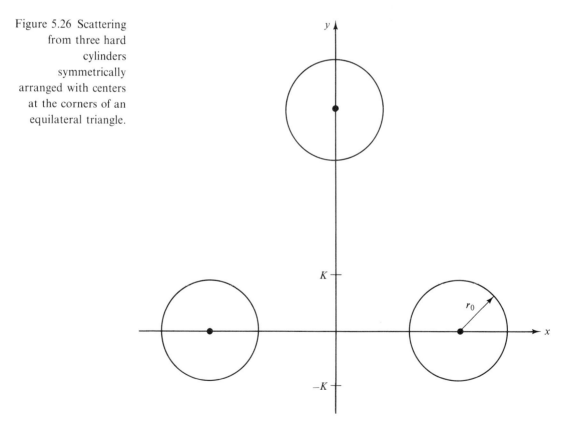

Notes

1. The horseshoe map specifies the dynamics of points in the square S. The fate of orbits that leave the square depends on the dynamics that orbits experience outside the square. For example, they may be attracted to some periodic attractor outside the square, or the dynamics may be such that orbits which leave the square are eventually fed back into the square. In the latter case the invariant set of the horseshoe map may be embedded in some larger chaotic invariant set that forms a chaotic attractor. In the former case initial conditions near the invariant set of the horseshoe will experience a chaotic transient before settling down to motion on the attracting periodic orbit.

2. In Eq. (3.46) we have stated that the dimension d_0 of the intersection of two smooth surfaces (these are nonfractal sets) of dimensions d_1 and d_2 is generically $d_0 = d_1 + d_2 - N$ where N is the dimension of the Cartesian space in which these sets lie. Mattila (1975) proves that this equation holds fairly generally for the case where set 1 is a plane of dimension d_1, set 2 is a fractal of Hausdorff dimension d_2, and d_0 is the Hausdorff dimension of the intersection. The concept of Hausdorff dimension is discussed in the appendix to Chapter 3, but we note here that it is very closely related to the box-counting dimension and, based on examples (Pelikan, 1985), it is believed that the box-counting dimension and the Hausdorff dimension of fractal basin boundaries are typically the same.

3. The construction for the one-dimensional map of Figure 5.6 (*a*) is analogous to the construction here. In particular, compare Figures 5.6(*b*) and 5.11.

4. For reviews dealing with chaotic scattering see Eckhardt (1988a) and Smilansky (1992).

5. Time reversal symmetry can be absent for other problems of Hamiltonian mechanics (e.g., for charged particle motion in a static magnetic field).

6. As in the case of fractal basin boundaries, the difficulty in making a determination (in this case, a determination of the scattering angle) increases with the fractal dimension. The worst case is attained when the scattering is nonhyperbolic. In that situation orbits entering the scattering region can stick for a long time near a hierarchy of bounding KAM surfaces (Meiss and Ott, 1985) and this leads to very complicated behavior. (KAM surfaces are discussed in Chapter 7.) Lau *et al.* (1991) show that in this case the dimension of the set of values on which the scattering function is singular is one, in spite of the fact that this set has zero Lebesgue measure.

7. Equation (5.11) holds only in the case of time reversible dynamics.[5]

CHAPTER SIX

Quasiperiodicity

6.1 Frequency spectrum and attractors

In Chapter 1 we introduced three kinds of dynamical motions for continuous time systems: steady states (as in Figure 1.10(a)), periodic motion (as in Figure 1.10(b)), and chaotic motion (as in Figure 1.2). In addition to these three, there is another type of dynamical motion that is common; namely, *quasiperiodic* motion. Quasiperiodic motion is especially important in Hamiltonian systems where it plays a central role (see Chapter 7). Furthermore, in dissipative systems quasiperiodic *attracting* motions frequently occur.

Let us contrast quasiperiodic motion with periodic motion. Say we have a system of differential equations with a limit cycle attractor (Figure 1.10(b)). For orbits on the attractor, a dynamical variable, call it $f(t)$, will vary periodically with time. This means that there is some smallest time $T > 0$ (the period) such that $f(t) = f(t + T)$. Correspondingly, the Fourier transform of $f(t)$,

$$\hat{f}(\omega) = \int_{-\infty}^{\infty} f(t) \exp(i\omega t) \, dt, \qquad (6.1)$$

consists of delta function spikes of varying strength located at integer multiples of the fundamental frequency $\Omega = 2\pi/T$,

$$\hat{f}(\omega) = 2\pi \sum_n a_n \delta(\omega - n\Omega). \qquad (6.2)$$

Basically, quasiperiodic motion can be thought of as a mixture of periodic motions of several different fundamental frequencies. We speak of N-frequency quasiperiodicity when the number of fundamental frequencies that are 'mixed' is N. In the case of N-frequency quasiperiodic motion a dynamical variable $f(t)$ can be represented in terms of a function of N independent variables, $G(t_1, t_2, \ldots, t_N)$, such that G is periodic in each of its N independent variables. That is,

$$G(t_1, t_2, \ldots, t_i + T_i, \ldots, t_N) = G(t_1, t_2, \ldots, t_i, \ldots, t_N), \qquad (6.3)$$

where, for each of the N variables, there is a period T_i. Furthermore, the N frequencies $\Omega_i \equiv 2\pi/T_i$ are *incommensurate*. This means that no one of the frequencies Ω_i can be expressed as a linear combination of the others using coefficients that are rational numbers. In particular, a relation of the form

$$m_1\Omega_1 + m_2\Omega_2 + \cdots + m_N\Omega_N = 0 \tag{6.4}$$

does not hold for *any* set of integers, $m_1, m_2, \ldots, m_N$ (negative integers are allowed), except for the trivial solution $m_1 = m_2 = \cdots = m_N = 0$. In terms of the function G, an N-frequency quasiperiodic dynamical variable $f(t)$ can be represented as

$$f(t) = G(t, t, t, \ldots, t). \tag{6.5}$$

That is, f is obtained from G by setting all its N variables equal to t; $t_1 = t_2 = \cdots = t_N = t$. Due to the periodicity of G, it can be represented as an N-tuple Fourier series of the form

$$G = \sum_{n_1, n_2, \ldots, n_N} a_{n_1 \ldots n_N} \exp[i(n_1\Omega_1 t_1 + n_2\Omega_2 t_2 + \cdots + n_N\Omega_N t_N)].$$

Thus setting $t = t_1 = t_2 = \cdots = t_N$ and taking the Fourier transform we obtain,

$$\hat{f}(\omega) = 2\pi \sum_{n_1, n_2, \ldots, n_N} a_{n_1 \ldots n_N} \delta(\omega - (n_1\Omega_1 + n_2\Omega_2 + \cdots + n_N\Omega_N)) \tag{6.6}$$

Hence the Fourier transform of a dynamical variable $\hat{f}(\omega)$ consists of delta functions at all integer linear combinations of the N fundamental frequencies $\Omega_1, \ldots, \Omega_N$.

Figure 6.1 shows the magnitude squared of the Fourier transform (i.e., the frequency power spectrum) of a dynamical variable for three experimental situations: (a) a case with a limit cycle attractor, (b) a case with a two frequency quasiperiodic attractor, and (c) a case with a chaotic attractor. These results (Swinney and Gollub, 1978) were obtained for an experiment on Couette–Taylor flow (see Figure 3.10(a)). The three spectra shown correspond to three values of the rotation rate of the inner cylinder in Figure 3.10(a), with (a) corresponding to the smallest rate and (c) corresponding to the largest rate. Note that for the quasiperiodic case the frequencies $n_1\Omega_1 + n_2\Omega_2$ are dense on the ω-axis, but, since their amplitudes decrease with increasing n_1 and n_2, peaks at frequencies corresponding to very large values of n_1 and n_2 are eventually below the overall noise level of the experiment. In the chaotic case, Figure 6.1(c), we see that peaks at the two basic frequencies Ω_1 and Ω_2 are present, but that the spectrum has also developed a broad continuous component. (Note that the broad continuous component in Figure 6.1(c) is far above the noise level of $\sim 10^{-4}$ evident in Figure 6.1(a).) The situation in Figure 6.1(c) is in contrast to that in Figure 6.1(b), where the only apparent frequency components are discrete (namely $n_1\Omega_1 + n_2\Omega_2$). The presence

Figure 6.1 Results for frequency power spectra for a Couette–Taylor experiment with increasing rotation rate of the inner cylinders (Gollub and Swinney, 1975).

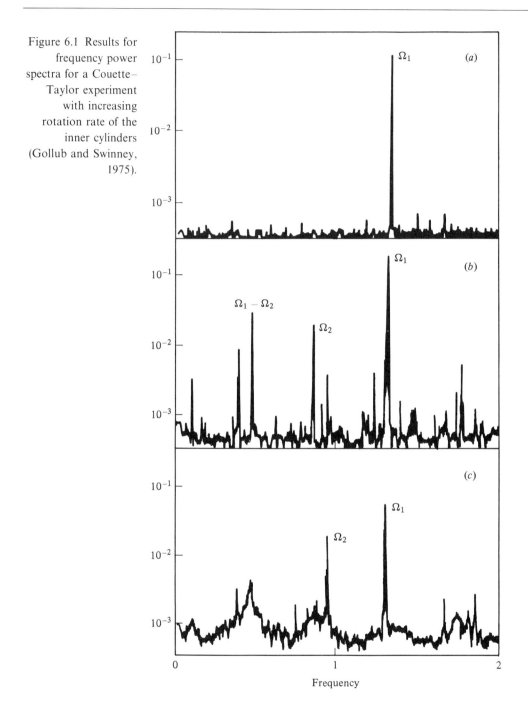

of a continuous component in a frequency power spectrum is a hallmark of chaotic dynamics.

A simple way to envision the creation of a quasiperiodic signal with a mixture of frequencies is illustrated in Figure 6.2, which shows two sinusoidal voltage oscillators in series with a nonlinear resistive element whose resistance R is a function of the voltage V across it, $R = R(V)$. Since the voltage sources are in series, we have $V = v_1 \sin(\Omega_1 t + \theta_0^{(1)}) + v_2 \sin(\Omega_2 t + \theta_0^{(2)})$. The current through the resistor, $I(t) = V/R(V)$, is a nonlinear function of V and hence will typically have all frequency components $n_1\Omega_1 + n_2\Omega_2$. Assuming that Ω_1 and Ω_2 are intercommensurate, the current $I(t)$ is two frequency quasiperiodic. The situation shown in Figure 6.2 is, in a sense, *too* simple to give very interesting behavior. If, for example, the value of the current I were to effect the dynamics of the voltage source oscillators, then a much richer range of behaviors would be possible, including *frequency locking* and chaos. Frequency locking refers to a situation where the interaction of two nonlinear oscillators causes them to self-synchronize in a coherent way so that their basic frequencies become commensurate (as we shall see, this implies that the motion is periodic) and remain locked in their commensurate relationship over a range of parameters. This will be discussed further shortly.

Let us now specialize to the case of attracting two frequency quasiperiodicity ($N = 2$) and ask, what is the geometrical shape of the attractor in phase space in such a case? To answer this, assume that we have a two frequency quasiperiodic solution of the dynamical system Eq. (1.3). In this case every component $x^{(i)}$ of the vector $\mathbf{x}$ giving the system state can be expressed as

$$x^{(i)}(t) = G^{(i)}(t_1, t_2)|_{t_1 = t_2 = t}.$$

Since $G^{(i)}$ is periodic in t_1 and t_2, we only need specify the value of t_1 and t_2 modulo T_1 and T_2 respectively. That is, we can regard the $G^{(i)}$ as being functions of two *angle* variables

$$\bar{\theta}_j = \Omega_j t_j \text{ modulo } 2\pi; j = 1, 2. \tag{6.7}$$

Figure 6.2 Two sinusoidal voltage sources driving a nonlinear resistor.

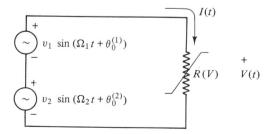

Thus the system state is specified by specifying two angles,

$$\mathbf{x} = \mathbf{G}(\bar{\theta}_1/\Omega_1, \bar{\theta}_2/\Omega_2),\qquad(6.8)$$

where $\mathbf{G}$ is periodic with period 2π in $\bar{\theta}_1$ and $\bar{\theta}_2$. Specification of one angle can be regarded geometrically as specifying a point on a circle. Specification of two angles can be regarded geometrically as specifying a point on a two-dimensional toroidal surface (cf. Figure 6.3). In the full phase space, the attractor is given by Eq. (6.8), which must hence be topologically equivalent to a two-dimensional torus (i.e., it is a distorted version of Figure 6.3). A two frequency quasiperiodic orbit on a toroidal surface in a three-dimensional $\mathbf{x}$ phase space is shown schematically in Figure 6.4. The orbit continually winds around the torus in the short direction (making an average of $\Omega_1/2\pi$ rotations per unit time) and simultaneously continually winds around the torus in the long direction (making an average of $\Omega_2/2\pi$ rotations per unit time). Provided that Ω_1 and Ω_2 are incommensurate, the orbit on the torus never closes on itself, and, as time goes to infinity the orbit will eventually come arbitrarily close to every point on the toroidal surface. If we consider the orbit originating from the initial condition $\mathbf{x}_0$ near (but not on) a *toroidal attractor*, as shown in Figure 6.4, then, as time progresses, the orbit circulates around the torus in the long and short directions and asymptotes to a two frequency quasiperiodic orbit on the torus.

We define the *rotation number* in the short direction as the average number of rotations executed by the orbit in the short direction for each rotation it makes in the long direction,

$$R = \Omega_1/\Omega_2.\qquad(6.9)$$

Figure 6.3 A point on a torus specifying the two angles $\bar{\theta}_1$ and $\bar{\theta}_2$.

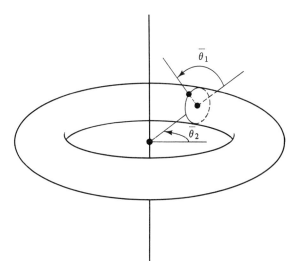

When R is irrational the orbit fills the torus, never closing on itself. When R is rational, $R = \tilde{p}/\tilde{q}$ with $\tilde{p}$ and $\tilde{q}$ integers that have no common factor, the orbit closes on itself after $\tilde{p}$ rotations the short way and $\tilde{q}$ rotations the long way. Such an orbit is periodic and has period $\tilde{p}T_1 = \tilde{q}T_2$. The case $R = 3$ is illustrated in Figure 6.5, where we see that the orbit closes on itself after three rotations the short way around and one rotation the long way around.

In Figures 6.2–6.4 we have restricted our considerations to two frequency quasiperiodicity. We emphasize, however, that the situation is essentially the same for N-frequency quasiperiodicity. In that case the orbit fills up an N-dimensional torus in the phase space. By an N-dimensional torus we mean an N-dimensional surface on which it is possible to specify uniquely any point by a smooth one to one relationship with the values of N angle variables. We denote the N-dimensional torus by the symbol T^N.

In some situations it is possible to rule out the possibility of quasiperiodicity. As an example, consider the system of equations studied by Lorenz, Eqs. (2.30). It was shown in the paper by Lorenz (1963) that all orbits eventually enter a spherical region, $X^2 + Y^2 + Z^2 < $ (constant), from which they never leave. Thus, X, Y and Z are bounded, and we may regard the phase space as Cartesian with axes X, Y and Z. A two frequency quasiperiodic orbit fills up a two-dimensional toroidal surface

Figure 6.4 Two frequency quasiperiodic orbit on a torus lying in a three-dimensional phase space $x = (x^{(1)}, x^{(2)}, x^{(3)})$.

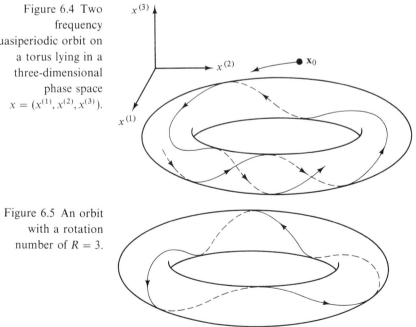

Figure 6.5 An orbit with a rotation number of $R = 3$.

in this space. Thus the toroidal surface is invariant under the flow. That is, evolving every point on the surface forward in time by any fixed amount maps the surface to itself. Furthermore, the volume inside the torus must also be invariant by the continuity of the flow. However, we have seen in Section 2.4.1 that following the points on a closed surface forward in time, the Lorenz equations contract the enclosed phase space volumes exponentially in time. Thus two frequency quasiperiodic motion is impossible for this system of equations.

6.2   The circle map

The system illustrated in Figure 6.2 is particularly simple. Since $\Omega_1 t$ and $\Omega_2 t$ appear only as the argument in sinusoids, we regard them as angles $\theta^{(1)}(t) = \Omega_1 t + \theta_0^{(1)}$ and $\theta^{(2)}(t) = \Omega_2 t + \theta_0^{(2)}$. In these terms the dynamical system reduces to

$$d\theta^{(1)}/dt = \Omega_1 \text{ and } d\theta^{(2)}/dt = \Omega_2.$$

Now taking a surface of section at $(\theta^{(2)} \text{ modulo } 2\pi) = (\text{const.})$, we obtain a one-dimensional map for $\theta_n = \theta^{(1)}(t_n)$ modulo 2π (where t_n denotes the time at the nth piercing of the surface of section),

$$\theta_{n+1} = (\theta_n + w) \text{ modulo } 2\pi, \qquad (6.10)$$

where $w = 2\pi\Omega_1/\Omega_2$. Geometrically, the map Eq. (6.10) can be thought of as a rigid rotation of the circle by the angle w. For incommensurate frequencies, Ω_1/Ω_2 is irrational, and for any initial condition, the orbit obtained from the map (6.10) densely fills the circle, creating a uniform invariant density of orbit points in the limit as time goes to infinity. On the other hand, if $\Omega_1/\Omega_2 = \tilde{p}/\tilde{q}$ is rational, then the orbit is periodic with period $\tilde{q}$ ($\theta_{n+\tilde{q}} = (\theta_n + \tilde{q}w)$ modulo $2\pi = \theta_n$). Thus there is only a zero Lebesgue measure set of w (namely, the rationals) for which periodic motion (as opposed to two frequency quasiperiodic motion) applies. Let us now ask, what would we expect to happen if the two voltage oscillators in Figure 6.2 were allowed to couple nonlinearly? Would the quasiperiodicity be destroyed and immediately be replaced by periodic orbits? Since the rationals are dense, and coupling is known to induce frequency locking, this question deserves some serious consideration. To answer this Arnold (1965) considered a model that addresses the main points. In particular, the effect of such coupling of the oscillator dynamics is to add nonlinearity to Eq. (6.10). Thus Arnold introduced the map,

$$\theta_{n+1} = (\theta_n + w + k\sin\theta_n) \text{ modulo } 2\pi, \qquad (6.11)$$

where the term $k\sin\theta$ models the effect of the nonlinear oscillator coupling. This map is called the *circle map*. In what follows we take w to lie in the range $[0, 2\pi]$.

Although deceptively simple in appearance, the circle map (like the logistic map) reveals a wealth of intricate behavior. It is of interest to understand the behavior of this map as a function of both w and the nonlinearity parameter k. A key role is played by the rotation number, which for this case is given by

$$R = \frac{1}{2\pi} \lim_{m \to \infty} \frac{1}{m} \sum_{n=0}^{m-1} \Delta\theta_n, \qquad (6.12)$$

where $\Delta\theta_n = w + k \sin \theta_n$. For $k = 0$, we have $R = w/2\pi$ and the periodic orbits (rational values of R) only occur for a set of w of Lebesgue measure zero (i.e., rational values of $w/2\pi$). What are the characters of the sets of w-values yielding rational and irrational R if $k > 0$? Arnold (1965) considered this problem for small k. Specifically, we ask whether the Lebesgue measure of w yielding irrational R (i.e., quasiperiodic motion) immediately becomes zero when k is made nonzero. Arnold proved the fundamental result that quasiperiodicity survives in the following sense. For small k the Lebesgue measure of $w/2\pi$ yielding quasiperiodicity is close to 1 and approaches 1 as $k \to 0$. The set of w-values yielding quasiperiodicity, however, is nontrivial because arbitrarily close to a w-value yielding quasiperiodicity (irrational R) there are *intervals* of w yielding attracting periodic motion (rational R). (The existence of intervals where R is rational is what we mean by the term frequency locking.) Thus, the periodic motions are dense in w. (This corresponds to the fact that rational numbers are dense.) The set of w-values yielding quasiperiodicity is a Cantor set of positive Lebesgue measure (in the terminology of Section 3.9, it is a 'fat fractal'). Arnold's result was an important advance and is closely related to the celebrated KAM theory (for Kolmogorov, Arnold and Moser) for Hamiltonian systems (see Chapter 7). Specifically, in dealing with the circle map, as well as the problem which KAM theory addresses, one has to confront the difficulty of the 'problem of small denominators.' To indicate briefly the nature of this problem, first note that Arnold was examining the case of small k. The natural approach is to do a perturbation expansion around the case $k = 0$ (i.e., the pure rotation, Eq. (6.10)). One problem is that at every stage of the expansion this results in infinite series terms of the form

$$\sum_m \frac{A_m}{1 - \exp(2\pi imR)} \exp(im\theta).$$

For R any irrational, the number $[(mR)$ modulo 1] can be made as small as we wish by a proper choice of the integer m (possibly very large). Hence, the denominator, $1 - \exp(2\pi imR)$, can become small, and thus there is the concern that the series might not converge. To estimate this effect say

$\exp(2\pi imR)$ is close to 1 so that the denominator is small. This occurs when mR is close to an integer; call it n. In this case

$$1 - \exp(2\pi imR) \simeq -2\pi i(mR - n).$$

Thus, the magnitude of a term in the sum is approximately

$$\frac{1}{2\pi m} \frac{|A_m|}{|R - n/m|}.$$

(Clearly, if R is rational, then $R = n/m$ for some n and m, and this expansion fails. But we are here interested in the case of quasiperiodic motion for which R is irrational.) The convergence of the sum will depend on the number R. In particular, R-values satisfying the inequality

$$\left| R - \frac{n}{m} \right| > \frac{K}{m^{(2+\varepsilon)}}$$

for some positive numbers K and ε and all values of the integers m and n ($m \neq 0$) are said to be 'badly approximated by rationals.' It is a basic fact of number theory that the set of numbers (R in our case) that are not badly approximated by rationals has Lebesgue measure zero. The coefficients A_m are obtained from Fourier expansion of an analytic function, and hence the A_m decay exponentially with m, i.e., for some positive numbers σ and c, we have $|A_m| < c \exp(-\sigma|m|)$. Thus,

$$\frac{1}{2\pi m} \frac{|A_m|}{|R - m/n|} < O(m^{(1+\varepsilon)} \exp(-\sigma|m|)).$$

The exponential decay $\exp(-\sigma|m|)$ is much stronger than the power law increase $m^{(1+\varepsilon)}$, and convergence of the sum is therefore obtained for all R-values that are badly approximated by rationals. This, however, is only the beginning of the story since, at each stage of the perturbation expansion, sums of this type appear. While these sums converge, it still remains to show convergence of the perturbation expansion itself. Arnold was able to prove convergence of his perturbation expansion. Thereby he showed that the Lebesgue measure of w in $[0, 2\pi]$ for which there is quasiperiodicity (i.e., irrational R) is not zero for small k and that this measure approaches 2π in the limit $k \to 0$. Thus, for small k, quasiperiodicity survives and occupies most of the Lebesgue measure.

Let us now address the issue of frequency locking. As an example, consider the rotation number $R = 0$. This corresponds to a fixed point of the map. Hence we look for solutions of

$$\theta = (\theta + w + k \sin \theta). \tag{6.13}$$

The solution of this equation is demonstrated graphically in Figure 6.6(a) for several values of w. Note that there are no solutions of the fixed point equation, Eq. (6.13), for the value of w labeled $w < -w_0$ in the figure. As w

is increased from $w < -w_0$, the graph of $(\theta + w + k \sin \theta)$ becomes tangent to the dashed $45°$ line at $w = -w_0$. Thus two fixed point orbits, one stable and one unstable, are born by a tangent bifurcation as w increases through $w = -w_0$. As w is increased further, the two solutions continue to exist, until, as w increases through w_0, they are destroyed in a backward tangent bifurcation. Figure 6.6(b) shows the corresponding bifurcation diagram. From Eq. (6.13) we have $w_0 = k$. Thus we see that, for $k > 0$, the stable fixed point ($R = 0$) exists in an *interval* of w values, $k > w > -k$, whereas at $k = 0$ we only have $R = 0$ at the single value $w = 0$. This is what we mean by frequency locking. Similarly, one can show that, for small k, an attracting period two orbit (corresponding to a rotation number $R = \frac{1}{2}$) exists in a range $w_{1/2}^- < w < w_{1/2}^+$, where

$$w_{1/2}^\pm = \pi \pm k^2/4 + O(k^3).\qquad(6.14)$$

In general, for any rational rotation number $R = \tilde{p}/\tilde{q}$ there is a frequency locking range of w in which the corresponding attracting period $\tilde{q}$ orbit exists, and this range $(w_{\tilde{p}/\tilde{q}}^-, w_{\tilde{p}/\tilde{q}}^+)$ has a width $\Delta w_{\tilde{p}/\tilde{q}} = w_{\tilde{p}/\tilde{q}}^+ - w_{\tilde{p}/\tilde{q}}^-$ which scales as

$$\Delta w_{\tilde{p}/\tilde{q}} = O(k^{\tilde{q}}).\qquad(6.15)$$

Figure 6.6(a) Fixed point solutions of the circle map denoted by dots. (b) Bifurcation diagram for the $R = 0$ stable (solid curve) and unstable (dashed curve) orbits.

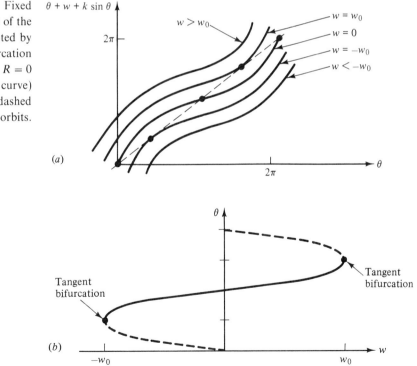

Furthermore, as w increases through the value $w_{\tilde{p}/\tilde{q}}^{-}$, the attracting period $\tilde{q}$ orbit with rotation number $\tilde{p}/\tilde{q}$ is born by a forward tangent bifurcation, and as w increases through the value $w_{\tilde{p}/\tilde{q}}^{+}$ the attracting period $\tilde{q}$ dies by a backward tangent bifurcation.

Note that, since the map function is monotonically increasing for $0 \leq k \leq 1$, its derivative and that of its n times composition are positive, $dM^n(\theta)/d\theta > 0$. Hence there can be no period doubling bifurcations of a period n orbit for any n in $0 \leq k \leq 1$, since the stability coefficient (slope of M^n) must be -1 at a period doubling bifurcation point.

Consider the total length in w (Lebesgue measure) of all frequency locked intervals in $[0, 2\pi]$,

$$\sum_{r=\tilde{p}/\tilde{q}} \Delta w_r.$$

Arnold's results show that this number is small for small k and decreases to zero in the limit $k \to 0$. Thus the set of w-values yielding quasiperiodic motion has most of the Lebesgue measure of w for small k. This set is a Cantor set of positive Lebesgue measure. (We have previously encountered such a set in Section 2.2 when we considered the set of r-values for which the logistic map yields attracting chaotic motion.) The situation can be illustrated schematically as in Figure 6.7 which shows regions of the $w-k$ plane (called Arnold tongues) in which the rotation numbers $R = 0, \frac{1}{2}, \frac{1}{3}$ and $\frac{2}{3}$ exist. We see that there are narrow frequency-locked tongues of rational R which extend down to $k = 0$. For higher periods (i.e., larger $\tilde{q}$ in $R = \tilde{p}/\tilde{q}$) the frequency-locked intervals becomes extremely small for small k. (This qualitative type of frequency-locking behavior occurring in tongues in parameter space has been found in numerical solutions of ordinary different equations, as well as in physical experiments.)

Figure 6.7 Arnold tongues for the circle map.

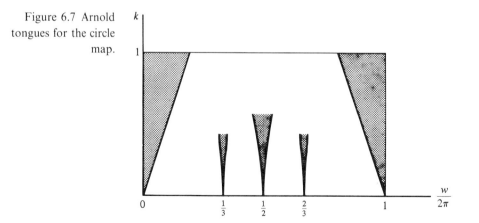

For $k < 1$ and a w-value yielding an irrational value of R, the orbit points on the resulting quasiperiodic orbit generate a smooth invariant density $\rho(\theta)$. In this case, by a smooth change of variables $\phi = f(\theta)$ the circle map can be transformed to the pure rotation

$$\phi_{n+1} = \phi_n + 2\pi R(w, k)$$

Since the pure rotation generates a uniform density $\tilde{\rho}(\phi) = 1/2\pi$, and the circle map is invertible for $k < 1$, we see by $\tilde{\rho}(\phi)\,d\phi = \rho(\theta)\,d\theta$ that the change of variables is

$$\phi = f(\theta) \equiv 2\pi \int_0^\theta \rho(\theta')\,d\theta'. \tag{6.16}$$

As k approaches 1 from below, the widths $\Delta w_{p/q}$ increase, and the sum $\Sigma \Delta w_r$, approaches 2π. That is, at $k = 1$, the entire Lebesgue measure in w is occupied by frequency-locked periodic orbits, and the quasiperiodic orbits occupy zero Lebesgue measure in w. Figure 6.8 shows a numerical plot of R versus w at $k = 1$. We see that R increases monotonically with w. The set of w values on which R increases is the Cantor set of zero Lebesgue measure on which R is irrational (i.e., the motion is quasiperiodic). The function R versus w at $k = 1$ is called a *complete devil's staircase*. At lower k we again obtain a monotonic function which increases only on the Cantor set of w-values where R is irrational, but now the Cantor set has positive Lebesgue measure (it is a fat fractal (Section 3.9)). We consequently say that R versus w is an *incomplete devil's staircase* for $1 > k > 0$.

The box-counting dimension of the set on which R increases for $k = 1$ (the complete devil's staircase case) has been calculated by Jensen, Bak

Figure 6.8 Complete devil's staircase at $k = 1$ (Jensen *et al.*, 1984).

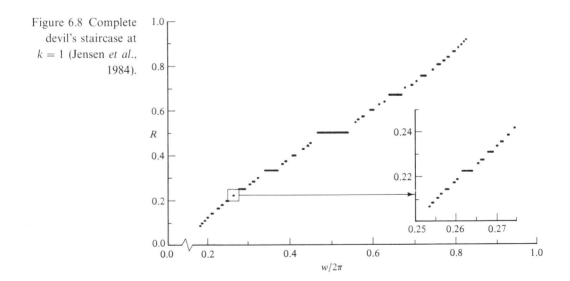

and Bohr (1983). They obtain a dimension value of $D_0 \simeq 0.87$. Furthermore, they claim that this value is universal in that it applies to a broad class of systems, not just the circle map. This contention is supported by the renormalization group theory of Cvitanović *et al.* (1985).

For $k > 1$, the circle map is noninvertible ($d\theta_{n+1}/d\theta_n$ changes sign as θ_n varies when $k > 1$). As a consequence of this, typical initial conditions can yield chaotic orbits but do not yield quasiperiodic orbits for $k > 1$. To see why quasiperiodic orbits do not result from typical initial conditions, we note that we have previously seen that a smooth change of variables Eq. (6.16) transforms the circle map to the pure rotation if there is a quasiperiodic orbit with a smooth invariant density $\rho(\theta)$. Since it is not possible to transform a noninvertible map to an invertible one (i.e., the pure rotation), we conclude that there can be no quasiperiodic orbits generating smooth invariant densities[1] for $k > 1$.

As an example of circle map type dynamics appearing in an experiment, we mention the paper of Brandstater and Swinney (1987) on Couette–Taylor flow (see Section 3.7). Under particular conditions the authors observe two frequency quasiperiodic motion on a two-dimensional toroidal surface. Figure 6.9(a) shows a delay coordinate plot of the orbit $V(t)$ versus $V(t - \tau)$ where $V(t)$ is the radial velocity component measured at a particular point in the flow. Taking a surface of section along the dashed line in Figure 6.9(a) one obtains a closed curve indicating that the orbit in Figure 6.9(a) lies on a two-dimensional torus. Figure 6.9(b) shows such a surface of section plot (for slightly different conditions from those in Figure 6.9(a)). Brandstater and Swinney then parameterize the location of orbit points in the surface of section by an angle θ measured from a point inside the closed curve. In Figure 6.9(c) they plot the value of θ at the $(n + 1)$th piercing of the surface of section versus its value at the nth piercing of the surface of section. We see that this map is indeed of a similar form to the circle map of Arnold:[2] it is invertible and is close to a pure rotation with an added nonlinear piece, $\theta_{n+1} = [\theta_n + w + P(\theta_n)]$ modulo 2π, where $P(\theta)$ is the periodic nonlinear piece, $P(\theta) = P(\theta + 2\pi)$. (In the absence of $P(\theta)$ the map would be two parallel straight lines at $45°$ (pure rotation), which Figure 6.9(c) would resemble if the wiggles due to $P(\theta)$ were absent.)

As an example of how circle map type phenomena can appear in a differential equation consider the equation,

$$d\theta/dt + \delta \sin \theta = V + W \cos(\Omega t), \qquad (6.17)$$

which may be viewed as a highly damped slowly forced pendulum such that the inertia term $d^2\theta/dt^2$ is negligible.[3] For small δ one may expand the solution as a power series in δ retaining only the first two terms,

$$\theta(t) = \theta^{(0)}(t) + \delta\theta^{(1)}(t) + O(\delta^2). \tag{6.18}$$

The lowest order solution is obtained by setting $\delta = 0$ in (6.17),

$$\theta^{(0)}(t) = \theta(t') + V(t - t') + (W/\Omega)[\sin(\Omega t) - \sin(\Omega t')], \tag{6.19}$$

and the next order solution is

$$\theta^{(1)}(t) = -\int_{t'}^{t} \sin\theta^{(0)}(t)\,dt. \tag{6.20}$$

Letting $\theta_n = \theta(t_n)$ modulo 2π, where $t_n = 2\pi n/\Omega$, we obtain a map from (6.18)–(6.20) by setting $t = t_{n+1}$ and $t' = t_n$,

$$\theta_{n+1} = \{\theta_n + w + kf(w, u)\sin[\theta_n + \phi(w, u) + \pi]\} \text{ modulo } 2\pi, \tag{6.21}$$

where (Problem 4)

$$f(w, u) = \left| \int_0^{2\pi} \exp\left[i\left(\frac{w\tau}{2\pi} + u\sin\tau\right) \right] d\tau \right|,$$

$w = 2\pi V/\Omega$, $k = \delta/\Omega$, $u = W/\Omega$, and $\phi(w, u)$ is the angle of the complex quantity whose magnitude appears above. Equation (6.21) with u fixed is, aside from the more complicated dependence on w, the same as the circle map. (Note, however, that Eq. (6.21) is only valid for small δ.) Thus frequency locking and Arnold tongues also occur here although the

Figure 6.9(a) Projection of the orbit onto the delay coordinate plane $V(t)$ versus $V(t - \tau)$. (b) The surface of section given by the dashed line in (a). (c) Experimental circle map obtained from (b) (Brandstater and Swinney, 1987).

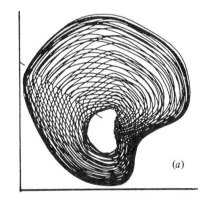

(a)

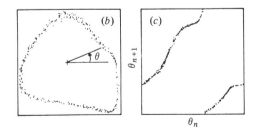

(b)   (c)

picture, Figure 6.7, will be distorted by the different parameter dependence of the map on w.

Equation (6.17) can be considered as a two-dimensional dynamical
system in the two variables $\theta^{(1)} = \theta$ and $\theta^{(2)} = \Omega t$ which are both angle
variables. Hence (6.17) describes a flow on a two-dimensional toroidal
surface. On this surface we can either have a quasiperiodic orbit, or an
attracting periodic orbit, the latter corresponding to a frequency-locked
situation. The attraction of orbits on the torus to a periodic orbit is
illustrated in Figure 6.10. As mentioned already, this behavior is
displayed by higher-dimensional systems. What happens in these higher-
dimensional systems is that there is an invariant two-dimensional torus
embedded in the phase space flow. On the torus, the flow can be either
quasiperiodic or else it can have an attracting periodic orbit (Figure 6.10).
When the flow is quasiperiodic, a surface of section yields a picture of the
attractor cross section which is either a closed curve, or several closed
curves, resulting from the intersection of the surface of section with the
attracting invariant torus. When the attractor is periodic, the surface of
section intersection with the attractor reveals a finite number of discrete
points (note, however, that there can still be an invariant torus on which
the attractor lies). We can think of the flow in the higher-dimensional
phase space as being attracted to a lower-dimensional (two-dimensional)
flow on the torus, on which, in turn, there can be quasiperiodic motion or
a periodic attractor.

A fairly common way that one sees chaos appearing as a system
parameter is varied is that first two frequency quasiperiodicity is seen,
then frequency locking to a periodic attractor, and then a chaotic
attractor. Since chaos is not possible for a two-dimensional flow, in order
for the chaos to appear, the orbit can no longer be on a two-dimensional
torus. Typically, as the parameter is increased toward the value yielding
chaos, the invariant two-dimensional torus is destroyed. When this
happens, it does so while in the parameter range in which the periodic

Figure 6.10 Attraction
of initial conditions on
a two-dimensional
torus to a periodic
orbit.

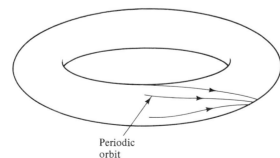

Periodic
orbit

attractor exists. In terms of the circle map, we can think of the destruction of the torus as analogous to the map becoming noninvertible as k increases through 1 (quasiperiodic orbits do not occur for typical initial conditions for $k > 1$). If we were to fix w and increase k, we might expect to see quasiperiodicity and then frequency locking as k is increased toward one, since the frequency locked regions have Lebesgue measure 2π in w at $k = 1$. The periodic solutions at $k = 1$ typically remain stable as k is increased past 1 into the region where chaos becomes possible. These periodic solutions can then become chaotic, for example, by going through a period doubling cascade.

In our discussion above of the onset of chaos for the circle map, we imagined a typically chosen variation along a path in parameter space; specifically, we imagined choosing a typical w and then increasing k. Another possibility is carefully to choose a path in parameter space such that we maintain the rotation number to be constant and irrational. Thus, as we increase k, we adjust w to keep $R(k, w)$ the same. Such a path threads between the frequency-locked Arnold tongues all the way up to $k = 1$. The same can be done in an experiment on a higher-dimensional system, in which case $k = 1$ corresponds to the point at which the torus is destroyed. Studies of this type of variation have revealed that there is a universal phenomonology in the behavior of systems approaching torus destruction along such a path in parameter space. The behavior depends on the rotation number R chosen but is essentially system-independent. Extensive work demonstrating this has been done for the case of the path on which the rotation number is held constant at the value given by the golden mean, $R = (\sqrt{5} - 1)/2 \equiv R_g$ (Shenker, 1982; Feigenbaum *et al.*, 1982; Ostlund *et al.*, 1983; Umberger *et al.*, 1986). This number is of particular significance because of its number theoretic properties. Specifically, in some sense (see Section 7.3.2), R_g is the most irrational of all irrational numbers in that it is most difficult to approximate by rational numbers of limited denominator size. These results for $R = R_g$ are obtained using the renormalization group technique, the same technique used to analyze the universal properties of the period doubling cascade (cf. Chapter 8). Perhaps the most striking of these results is that for the low frequency power spectrum of a process which is quasiperiodic with rotation number equal to the golden mean and parameters corresponding to the critical point at which the torus is about to be destroyed ($k = 1$ for the circle map). As illustrated in Figure 6.11, if one plots the frequency power spectrum $\hat{P}(\omega)$ divided by ω^2 versus the frequency ω on a log–log plot, then the result is predicted to be universal and periodic in $\log \omega$ for small ω. Furthermore, the periodicity length in $\log \omega$ is just the logarithm of the golden mean.

6.3   *N* frequency quasiperiodicity with *N* > 2

For N frequency quasiperiodicity we imagine a flow analogous to that pictured in Figure 6.4, but on an N-dimensional toroidal surface T^N. Figures 6.12 (a) and (b) illustrate the 'unwrapping' of quasiperiodic flows on T^2 and T^3. Since the N frequency quasiperiodic flow can be put in the form of a dynamical system on T^N given by

$$d\theta_i/dt = \Omega_i, \; i = 1, 2, \ldots, N,$$

via a suitable change of variables (cf. Eq. (6.8)), we see that an N frequency quasiperiodic attractor has N Lyapunov exponents that are zero. (For an N frequency quasiperiodic attractor, the other exponents are negative corresponding to attraction to the torus.)

It is useful to imagine the creation of quasiperiodic motion on an N torus for a continuous time dynamical system as arising via the successive addition of new active 'modes' of oscillation, each with its own frequency. Thus, say we start with a situation where the attracting motion is a steady state. We then increase some system parameter p. As p increases past $p_1, p_2, \ldots, p_N$ new active modes of oscillation, each with their own fundamental frequency, $\Omega_1, \Omega_2, \ldots, \Omega_N$, are introduced as p passes each p_i, and this leads the attractor to make transitions as follows: (steady

Figure 6.11 Universal behavior of the low frequency power spectrum for $R = R_0$ at criticality.

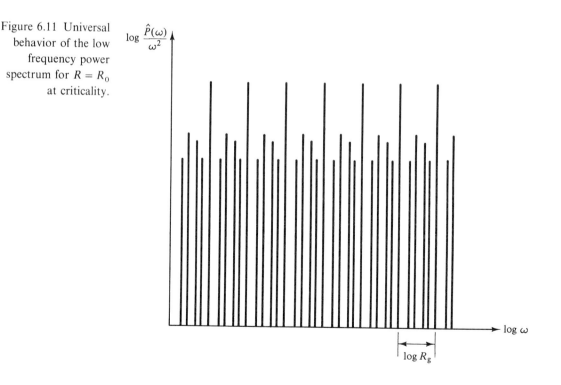

state) → (periodic) → (2 torus) → ··· → (N torus). The mechanism whereby new active modes of oscillation are added is the *Hopf bifurcation* which we now briefly describe with reference to the first transition, (steady state) → (periodic).

Consider the case where the linearized equations about a fixed point $\mathbf{x} = \mathbf{x}_*$ have a solution of Eq. (4.6) such that there is a pair of complex conjugate eigenvalues, $s = \sigma(p) \pm i\omega(p)$, and its real part $\sigma(p)$ increases with p, being negative for $p < p_1$, zero for $p = p_1$, and positive for $p > p_1$. Furthermore, we assume that all other eigenvalues s have negative real parts when p increases through p_1. The essential dynamics for p close to p_1 and $\mathbf{x}$ close to $\mathbf{x}_*$ can be studied by restricting attention to a two-dimensional subspace of the phase space. For $p < p_1$ an appropriate reduced set of two-dimensionalized linearized equations yield spiraling in toward $\mathbf{x}_*$, while for $p > p_1$ an initial condition spirals out from $\mathbf{x}_*$ (see Figure 4.6). As it spirals out, increasing its distance from $\mathbf{x}_*$ the linear approximation to the dynamics becomes less well satisfied, and nonlinear terms in the equations of motion must be considered. Including the nonlinearity to lowest order by making a Taylor series expansion of the dynamical equations about the point $\mathbf{x} = \mathbf{x}_*$, the essential two-dimensional dynamics can be cast in a *normal form* (for example, Guckenheimer and Holmes (1983)),

$$\frac{dr}{dt} = [(p - p_1)\sigma' - ar^2]r, \tag{6.22a}$$

$$\frac{d\theta}{dt} = [\Omega_1 + (p - p_1)\omega' + br^2], \tag{6.22b}$$

where $\sigma' = (d\sigma(p)/dp) > 0$, $\omega' = d\omega(p)/dp$ both evaluated at $p = p_1$, and $\Omega_1 = \omega(p_1)$. Here r and θ are scaled polar coordinates centered at $\mathbf{x}_*$.

Figure 6.12 Unwrapping of quasiperiodic flows on (a) T^2 and (b) T^3. (After Bergé *et al.*, 1986).

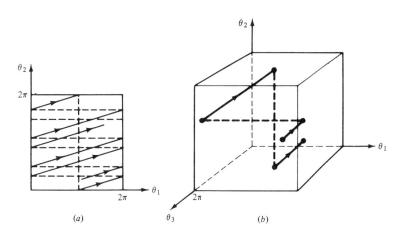

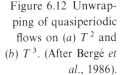

Thus, for $p < p_1$ the trajectory spirals in, approaching $\mathbf{x}_*$, while for $p > p_1$ it spirals out approaching the limit cycle circle $r = \bar{r}(p) \equiv [(\sigma'/a)(p - p_1)]^{1/2}$ on which it circulates with angular frequency $\omega_1 + (p - p_1)\omega' + b\bar{r}^2(p)$ (we assume here that[4] $a > 0$ in which case the bifurcation is said to be 'supercritical'). (See Figure 6.13.) Thus the Hopf bifurcation creates a periodic orbit whose frequency at the bifurcation is Ω_1. New frequencies are added by successive additional Hopf bifurcations at the parameter values $p_2, p_3, \ldots, p_N$, thus leading to motion on a torus T^N.

In an early paper, Ruelle and Takens (1971) considered four frequency quasiperiodic flows on the torus T^4. They showed that it was possible to make *arbitrarily small* (but very carefully chosen), smooth perturbations to the flow such that the quasiperiodic flow was converted to a chaotic flow on a strange attractor lying in the torus T^4. Furthermore, these chaotic attractors, once created, cannot be destroyed by arbitrarily small perturbations of the flow. Subsequently, Newhouse, Ruelle and Takens (1978) showed that the same could be said for a three frequency quasiperiodic flow[5] on the torus T^3. It had been tentatively conjectured that these results meant that three and four frequency quasiperiodicities are unlikely to occur because they are 'easily' destroyed and surplanted by chaos; and furthermore that, if there was two frequency quasiperiodicity and a third frequency was destabilized as some stress parameter of the system is increased, then the flow would immediately become chaotic. In particular, the relevance of this to the onset of turbulence in fluids was discussed. While these speculations turned out not to be true in detail, these papers played an important role in providing early motivation for

Figure 6.13 Illustration of a supercritical Hopf bifurcation for $a > 0$.

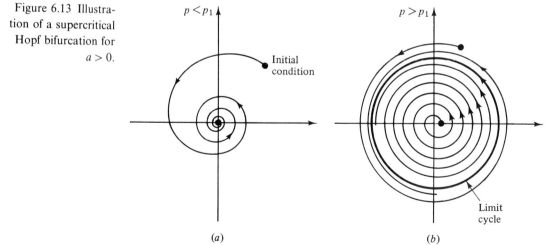

(a) (b)

the study of chaos, especially in fluids. Furthermore, they pointed out that broad frequency spectra need not be the result of the successive addition of a great many discrete new frequency components as a system parameter is increased, as had been previously proposed, but could instead appear more abruptly as the result of the onset of a chaotic attractor.

Numerical experiments were performed by Grebogi *et al.* (1985c) to see if three frequency quasiperiodicity would occur and, if so, to obtain an idea of how often. They, like Newhouse *et al.*, assumed a flow on a three torus T^3. Then, taking a surface of section at times corresponding to one of the flow periods, a *map* on a two torus results, which Grebogi *et al.* took to be of the form

$$\left.\begin{array}{l} \theta_{n+1} = [\theta_n + w_1 + kP_1(\theta_n, \phi_n)] \text{ modulo } 2\pi, \\ \phi_{n+1} = [\phi_n + w_2 + kP_2(\theta_n, \phi_n)] \text{ modulo } 2\pi, \end{array}\right\} \tag{6.23}$$

where θ and ϕ are angles and $P_{1,2}$ are 2π periodic in both θ and ϕ. Equations (6.23) may be thought of as the extension to three frequency quasiperiodicity of the circle map model of two frequency quasiperiodicity. For $k = 0$, three frequency quasiperiodic flows correspond to w_1 and w_2 being incommensurate with themselves and 2π. That is, the only solution of

$$m_1 w_1 + m_2 w_2 + 2\pi m_3 = 0$$

for integer $m_{1,2,3}$ is the trivial solution $m_1 = m_2 = m_3 = 0$. Grebogi *et al.* then arbitrarily chose particular sinusoidal forms for P_1 and P_2 and tested to see what fraction of the measure of (w_1, w_2) was occupied by different types of attractors for various sizes of the nonlinearity parameter k. This was done by choosing many pairs (w_1, w_2) randomly in $[0, 2\pi] \times [0, 2\pi]$ and calculating the two Lyapunov exponents h_1 and h_2 on the attractor. (By convention $h_1 \geq h_2$.) If $h_1 = h_2 = 0$, the flow is three frequency quasiperiodic; if $h_1 = 0$ and $h_2 < 0$, the flow is two frequency quasiperiodic; if h_1 and h_2 are both less than zero the flow is periodic; and if $h_1 > 0$ the flow is chaotic. The results are shown in Table 6.1. The value k_c is the value of k past which the map (6.23) becomes noninvertible. Three frequency quasiperiodicity is not possible for $k > k_c$ (analogous to $k > 1$ for the circle map). As is evident, three frequency quasiperiodicity is very common at moderate values of the nonlinearity parameter k. (Grebogi *et al.* also obtained similar results for flows on T^4.)

The situation is roughly analogous to that which occurs in the circle map: Two frequency quasiperiodicity motion can be converted to periodic motion by an arbitrarily small (but carefully chosen) change in w (the locked regions are dense in w and the quasiperiodicity exists on a Cantor set). Once w has been changed so that it lies in the interior of a phase locked interval, perturbations of w that are too small will be

Table 6.1 *Fraction of attractors of various types.*

Attractor	Lyapunov Exponents	$\dfrac{k}{k_c} = \dfrac{3}{8}$	$\dfrac{k}{k_c} = \dfrac{3}{4}$	$\dfrac{k}{k_c} = \dfrac{9}{8}$
Three frequency quasiperiodic	$h_1 = h_2 = 0$	82%	44%	0%
Two frequency quasiperiodic	$h_1 = 0$ $h_2 < 0$	16%	38%	33%
Periodic	$h_{1,2} < 0$	2%	11%	31%
Chaotic	$h_1 > 0$	$\approx 0\%$	7%	36%

insufficient to move it out of the phase locked interval. Nevertheless, the measure of w corresponding to quasiperiodicity is positive, and quasiperiodicity is, therefore, expected to occur. The key point is that, in deciding whether a phenomenon can occur in practice, one should ask whether the measure in parameter space over which the phenomenon occurs is zero or positive, not whether carefully chosen arbitrarily small perturbations of the system can destroy the phenomenon. If the measure in parameter space yielding a particular phenomenon is positive, then a random choice of parameters has a nonzero probability of yielding that phenomenon, and we can expect that sometimes it will occur.

In the works of Ruelle and Takens and Newhouse *et al.*, the flow was on a torus. As mentioned in Section 6.2, however, it is possible for invariant tori to be destroyed as a system parameter is varied (in the context of Eqs. (6.23), this corresponds to $k > k_c$). Thus another question that naturally arises is the following. Say that one sees a transition in which, at some value of a parameter, there is three frequency quasiperiodicity, while at another there is chaos. Is the chaotic motion on an invariant three-dimensional toroidal surface embedded in the phase space, or has the surface T^3 been destroyed as the parameter is varied from the quasiperiodic value to the chaotic value? Both possibilities should be possible depending on the specific system considered. An example has been considered numerically by Battelino *et al.* (1989) who formulated a numerical technique for testing whether a chaotic attractor lies on a three torus. They found, for their example (which involved coupled van der Pol oscillators), that destruction of the three torus apparently preceded the occurance of a chaotic attractor.

6.4 Strange nonchaotic attractors of quasiperiodically forced systems

In Chapter 1 we have defined a strange attractor as one which has fractal phase space structure, while we have defined a chaotic attractor as one on which typical orbits exhibit sensitive dependence on initial conditions. The logistic map at $r = 4$ has a chaotic attractor with Lyapunov exponent $h = \ln 2 > 1$, but this attractor is not strange; it is simply the interval $[0, 1]$. Strange attractors that are not chaotic are also possible. For example, the logistic map at $r = r_\infty$ (the accumulation point for period doublings) has an attractor which has a Lyapunov exponent $h = 0$ but is a Cantor set with fractal dimension $d \approx 0.54$ (Grassberger, 1981). Hence, it is a strange nonchaotic attractor. Furthermore, there is a countably infinite set of r-values corresponding to the accumulation points of period doublings experienced by each period p orbit born in a tangent bifurcation (at the beginning of a window). At each such r the attractor is fractal (with $d \approx 0.54$) and $h = 0$. Note, however, that the set of parameter values r which for the logistic map yield strange nonchaotic attractors has zero Lebesgue measure in r because these r-values are countable. Hence we say these attractors are not typical. A natural question that arises is whether there are any systems for which strange nonchaotic attractors are typical in the sense that they occupy a positive Lebesgue measure of the parameter space. This question has been considered in a series of papers[6] where the authors demonstrated that strange nonchaotic attractors are indeed typical in systems that are driven by a two frequency quasiperiodic forcing function. For example, Romeiras and Ott (1987) consider strange nonchaotic attractors for the quasiperiodicly forced damped pendulum equation

$$\frac{\mathrm{d}^2\theta}{\mathrm{d}t^2} + \delta\frac{\mathrm{d}\theta}{\mathrm{d}t} + \Omega^2 \sin\theta = T_1 \sin(\Omega_1 t) + T_2 \sin(\Omega_2 t) + K, \quad (6.24)$$

where Ω_1/Ω_2 is irrational and δ, Ω^2, K and $T_{1,2}$ are parameters. They find that as $T_{1,2}$ are increased there is a transition to a situation where strange nonchaotic attractors are observed on a Cantor set of positive Lebesgue measure in the parameter space. Further increase of $T_{1,2}$ then produces a transition to chaos. In the theoretical studies[6] it was shown that strange nonchaotic attractors have a distinctive signature in their frequency spectra, and this has been observed in experiments on a quasiperiodicly forced magnetoelastic ribbon by Ditto *et al.* (1990b).

Problems

1. Assuming that in Figure 6.2 the nonlinear resistor has a resistance

$$R(V) = (V/V_0)R_0 \exp(-V/V_0)$$

(where R_0 and V_0 are constants), find the Fourier transform of the current. What is the coefficient of the delta function $\delta[\omega - (m_1\Omega_1 + m_2\Omega_2)]$? *Hint*: The modified Bessel function of order n can be expressed as

$$I_n(x) = \frac{1}{2\pi} \int_0^{2\pi} \exp(in\theta)\exp(x\sin\theta)\,d\theta.$$

2. Consider the damped pendulum equation with a forcing on the right-hand side which consists of a sinusoidal part, $T\sin(\Omega t)$, and a constant part, K,

$$d^2\theta/dt^2 + \delta\,d\theta/dt + \sin\theta = K + T\sin(\Omega t).$$

Define phase space variables $x^{(1)} = d\theta/dt$, $x^{(2)} = \theta$ modulo 2π, and $x^{(3)} = \Omega t$ modulo 2π.

(*a*) Show that volumes in phase space shrink exponentially with time.

(*b*) If a surface of section $x^{(3)} = (\text{constant})$ is taken, show that areas shrink with each iterate of the surface of section map.

(*c*) Say that there is a solution of the equation denoted $\theta = \tilde{\theta}(t)$. Show that, if $K = 0$, then $\theta = -\tilde{\theta}(t + \pi/\Omega)$ is also a solution.

(*d*) Using the results above show that quasiperiodic solutions filling a toroidal surface are not possible for $K = 0$. (Remark: Quasiperiodic solutions for $K \neq 0$ do exist and have been investigated in a number of papers; see for example D'Humieres *et al.* (1982).)

3. Derive Eq. (6.14).

4. Derive Eq. (6.21) and the expression given for $f(w, u)$.

5. Write a program to calculate the rotation number R of the circle map Eq. (6.11). Plot R versus w for $k = 0.4$, for $k = 0.8$, and for $k = 1.0$.

6. Show that the van der Pol equation, Eq. (1.13), undergoes a Hopf bifurcation of the steady state $(x, dx/dt) = (0, 0)$ as the parameter η is increased from -1 to $+1$. At what value of η does the bifurcation occur?

Notes

1. There are quasiperiodic orbits in $k > 1$, but these only occur on a zero Lebesgue measure Cantor set of θ-values (Kadanoff, 1983). Consequently these orbits do not generate a smooth density. Also, since a randomly chosen initial condition clearly has zero probability of falling on such an orbit, and, since these orbits are nonattracting, we conclude that these orbits are not realized for typical initial conditions.

2. We note, however, that although quasiperiodicity is present in Brandstadter and Swinney's experiment, frequency locking does not seem to occur. The reason for this appears to be connected with the circular symmetry of the Couette–Taylor configuration.

3. This equation with the $d^2\theta/dt^2$ term retained has been extensively studied in the context of transitions from quasiperiodic solutions to chaotic solutions. See, for example, D'Humieres *et al.* (1982). With the $d^2\theta/dt^2$ term deleted chaos is necessarily ruled out.

4. The case $a < 0$ is referred to as 'subcritical'. In this case, when p increases through p_1, an orbit initially near $\mathbf{x}_*$ will typically move far from $\mathbf{x}_*$ (this is unlike the case where $a > 0$).

5. In the paper of Newhouse *et al.* (1978) the small perturbations to the flow were required to have small first- and second-order derivatives but could have large higher-order derivatives. In contrast, in the case of the torus T^4 treated by Ruelle and Takens (1971), the perturbations could be such that derivatives of all orders were small.

6. See Grebogi *et al.* (1984), Bondeson *et al.* (1985) and Romeiras *et al.* (1989) and references therein. Bondeson *et al.* (1985) shows that strange nonchaotic behavior in their system can be shown on the basis of Anderson localization (see Chapter 10) in a quasiperiodic potential.

CHAPTER SEVEN

Chaos in Hamiltonian systems

Hamiltonian systems are a class of dynamical systems that occur in a wide variety of circumstances.[1] The special properties of Hamilton's equations endow these systems with attributes that differ qualitatively and fundamentally from other systems. (For example, Hamilton's equations do not possess attractors.)

Examples of Hamiltonian dynamics include not only the well-known case of mechanical systems in the absence of friction, but also a variety of other problems such as the paths followed by magnetic field lines in a plasma, the mixing of fluids, and the ray equations describing the trajectories of propagating waves. In all of these situations chaos can be an important issue. Furthermore, chaos in Hamiltonian systems is at the heart of such fundamental questions as the foundations of statistical mechanics and the stability of the solar system. In addition, Hamiltonian mechanics and its structure are reflected in quantum mechanics. Thus, in Chapter 10 we shall treat the connection between chaos in Hamiltonian systems and related quantum phenomena. The present chapter will be devoted to a discussion of Hamiltonian dynamics and the role that chaos plays in these systems. We begin by presenting a summary of some basic concepts in Hamiltonian mechanics.[2,3]

7.1 Hamiltonian systems

The dynamics of a Hamiltonian system is completely specified by a single function, the Hamiltonian, $H(\mathbf{p}, \mathbf{q}, t)$. The state of the system is specified by its 'momentum' $\mathbf{p}$ and 'position' $\mathbf{q}$. Here the vectors $\mathbf{p}$ and $\mathbf{q}$ have the same dimensionality which we denote N. We call N the number of *degrees of freedom* of the system. For example, Hamilton's equations for the motion of K point masses interacting in three-dimensional space via gravitational attraction has $N = 3K$ degrees of freedom, corresponding to the three spacial coordinates needed to specify the location of each mass. Hamilton's equations determine the trajectory $(\mathbf{p}(t), \mathbf{q}(t))$ that the system follows in the $2N$-dimensional phase space, and are given by

$$d\mathbf{p}/dt = -\partial H(\mathbf{p}, \mathbf{q}, t)/\partial \mathbf{q}, \tag{7.1a}$$

$$d\mathbf{q}/dt = \partial H(\mathbf{p}, \mathbf{q}, t)/\partial \mathbf{p}. \tag{7.1b}$$

In the special case that the Hamiltonian has no explicit time dependence, $H = H(\mathbf{p}, \mathbf{q})$, we can use Hamilton's equations to show that, as $\mathbf{p}$ and $\mathbf{q}$ vary with time, the value of $H(\mathbf{p}(t), \mathbf{q}(t))$ remains a constant:

$$\frac{dH}{dt} = \frac{d\mathbf{q}}{dt} \cdot \frac{\partial H}{\partial \mathbf{q}} + \frac{d\mathbf{p}}{dt} \cdot \frac{\partial H}{\partial \mathbf{p}} = \frac{\partial H}{\partial \mathbf{p}} \cdot \frac{\partial H}{\partial \mathbf{q}} - \frac{\partial H}{\partial \mathbf{q}} \cdot \frac{\partial H}{\partial \mathbf{p}} = 0.$$

Thus, identifying the value of the Hamiltonian with the energy E of the system, we see that the energy is conserved for time-independent systems, $E = H(\mathbf{p}, \mathbf{q}) = $ (constant).

7.1.1 Symplectic structure

We can write Eqs. (7.1) in the form

$$d\tilde{\mathbf{x}}/dt = \mathbf{F}(\tilde{\mathbf{x}}, t), \tag{7.2}$$

by taking $\tilde{\mathbf{x}}$ to be the $2N$-dimensional vector

$$\tilde{\mathbf{x}} = \begin{pmatrix} \mathbf{p} \\ \mathbf{q} \end{pmatrix},$$

and by taking $\mathbf{F}(\tilde{\mathbf{x}})$ to be

$$\mathbf{F}(\tilde{\mathbf{x}}, t) = \mathbf{S}_N \cdot \partial H/\partial \tilde{\mathbf{x}}, \tag{7.3}$$

with

$$\mathbf{S}_N = \begin{bmatrix} \mathbf{O}_N & -\mathbf{I}_N \\ \mathbf{I}_N & \mathbf{O}_N \end{bmatrix} \tag{7.4}$$

where $\mathbf{I}_N$ is the N-dimensional identity matrix, $\mathbf{O}_N$ is the $N \times N$ matrix of zeros, and

$$\frac{\partial H}{\partial \tilde{\mathbf{x}}} = \begin{bmatrix} \partial H/\partial \mathbf{p} \\ \partial H/\partial \mathbf{q} \end{bmatrix} \tag{7.5}$$

From this we see how restricted the class of Hamiltonian systems is. In particular, a general system of the form (7.2) requires the specification of all the components of the *vector* function $\mathbf{F}(\tilde{\mathbf{x}}, t)$, while by (7.3), if the system is Hamiltonian, it is specified by a single scalar function of $\mathbf{p}, \mathbf{q}$ and t (the Hamiltonian).

One of the basic properties of Hamilton's equations is that they preserve $2N$-dimensional volumes in the phase space. This follows by taking the divergence of $\mathbf{F}(\tilde{\mathbf{x}})$ in Eq. (7.2), which gives

$$\frac{\partial}{\partial \tilde{\mathbf{x}}} \cdot \mathbf{F} = \frac{\partial}{\partial \mathbf{p}} \cdot \left(-\frac{\partial H}{\partial \mathbf{q}} \right) + \frac{\partial}{\partial \mathbf{q}} \cdot \left(\frac{\partial H}{\partial \mathbf{p}} \right) = 0. \tag{7.6}$$

Thus, if we consider an initial closed surface S_0 in the $2N$-dimensional phase space and evolve each point on the surface forward with time, we obtain at each instant of time t a new closed surface S_t which contains within it precisely the same $2N$-dimensional volume as does S_0. This follows from

$$\frac{\mathrm{d}}{\mathrm{d}t} \int_{S_t} \mathrm{d}^{2N} \tilde{\mathbf{x}} = \oint_{S_t} \frac{\mathrm{d}\tilde{\mathbf{x}}}{\mathrm{d}t} \cdot \mathrm{d}\mathbf{S} = \oint_{S_t} \mathbf{F} \cdot \mathrm{d}\mathbf{S} = \int_{S_t} \frac{\partial}{\partial \tilde{\mathbf{x}}} \cdot \mathbf{F} \, \mathrm{d}^{2N} \tilde{\mathbf{x}} = 0,$$

where $\int_{S_t} \ldots$ denotes integration over the volume enclosed by S_t, $\oint_{S_t} \ldots$ denotes a surface integral over the closed surface S_t, and the third equality is from the divergence theorem (cf. Eq. (1.12)). As a consequence of this result, Hamiltonian systems do not have attractors in the usual sense. This incompressibility of phase space volumes for Hamiltonian systems is called Liouville's theorem.

Perhaps the most basic structural property of Hamilton's equations is that they are *symplectic*. That is, if we consider three orbits that are infinitesimally displaced from each other, $(\mathbf{p}(t), \mathbf{q}(t))$, $(\mathbf{p}(t) + \delta\mathbf{p}(t), \mathbf{q}(t) + \delta\mathbf{q}(t))$ and $(\mathbf{p}(t) + \delta\mathbf{p}'(t), \mathbf{q}(t) + \delta\mathbf{q}'(t))$, where $\delta\mathbf{p}$, $\delta\mathbf{q}$, $\delta\mathbf{p}'$ and $\delta\mathbf{q}'$ are infinitesimal N vectors, then the quantity,

$$\delta\mathbf{p} \cdot \delta\mathbf{q}' - \delta\mathbf{q} \cdot \delta\mathbf{p}',$$

which we call the differential symplectic area, is independent of time,

$$\frac{\mathrm{d}}{\mathrm{d}t} (\delta\mathbf{p} \cdot \delta\mathbf{q}' - \delta\mathbf{q} \cdot \delta\mathbf{p}') = 0. \tag{7.7}$$

The differential symplectic area can also be written as

$$\delta\mathbf{p} \cdot \delta\mathbf{q}' - \delta\mathbf{q} \cdot \delta\mathbf{p}' = \delta\tilde{\mathbf{x}}^\dagger \cdot \mathbf{S}_N \cdot \delta\tilde{\mathbf{x}}'. \tag{7.8}$$

where $\dagger$ denotes transpose.

To derive (7.7) we differentiate (7.8) with respect to time and use Eqs. (7.2)–(7.5):

$$\frac{\mathrm{d}}{\mathrm{d}t} \delta\tilde{\mathbf{x}}^\dagger \cdot \mathbf{S}_N \cdot \delta\tilde{\mathbf{x}}' = \frac{\mathrm{d}\delta\tilde{\mathbf{x}}^\dagger}{\mathrm{d}t} \cdot \mathbf{S}_N \cdot \delta\tilde{\mathbf{x}}' + \delta\tilde{\mathbf{x}}^\dagger \cdot \mathbf{S}_N \cdot \frac{\mathrm{d}\delta\tilde{\mathbf{x}}'}{\mathrm{d}t}$$

$$= \left(\frac{\partial \mathbf{F}}{\partial \tilde{\mathbf{x}}} \cdot \delta\tilde{\mathbf{x}} \right)^\dagger \cdot \mathbf{S}_N \cdot \delta\tilde{\mathbf{x}}' + \delta\tilde{\mathbf{x}}^\dagger \cdot \mathbf{S}_N \cdot \frac{\partial \mathbf{F}}{\partial \tilde{\mathbf{x}}} \cdot \delta\tilde{\mathbf{x}}$$

$$= \delta\tilde{\mathbf{x}}^\dagger \cdot \left[\left(\frac{\partial \mathbf{F}}{\partial \tilde{\mathbf{x}}} \right)^\dagger \cdot \mathbf{S}_N + \mathbf{S}_N \cdot \frac{\partial \mathbf{F}}{\partial \tilde{\mathbf{x}}} \right] \cdot \delta\tilde{\mathbf{x}}'$$

$$= \delta\tilde{\mathbf{x}}^\dagger \cdot \left[\left(\mathbf{S}_N \cdot \frac{\partial^2 H}{\partial\tilde{\mathbf{x}}\,\partial\tilde{\mathbf{x}}} \right)^\dagger \cdot \mathbf{S}_N + \mathbf{S}_N \cdot \mathbf{S}_N \cdot \frac{\partial^2 H}{\partial\tilde{\mathbf{x}}\,\partial\tilde{\mathbf{x}}} \right] \cdot \delta\tilde{\mathbf{x}}'$$

$$= \delta\tilde{\mathbf{x}}^\dagger \cdot \left[\left(\frac{\partial^2 H}{\partial\tilde{\mathbf{x}}\,\partial\tilde{\mathbf{x}}} \right)^\dagger \cdot \mathbf{S}_N^\dagger \cdot \mathbf{S}_N + \mathbf{S}_N \cdot \mathbf{S}_N \cdot \frac{\partial^2 H}{\partial\tilde{\mathbf{x}}\,\partial\tilde{\mathbf{x}}} \right] \cdot \delta\tilde{\mathbf{x}}'$$

$$= 0,$$

where we have used $\mathbf{S}_N \cdot \mathbf{S}_N = -\mathbf{I}_{2N}$ (where $\mathbf{I}_{2N}$ is the $2N$-dimensional identity matrix), $\mathbf{S}_N^\dagger = -\mathbf{S}_N$ and noted that $\partial^2 H/\partial\tilde{\mathbf{x}}\,\partial\tilde{\mathbf{x}}$ is a symmetric matrix. (In terms of the notation of Chapter 4, $\partial\mathbf{F}/\partial\tilde{\mathbf{x}} = \mathbf{DF}$.) For the case of one-degree-of-freedom systems ($N = 1$), Eq. (7.7) says that infinitesimal areas are preserved following the flow. (Figure 7.1 shows two infinitesimal vectors defining an infinitesimal parallelogram. The parallelogram area is $\delta p'\delta q - \delta q'\delta p$.) Since infinitesimal areas are preserved by a Hamiltonian flow with $N = 1$, so are noninfinitesimal areas. Thus for $N = 1$ Liouville's theorem and the symplectic condition are the same condition. For $N > 1$ the symplectic condition is not implied by Liouville's theorem. It can be shown,[2] however, that the symplectic condition implies volume conservation; so the symplectic condition is the more fundamental requirement for Hamiltonian mechanics. We interpret (7.7) as saying that the algebraic sum of the parallelogram areas formed by projecting the vectors $\delta\mathbf{p}$, $\delta\mathbf{q}$, $\delta\mathbf{p}'$, $\delta\mathbf{q}'$ to the N coordinate planes (p_i, q_i) is conserved,

$$\delta\mathbf{p} \cdot \delta\mathbf{q}' - \delta\mathbf{q} \cdot \delta\mathbf{p}' = \sum_{i=1}^{N} (\delta p_i \delta q_i' - \delta q_i \delta p_i').$$

The quantity $\delta\mathbf{p} \cdot \delta\mathbf{q}' - \delta\mathbf{q} \cdot \delta\mathbf{p}'$ is the differential form of *Poincaré's integral invariant*,

$$\oint_\gamma \mathbf{p} \cdot d\mathbf{q} = \sum_{i=1}^{N} \oint_\gamma p_i \, dq_i, \qquad (7.9a)$$

Figure 7.1 Infinitesimal area defined by the two infinitesimal vectors $(\delta p, \delta q)$ and $(\delta p', \delta q')$ for $N = 1$.

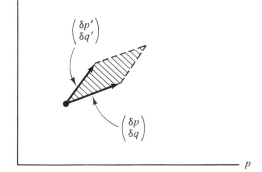

where the integral is taken around a closed path γ in $(\mathbf{p},\mathbf{q})$-space. We also refer to the quantity $\int_\gamma \mathbf{p} \cdot d\mathbf{q}$ as the *symplectic area*. Poincaré's integral invariant is independent of time if the closed path γ is taken following the flow in phase space.[2] That is, $\gamma(t)$ is the path obtained from $\gamma(0)$ by evolving all the points on $\gamma(0)$ forward in time by the amount t via Hamilton's equations.

A useful generalization of the above statement of the invariance of $\oint \mathbf{p} \cdot d\mathbf{q}$ following the flow is the *Poincaré–Cartan integral theorem*. Consider the $(2N + 1)$-dimensional extended phase space $(\mathbf{p},\mathbf{q},t)$. Let Γ_1 be a closed curve in this space and consider the tube of trajectories through points on Γ_1 as shown in Figure 7.2 for $N = 1$. The Poincare–Cartan integral theorem states that the 'action integral' around the path Γ_1, $\oint_{\Gamma_1} (\mathbf{p} \cdot d\mathbf{q} - H\,dt)$, is the same value for any other path Γ_2 encircling the same tube of trajectories,

$$\oint_{\Gamma_1} (\mathbf{p} \cdot d\mathbf{q} - H\,dt) = \oint_{\Gamma_2} (\mathbf{p} \cdot d\mathbf{q} - H\,dt), \qquad (7.9b)$$

where Γ_1 and Γ_2 are illustrated in Figure 7.2. If Γ_1 and Γ_2 are taken at two different (but constant) times, then $dt = 0$ in the above, and we recover the invariant (7.9a). Another important case is where H has no explicit t dependence, $H = H(\mathbf{p},\mathbf{q})$. In this case the Hamiltonian is a constant of the motion, and, if we restrict the points on Γ_1 to all have the same value of H

Figure 7.2 Trajectory tube through Γ_1.

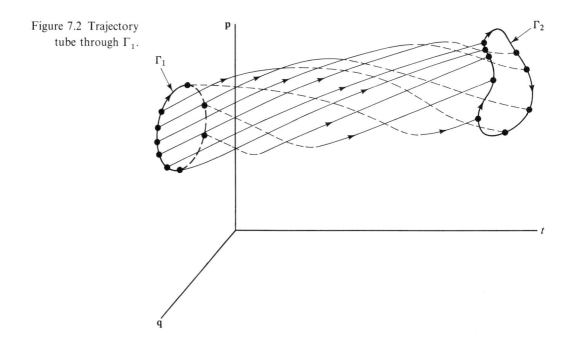

(i.e., Γ_1 lies on the $2N$-dimensional surface $H(\mathbf{p}, \mathbf{q}) = $ (constant)), then $\oint H \, dt = 0$ for any closed path on the $2N$-dimensional constant H surface in $(\mathbf{p}, \mathbf{q}, t)$ space. In this case we obtain

$$\oint_{\Gamma_1} \mathbf{p} \cdot d\mathbf{q} = \oint_{\Gamma_2} \mathbf{p} \cdot d\mathbf{q}. \tag{7.9c}$$

Note that, in contrast to (7.9a), the closed paths Γ_1 and Γ_2 in (7.9c) are in $(\mathbf{p}, \mathbf{q}, t)$ space and thus need not be taken at constant time.

7.1.2 Canonical changes of variables

If we introduce some arbitrary change of variables, $\tilde{\mathbf{x}}' = \mathbf{g}(\tilde{\mathbf{x}})$, the Hamiltonian form of the equations may not be preserved. Changes of variables which preserve the Hamiltonian form of the equations are said to be *canonical*, and the momentum and position vectors in terms of which one has a system in the Hamiltonian form (7.1) are said to be *canonically conjugate*. Specifically, if $\mathbf{p}$ and $\mathbf{q}$ satisfy (7.1), then a canonical change of variables to a new set of canonically conjugate variables $\bar{\mathbf{p}}$ and $\bar{\mathbf{q}}$,

$$\bar{\mathbf{p}} = \mathbf{g}_1(\mathbf{p}, \mathbf{q}, t),$$

$$\bar{\mathbf{q}} = \mathbf{g}_2(\mathbf{p}, \mathbf{q}, t),$$

leads to evolution equations for $\bar{\mathbf{p}}$ and $\bar{\mathbf{q}}$ of the form,

$$d\bar{\mathbf{p}}/dt = -\partial \bar{H}(\bar{\mathbf{p}}, \bar{\mathbf{q}}, t)/\partial \bar{\mathbf{q}},$$

$$d\bar{\mathbf{q}}/dt = \partial \bar{H}(\bar{\mathbf{p}}, \bar{\mathbf{q}}, t)/\partial \bar{\mathbf{p}},$$

where $\bar{H}$ is a new transformed Hamiltonian for the system. One way to specify a canonical change of variables is to introduce a *generating function*,[2] $S(\bar{\mathbf{p}}, \mathbf{q}, t)$ which is a function of the 'old' position coordinates $\mathbf{q}$ and the 'new' momentum coordinates $\bar{\mathbf{p}}$. In terms of $S(\bar{\mathbf{p}}, \mathbf{q}, t)$, the change of variables is specified by

$$\bar{\mathbf{q}} = \frac{\partial S(\bar{\mathbf{p}}, \mathbf{q}, t)}{\partial \bar{\mathbf{p}}}, \, \mathbf{p} = \frac{\partial S(\bar{\mathbf{p}}, \mathbf{q}, t)}{\partial \mathbf{q}}. \tag{7.10}$$

Thus the change of variables is given implicitly: to obtain $\bar{\mathbf{p}}$ in terms of $\mathbf{p}$ and $\mathbf{q}$ solve $\mathbf{p} = \partial S/\partial \mathbf{q}$ for $\bar{\mathbf{p}}$; to obtain $\bar{\mathbf{q}}$ in terms of $\mathbf{p}$ and $\mathbf{q}$ substitute the solution for $\bar{\mathbf{p}}$ into $\bar{\mathbf{q}} = \partial S/\partial \mathbf{p}$. Note that the change of variables specified by Eq. (7.10) is guaranteed to be symplectic. That is

$$\delta \mathbf{p} \cdot \delta \mathbf{q}' - \delta \mathbf{q} \cdot \delta \mathbf{p}' = \delta \bar{\mathbf{p}} \cdot \delta \bar{\mathbf{q}}' - \delta \bar{\mathbf{q}} \cdot \delta \bar{\mathbf{p}}'.$$

This can be checked by differentiating Eq. (7.10),

$$\delta \mathbf{q} = \frac{\partial^2 S}{\partial \bar{\mathbf{p}} \, \partial \bar{\mathbf{p}}} \cdot \delta \bar{\mathbf{p}} + \frac{\partial^2 S}{\partial \bar{\mathbf{p}} \, \partial \mathbf{q}} \cdot \delta \mathbf{q},$$

$$\delta \mathbf{p} = \frac{\partial^2 S}{\partial \mathbf{q} \, \partial \bar{\mathbf{p}}} \cdot \delta \bar{\mathbf{p}} + \frac{\partial^2 S}{\partial \mathbf{q} \, \partial \mathbf{q}} \cdot \delta \mathbf{q},$$

and substituting into the symplectic condition given above. In terms of the generating function the new Hamiltonian is given by[2]

$$\bar{H}(\bar{\mathbf{p}}, \bar{\mathbf{q}}, t) = H(\mathbf{p}, \mathbf{q}, t) + \partial S/\partial t. \tag{7.11}$$

7.1.3 Hamiltonian maps

Say we consider a Hamiltonian system and define the 'time T map' $\mathscr{M}_T$ for the system as

$$\mathscr{M}_T(\tilde{\mathbf{x}}(t), t) = \tilde{\mathbf{x}}(t + T). \tag{7.12}$$

(The explicit dependence on t in the second argument of $\mathscr{M}_T$ is absent if the Hamiltonian is time-independent.)

Taking a differential variation of Eq. (7.12) with respect to $\tilde{\mathbf{x}}$, we have

$$\frac{\partial \mathscr{M}_T}{\partial \tilde{\mathbf{x}}} \cdot \delta \tilde{\mathbf{x}}(t) = \delta \tilde{\mathbf{x}}(t + T).$$

The symplectic condition for a Hamiltonian flow

$$\delta \tilde{\mathbf{x}}^\dagger(t + T) \cdot \mathbf{S}_N \cdot \delta \tilde{\mathbf{x}}'(t + T) = \delta \tilde{\mathbf{x}}(t) \cdot \mathbf{S}_N \cdot \delta \tilde{\mathbf{x}}'(t)$$

yields

$$\delta \tilde{\mathbf{x}}^\dagger(t) \cdot \mathbf{S}_N \cdot \delta \tilde{\mathbf{x}}'(t) = \left(\frac{\partial \mathscr{M}_T}{\partial \tilde{\mathbf{x}}} \cdot \delta \tilde{\mathbf{x}}(t)\right)^\dagger \cdot \mathbf{S}_N \cdot \left(\frac{\partial \mathscr{M}_T}{\partial \tilde{\mathbf{x}}} \cdot \delta \tilde{\mathbf{x}}'(t)\right)$$

which (since $\delta \tilde{\mathbf{x}}(t)$ and $\delta \tilde{\mathbf{x}}'(t)$ are arbitrary) implies that the matrix $\partial \mathscr{M}_T/\partial \tilde{\mathbf{x}}$ satisfies

$$\mathbf{S}_N = \left(\frac{\partial \mathscr{M}_T}{\partial \tilde{\mathbf{x}}}\right)^\dagger \cdot \mathbf{S}_N \cdot \left(\frac{\partial \mathscr{M}_T}{\partial \tilde{\mathbf{x}}}\right).$$

The matrix $\partial \mathscr{M}_T/\partial \tilde{\mathbf{x}}$ is said to be a symplectic matrix, and we define a general $2N \times 2N$ matrix $\mathbf{A}$ to be symplectic if it satisfies

$$\mathbf{S}_N = \mathbf{A}^\dagger \cdot \mathbf{S}_N \cdot \mathbf{A}. \tag{7.13}$$

The product of symplectic matrices is also symplectic. To see this, suppose that $\mathbf{A}$ and $\mathbf{B}$ are symplectic. Then

$$(\mathbf{AB})^\dagger \cdot \mathbf{S}_N \cdot (\mathbf{AB}) = \mathbf{B}^\dagger \cdot (\mathbf{A}^\dagger \cdot \mathbf{S}_N \cdot \mathbf{A}) \cdot \mathbf{B} = \mathbf{B}^\dagger \cdot \mathbf{S}_N \cdot \mathbf{B} = \mathbf{S}_N.$$

So $\mathbf{AB}$ is symplectic.

One consequence of the conservation of phase space volumes for Hamiltonian systems is the *Poincaré recurrence theorem*. Say we consider a time-independent Hamiltonian $H = H(\mathbf{p}, \mathbf{q})$ for the case where all orbits are bounded. (This occurs if the energy surface is bounded; i.e., there are no solutions of $E = H(\mathbf{p}, \mathbf{q})$ with $|\mathbf{p}| \to \infty$ or $|\mathbf{q}| \to \infty$.) Now pick *any* initial point in phase space, and surround it with a ball R_0 of small radius $\varepsilon > 0$. Poincaré's recurrence theorem states that, if there are points which leave the initial ball, there are always some of these which will return to it if we wait long enough, and this is true no matter how small we choose ε to be.

In order to see that this is so consider the time T map, Eq. (7.12), which evolves points forward in time by an amount T. Say that under the time T map the initial ball R_0 is mapped to a region R_1 outside the initial ball ($R_1 \cap R_0$ is empty). Continue mapping so as to obtain regions $R_2, R_3, \ldots$. By Liouville's theorem all these regions have the same volume, equal to the volume of the initial ball R_0. Since the orbits are bounded, they are confined to a finite volume region of phase space. Thus, as we repeatedly apply the time T map, we must eventually find that we produce a region R_r which overlaps a previously produced region R_s, $r > s$. (If this is not so, then we would eventually come to the impossible situation where the sum of the volumes of the nonoverlapping Rs would eventually be larger than the volume of the bounded region that they are confined to.) Now apply the inverse of the time T map to R_r and R_s. This inverse must produce intersecting regions (namely R_{r-1} and R_{s-1}). Applying the inverse map s times we conclude that R_{r-s} (recall that $r - s > 0$) intersects the original ball R_0. Thus, as originally claimed, there are points in R_0 which return to R_0 after some time $(r - s)T$.

As in the case of general non-Hamiltonian systems, the surface of section technique also provides an extremely useful tool for analysis in Hamiltonian systems. There are two cases that are of interest.

(*a*) The Hamiltonian depends periodically on time: $H = H(\mathbf{p}, \mathbf{q}, t) = H(\mathbf{p}, \mathbf{q}, t + \tau)$, where τ is the period.

(*b*) The Hamiltonian has no explicit dependence on time: $H = H(\mathbf{p}, \mathbf{q})$.

First, we consider the case of a time periodic Hamiltonian. In that case, we can consider the phase space as having dimension $2N + 1$ by replacing the argument t in H by a dependent variable ξ, taking the phase space variables to be $(\mathbf{p}, \mathbf{q}, \xi)$, and supplementing Hamilton's equations by the addition of the equation $d\xi/dt = 1$. Since the Hamiltonian is periodic in ξ with period τ, we can consider ξ as an angle variable and replace its value in the Hamiltonian by

$$\bar{\xi} = \xi \text{ modulo } \tau.$$

We then use for our surface of section the surface, $\bar{\xi} = t_0$, where t_0 is a constant between zero and τ. (This is the same construction we used for the periodically driven damped pendulum equation in Chapter 1.) Since the Hamiltonian is time periodic, the time T map defined by (7.12) satisfies

$$\mathcal{M}_\tau(\tilde{\mathbf{x}}, t_0) = \mathcal{M}_\tau(\tilde{\mathbf{x}}, t_0 + n\tau),$$

where n is an integer and we have taken $T = \tau$. Hence the surface of section map, which we denote $\mathbf{M}(\tilde{\mathbf{x}})$, is

$$\mathbf{M}(\mathbf{x}) = \mathcal{M}_\tau(\tilde{\mathbf{x}}, t_0),$$

and $\mathbf{M}$ is endowed with the same symplectic properties as $\mathcal{M}_\tau$ (i.e., the matrix $\partial \mathbf{M}/\partial \tilde{\mathbf{x}}$ satisfies (7.13)).

Example: Consider the 'kicked rotor' illustrated in Figure 7.3. There is a bar of moment of inertia $\bar{I}$ and length l, which is fastened at one end to a frictionless pivot. The other end is subjected to a vertical periodic impulsive force of impulse strength K/l applied at times $t = 0, \tau, 2\tau, \ldots$. (There is no gravity.) Using canonically conjugate variables p_θ (representing the angular momentum) and θ (the angular position of the rotor), we have that the Hamiltonian for this system and the corresponding equations of motion obtained from (7.1) are given by

$$H(p_\theta, \theta, t) = p_\theta^2/(2\bar{I}) + K \cos\theta \sum_n \delta(t - n\tau),$$

$$\frac{dp_\theta}{dt} = K \sin\theta \sum_n \delta(t - n\tau). \tag{7.14a}$$

$$\frac{d\theta}{dt} = p_\theta/\bar{I}, \tag{7.14b}$$

where $\delta(\ldots)$ denotes the Dirac delta function. From Eqs. (7.14) we see that p_θ is constant between the kicks but changes discontinuously at each kick. The position variable θ varies linearly with t between kicks (because p_θ is constant) and is continuous at each kick. For our surface of section we examine the values of p_θ and θ just after each kick. Let p_n and θ_n denote the values of p_θ and θ at times $t = n\tau + 0^+$, where 0^+ denotes a positive infinitesimal. By integrating (7.14a) through the delta function at $t = (n+1)\tau$, we then obtain

$$p_{n+1} - p_n = K \sin\theta_{n+1},$$

and from (7.14b), $\theta_{n+1} - \theta_n = p_n\tau/\bar{I}$. Without loss of generality we can take $\tau/\bar{I} = 1$ to obtain the map

$$\theta_{n+1} = (\theta_n + p_n) \text{ modulo } 2\pi, \tag{7.15a}$$

Figure 7.3 The kicked rotor. There is no gravity and no friction at the pivot.

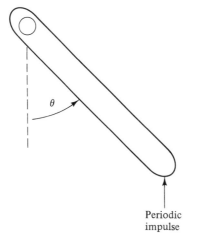

θ

Periodic
impulse

$$p_{n+1} = p_n + K \sin \theta_{n+1}, \tag{7.15b}$$

where we have added a modulo 2π to Eq. (7.15a) since θ is an angle, and we wish to restrict its value to be between zero and 2π. The map given by Eqs. (7.15) is often called the 'standard map' and has proven to be a very convenient model for the study of the typical chaotic behavior of Hamiltonian systems that yield a two-dimensional map. It is a simple matter to check that Eqs. (7.15) preserve area in (p, θ)-space. To do this we calculate the determinant of the Jacobian of the map and verify that it is 1:

$$\det \begin{bmatrix} \partial \theta_{n+1}/\partial \theta_n & \partial \theta_{n+1}/\partial p_n \\ \partial p_{n+1}/\partial \theta_n & \partial p_{n+1}/\partial p_n \end{bmatrix} = \det \begin{bmatrix} 1 & 1 \\ K \cos \theta_{n+1} & 1 + K \cos \theta_{n+1} \end{bmatrix} = 1.$$

Since $N = 1$ this also implies that the map is symplectic, as required.

We now consider the second class of surface of sections that we have mentioned, namely the case where the Hamiltonian has no explicit time dependence. In this case, since the energy is conserved, the motion of the system is restricted to the $(2N - 1)$-dimensional surface given by $E = H(\mathbf{p}, \mathbf{q})$. Taking a surface of section we would then obtain a $(2N - 2)$-dimensional map. Say we choose for our surface of section the plane $q_1 = K_0$ (where K_0 is a constant), and say we give the values of the $2N - 2$ quantities, $p_2, p_3, \ldots, p_N, q_2, q_3, \ldots, q_N$, on this plane. Let $\hat{\mathbf{x}}$ denote the vector specifying these coordinate values on the surface of section $\hat{\mathbf{x}} \equiv (p_2, p_3, \ldots, p_N, q_2, q_3, \ldots, q_N)$. Is there a map, $\hat{\mathbf{x}}_{n+1} = \mathbf{M}(\hat{\mathbf{x}}_n)$, evolving points forward on the surface of section; i.e., does a knowledge of $\hat{\mathbf{x}}_n$ uniquely determine the location of the next point on the surface of section? Given $\hat{\mathbf{x}}$ on the surface of section, the only unknown is p_1 (q_1 is known since we are on the surface of section $q_1 = K_0$). If we can determine p_1 then the phase space position $\tilde{\mathbf{x}}$ is known, and this uniquely determines the system's future evolution, and hence $\hat{\mathbf{x}}$ at the next piercing of the surface of section. To find p_1 we attempt to solve the equation $E = H(\mathbf{p}, \mathbf{q})$ for the single unknown p_1. The problem is that this equation will in general have multiple solutions for p_1. For example, for the commonly encountered case where the Hamiltonian is in the form of a kinetic energy $p^2/2m$, plus a potential energy,

$$H(\mathbf{p}, \mathbf{q}) = p^2/2m + V(\mathbf{q}),$$

for given $\hat{\mathbf{x}}$ there are two roots for p_1,

$$p_1 = \pm \{2m[E - V(q)] - (p_2^2 + p_3^2 + \cdots + p_N^2)\}^{1/2}. \tag{7.16}$$

To make our determination of p_1 unique we adopt the following procedure. We specify $\hat{\mathbf{x}}_n$ to be the coordinates $(p_2, \ldots, p_N, q_2, \ldots, q_N)$ at the nth time at which $q_1(t) = K_0$ *and* $p_1 > 0$. That is, we only count surface of section piercings which cross $q_1 = K_0$ from $q_1 < K_0$ to $q_1 > K_0$ and not vice versa (for the Hamiltonian under consideration $dq_1/dt = p_1$).

Hence, we *define* the surface of section so that we always take the positive root in (7.16) for p_1 (we could equally well have chosen $p_1 < 0$, rather than $p_1 > 0$, in our definition). With this definition, specification of $\hat{\mathbf{x}}_n$ uniquely determines a point in phase space. This point can be advanced by Hamilton's equations to the next time that $q_1(t) = K_0$ with $p_1 > 0$ thus determining $\hat{\mathbf{x}}_{n+1}$. In this way we determine a map,

$$\hat{\mathbf{x}}_{n+1} = \mathbf{M}(\hat{\mathbf{x}}_n).$$

This $(2N - 2)$-dimensional map is symplectic in the remaining canonical variables $\hat{\mathbf{p}} = (p_2, \ldots, p_N)$ and $\hat{\mathbf{q}} = (q_2, \ldots, q_N)$. (This also implies that the map conserves $(2N - 2)$-dimensional volumes.) To show that the map is symplectic, we need to demonstrate that the symplectic area,

$$\oint_\Gamma \hat{\mathbf{p}} \cdot \mathrm{d}\hat{\mathbf{q}},$$

is invariant when the closed path Γ around which the integral is taken is acted on by the map $\mathbf{M}$. This follows immediately from the Poincaré–Cartan theorem in the form of Eq. (7.9c). Writing $\mathbf{p} \cdot \mathrm{d}\mathbf{q} = p_1 \, \mathrm{d}q_1 + \hat{\mathbf{p}} \cdot \mathrm{d}\hat{\mathbf{q}}$ and noting that $q_1 = K_0$ on the surface of section, we have $\mathrm{d}q_1 = 0$, and the desired result follows. (Note that use of the Poincaré–Cartan theorem (rather than the integral invariant (7.9a)) is necessary here because two differential initial conditions starting in the surface of section take different amounts of time to return to it.)

Thus, we see that, in the cases of both a time-periodic Hamiltonian and a time-independent Hamiltonian, the resulting maps are symplectic. For this reason symplectic maps have played an important role, especially with respect to numerical experiments, in elucidating possible types of chaotic behavior in Hamiltonian systems.

One consequence of the symplectic nature of these maps is that the Lyapunov exponents occur in pairs $\pm h_1, \pm h_2, \pm h_3, \ldots$. Thus for each positive exponent there is a negative exponent of equal magnitude, and the number of zero exponents is even. To see why this is so we recall that the Lyapunov exponents are obtained from the product of the matrices $\mathbf{DM}(\tilde{\mathbf{x}}_n)\mathbf{DM}(\tilde{\mathbf{x}}_{n-1}) \ldots \mathbf{DM}(\tilde{\mathbf{x}}_0)$; see Section 3.4. In the Hamiltonian case the matrices $\mathbf{DM}(\tilde{\mathbf{x}}_j)$ are symplectic. Since the product of symplectic matrices is also symplectic, the overall matrix, $\mathbf{DM}(\tilde{\mathbf{x}}_n) \ldots \mathbf{DM}(\tilde{\mathbf{x}}_0)$, is symplectic. Now let us examine what the symplectic condition implies for the eigenvalues of a matrix. The eigenvalues λ of a symplectic matrix $\mathbf{A}$ are the roots of its characteristic polynomial

$$D(\lambda) = \det[\mathbf{A} - \lambda\mathbf{I}].$$

Multiplying Eq. (7.13) on the left by $\mathbf{S}_N^{-1}(\mathbf{A}^\dagger)^{-1}$ we have

$$\mathbf{A} = \mathbf{S}_N^{-1}(\mathbf{A}^\dagger)^{-1}\mathbf{S}_N.$$

The characteristic polynomial then becomes

$$D(\lambda) = \det[\mathbf{S}_N^{-1}(\mathbf{A}^\dagger)^{-1}\mathbf{S}_N - \lambda\mathbf{I}]$$
$$= \det[(\mathbf{A}^\dagger)^{-1} - \lambda\mathbf{I}]$$
$$= \det[\mathbf{A}^{-1} - \lambda\mathbf{I}].$$

Thus, the eigenvalues of $\mathbf{A}$ and $\mathbf{A}^{-1}$ are the same. Since the eigenvalues of $\mathbf{A}^{-1}$ and $\mathbf{A}$ are also inverses of each other, we see that the eigenvalues must occur in pairs, (λ, λ^{-1}). Because the Lyapunov exponents are obtained from the logarithms of the magnitudes of the eigenvalues, $(h = \ln|\lambda|)$ we conclude that they occur in pairs $\pm h$.

As an example, we consider the stability of a periodic orbit of a symplectic two-dimensional map. If the period of the orbit is r, then the problem reduces to considering the stability of the fixed points of $\mathbf{M}^r(\tilde{\mathbf{x}})$ which is also area preserving. Hence, it suffices to examine the stability of the fixed points of symplectic two-dimensional maps. Let $\mathbf{J} = \mathbf{DM}^r$ denote the Jacobian matrix of the map at such a fixed point. Since the map is symplectic, we have $\det\mathbf{J} = 1$. The eigenvalues of $\mathbf{J}$ are given by

$$\det\begin{bmatrix} J_{11} - \lambda & J_{12} \\ J_{21} & J_{22} - \lambda \end{bmatrix} = \lambda^2 - \hat{T}\lambda + 1 = 0,$$

where $\hat{T} \equiv J_{11} + J_{22}$ is the trace of $\mathbf{J}$, and the last term in the quadratic is one by virtue of $\det\mathbf{J} = 1$. The solutions of the quadratic are

$$\lambda = [\hat{T} \pm (\hat{T}^2 - 4)^{1/2}]/2.$$

Since $\{[\hat{T} + (\hat{T}^2 - 4)^{1/2}/2\}\{\hat{T} - (\hat{T}^2 - 4)^{1/2}]/2\} = 1$, the roots are reciprocals of each other as required for the symplectic map. There are three cases:

(a) $\hat{T} > 2$; the roots are real and positive $(\lambda, 1/\lambda > 0)$;
(b) $2 > \hat{T} > -2$; the roots are complex and of magnitude one $(\lambda, 1/\lambda = \exp(\pm i\theta))$;
(c) $\hat{T} < -2$; the roots are real and negative $(\lambda, 1/\lambda < 0)$.

In case (a) we say that the periodic orbit is *hyperbolic*; in case (b) we say the periodic orbit is *elliptic*; and in case (c) we say that the periodic orbit is *hyperbolic with reflection*. Note that, in the linear approximation, cases (a) and (c) lead typical nearby orbits to diverge exponentially from the periodic orbit (linear instability); while in case (b), in the linear approximation a nearby orbit remains nearby forever (linear stability). In the latter case, the nearby linearized orbit remains on an ellipse encircling the periodic orbit and circles around it at a rate $\theta/2\pi$ per iterate of $\mathbf{M}$. (Because the product of the two roots is one, in no case can the periodic orbit be an attractor, since that requires that the magnitude of *both* roots be less than one.)

7.1.4 Integrable systems

In the case where the Hamiltonian has no explicit time dependence, $H = H(\mathbf{p}, \mathbf{q})$, we have seen that Hamilton's equations imply that $dH/dt = 0$, and the energy $E = H(\mathbf{p}, \mathbf{q})$ is a conserved quantity. Thus, orbits with a given energy E are restricted to lie on the $(2N - 1)$-dimensional energy surface $E = H(\mathbf{p}, \mathbf{q})$. A function $f(\mathbf{p}, \mathbf{q})$ is said to be a *constant of the motion* for a system with Hamiltonian H, if, as $\mathbf{p}(t)$ and $\mathbf{q}(t)$ evolve with time in accordance with Hamilton's equations, the value of the function f does not change, $f(\mathbf{p}, \mathbf{q}) = $ (constant). For example, for time-independent Hamiltonians, H is a constant of the motion. More generally, differentiating $f(\mathbf{p}(t), \mathbf{q}(t))$ with respect to time, and assuming that there is no explicit time dependence of the Hamiltonian, we have

$$\frac{df}{dt} = \frac{d\mathbf{p}}{dt} \cdot \frac{\partial f}{\partial \mathbf{p}} + \frac{d\mathbf{q}}{dt} \cdot \frac{\partial f}{\partial \mathbf{q}} = \frac{\partial H}{\partial \mathbf{p}} \cdot \frac{\partial f}{\partial \mathbf{q}} - \frac{\partial H}{\partial \mathbf{q}} \cdot \frac{\partial f}{\partial \mathbf{p}}.$$

We call the expression appearing on the right-hand side of the second equality the *Poisson bracket* of f and H, and we abbreviate it as $[f, H]$, where

$$[g_1, g_2] \equiv \frac{\partial g_1}{\partial \mathbf{q}} \cdot \frac{\partial g_2}{\partial \mathbf{p}} - \frac{\partial g_1}{\partial \mathbf{p}} \cdot \frac{\partial g_2}{\partial \mathbf{q}}. \tag{7.17}$$

Note that $[g_1, g_2] = -[g_2, g_1]$. Thus the condition that f be a constant of the motion for a time-independent Hamiltonian is that its Poisson bracket with H be zero,

$$[f, H] = 0. \tag{7.18}$$

(The Hamiltonian is a constant of the motion since $[H, H] = 0$.)

A time-independent Hamiltonian system is said to be *integrable* if it has N *independent* global constants of the motion $f_i(\mathbf{p}, \mathbf{q}), i = 1, 2, \ldots, N$ (one of these is the Hamiltonian itself; we choose this to be the $i = 1$ constant, $f_1(\mathbf{p}, \mathbf{q}) \equiv H(\mathbf{p}, \mathbf{q})$), and, furthermore, if

$$[f_i, f_j] = 0, \tag{7.19}$$

for all i and j.

We already know that the Poisson bracket of f_i with f_1 is zero for all $i = 1, 2, \ldots, N$, since the f_i are constants of the motion (see Eq. (7.18)). If the condition (7.19) holds for all i and j, then we say that the N constants of the motion f_i are *in involution*. The constants of the motion f_i are 'independent' if no one of them can be expressed as a function of the $(N - 1)$ other constants. The requirement that an integrable system has N independent constants of the motion implies that the trajectory of the system in the phase space is restricted to lie on the N-dimensional surface

$$f_i(\mathbf{p}, \mathbf{q}) = k_i, \tag{7.20}$$

$i = 1, 2, \ldots, N$, where k_i are N constants. The requirement that the N independent constants f_i be in involution (Eq. (7.19)) restricts the topology of the surface, Eq. (7.20), to be of a certain type: it must be an N-dimensional torus (as defined in Section 6.3). This is demonstrated in standard texts[2] and will not be shown here. For the case $N = 2$, an orbit on the torus is as shown in Figure 7.4(a).

Given an integrable system it is possible to introduce a canonical change of variables $(\mathbf{p}, \mathbf{q}) \rightarrow (\bar{\mathbf{p}}, \bar{\mathbf{q}})$ such that the new Hamiltonian $\bar{H}$ depends only on $\bar{\mathbf{p}}$ and not on $\bar{\mathbf{q}}$. One possibility is to choose the constants of the motion themselves as the N components of $\bar{\mathbf{p}}$, $\bar{p}_i = f_i(\mathbf{p}, \mathbf{q})$. Since the f_i are constants, $d\bar{\mathbf{p}}/dt = \partial \bar{H}/\partial \bar{\mathbf{q}} = 0$ and hence $\bar{H} = \bar{H}(\bar{\mathbf{p}})$. In fact, we can construct many equivalent sets of constants of the motion by noting that any N independent functions of the N constants f_i could be used for the components of $\bar{\mathbf{p}}$ with the same result (namely, $d\bar{\mathbf{p}}/dt = \partial \bar{H}/\partial \bar{\mathbf{q}} = 0$). Of all these choices, one particular choice is especially convenient. This choice is the *action-angle variables* which we denote

$$(\bar{\mathbf{p}}, \bar{\mathbf{q}}) = (\mathbf{I}, \boldsymbol{\theta}),$$

where $\mathbf{I}$ is defined by

$$I_i = \frac{1}{2\pi} \oint_{\gamma_i} \mathbf{p} \cdot d\mathbf{q}, \tag{7.21}$$

Figure 7.4(a) Orbit on a 2-torus. (b) Two irreducible paths on a 2-torus.

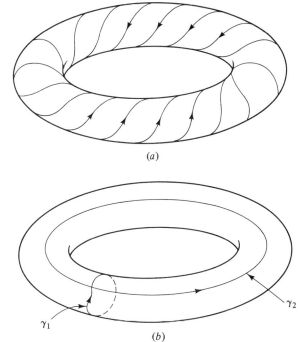

(a)

(b)

$i = 1, 2, \ldots, N$. (In (7.21) the γ_i denote N irreducible paths on the N-torus, each of which wrap around the torus in N angle directions that can be used to parameterize points on the torus (see Figure 7.4(b)). Note that deformations of the paths γ_i on the torus do not change the values of the integrals in (7.21) by virtue of the Poincaré–Cartan theorem, Eq. (7.9c).

The position coordinate $\boldsymbol{\theta}$, canonically conjugate to the momentum coordinate $\mathbf{I}$, is angle-like because, on one circuit following one of the irreducible paths γ_i around the torus, the variable θ_i increases by 2π, while the other variables θ_j with $j \neq i$ return to their original values. In order to see how this result is obtained, we first write the change of variables $(\mathbf{p}, \mathbf{q}) \to (\mathbf{I}, \boldsymbol{\theta})$ in terms of the generating function (Eq. (7.10)).

$$\boldsymbol{\theta} = \partial S(\mathbf{I}, \mathbf{q})/\partial \mathbf{I}, \tag{7.22a}$$

$$\mathbf{p} = \partial S(\mathbf{I}, \mathbf{q})/\partial \mathbf{q}. \tag{7.22b}$$

Let $\Delta_i \boldsymbol{\theta}$ denote the change of $\boldsymbol{\theta}$ on one circuit around the irreducible path γ_i, and $\Delta_i S$ denote the corresponding change of the generating function on one circuit. From (7.22b)

$$\Delta_i S = \oint_{\gamma_i} \mathbf{p} \cdot d\mathbf{q} = 2\pi I_i.$$

From (7.22a)

$$\Delta_i \boldsymbol{\theta} = \frac{\partial}{\partial \mathbf{I}} \Delta_i S = 2\pi \frac{\partial}{\partial \mathbf{I}} I_i$$

or

$$\Delta_i \theta_j = 2\pi \delta_{ij} \tag{7.23}$$

which is the desired result (here $\delta_{ij} \equiv 1$ if $i = j$ and $\delta_{ij} \equiv 0$ if $i \neq j$).

The new Hamiltonian in action–angle coordinates is, by construction, independent of $\boldsymbol{\theta}$, and hence Hamilton's equations reduce to

$$d\mathbf{I}/dt = 0,$$

$$d\boldsymbol{\theta}/dt = \partial \bar{H}(\mathbf{I})/\partial \mathbf{I} \equiv \boldsymbol{\omega}(\mathbf{I}).$$

The solution of these equations is $\mathbf{I}(t) = \mathbf{I}(0)$ and

$$\boldsymbol{\theta}(t) = \boldsymbol{\theta}(0) + \boldsymbol{\omega}(\mathbf{I})t. \tag{7.24}$$

Thus we can interpret $\boldsymbol{\omega}(\mathbf{I}) = \partial \bar{H}(\mathbf{I})/\partial \mathbf{I}$ as an angular velocity vector specifying trajectories on the N-torus. As in our discussion in Chapter 6, trajectories on a torus are N frequency quasiperiodic if there is no vector of integers $\mathbf{m} = (m_1, m_2, \ldots, m_N)$ such that

$$\mathbf{m} \cdot \boldsymbol{\omega} = 0, \tag{7.25}$$

except when $\mathbf{m}$ is the trivial vector all of whose components are zero. Assuming a typical smooth variation of $\bar{H}$ with $\mathbf{I}$ the condition $\mathbf{m} \cdot \boldsymbol{\omega} = 0$ with nonzero $\mathbf{m}$ is only satisfied for a *countable* set of $\mathbf{I}$. Thus, if one picks a point randomly with uniform probability in phase space, the probability is 1 that the point chosen will be on a torus for which the orbits are N frequency quasiperiodic and *fill up the torus*. Thus, for integrable systems, we can view the phase space as being completely occupied by N-tori almost all of which are in turn filled by N frequency quasiperiodic orbits. In contrast with the case of N frequency quasiperiodicity is the case of periodic motion, where orbits on the N-torus close on themselves (Figure 6.5). In this case

$$\boldsymbol{\omega} = \mathbf{m}\omega_0, \qquad (7.26)$$

where $\mathbf{m}$ is again a vector of integers and ω_0 is a scalar. The orbit in this case closes on itself after m_1 circuits in θ_1, m_2 circuits in $\theta_2, \ldots$. (Alternatively to (7.26), the condition for a periodic orbit can also be stated as requiring that $(N - 1)$-independent relations of the form (7.25) hold.[4]) Again assuming typical smooth variation of $\bar{H}$ with $\mathbf{I}$, we have that for integrable systems the set of tori which satisfy (7.26) and hence have periodic orbits, while having zero Lebesgue measure (i.e., zero phase space volume), is *dense in the phase space*. Thus, arbitrarily near any torus on which there is N frequency quasiperiodicity there are tori on which the orbits are periodic.

We now give an example of the procedure used for the reduction of an integrable system to action-angle variables. This procedure is based on the *Hamilton–Jacobi equation* obtained by combining (7.10) and (7.11),

$$H\left(\frac{\partial S(\mathbf{I}, \mathbf{q})}{\partial \mathbf{q}}, \mathbf{q}\right) = \bar{H}(\mathbf{I}). \qquad (7.27)$$

This equation may be regarded as a first-order partial differential equation for the generating function $S(\mathbf{I}, \mathbf{q})$.

Example: We consider a one degree of freedom Hamiltonian,

$$H(p, q) = p^2/(2m) + V(q),$$

where $V(q)$ is of the form shown in Figure 7.5. From Eq. (7.21) we have

$$I = \frac{1}{\pi} \int_{q_1}^{q_2} \{2m[E - V(q)]\}^{1/2} \, \mathrm{d}q. \qquad (7.28)$$

In the case of a harmonic oscillator, $V(q) = \frac{1}{2}m\Omega^2 q^2$, we have $q_2 = -q_1 = [2E/(m\Omega^2)]^{1/2}$, and the integral for I yields $I = E/\Omega$. Thus we have

$$\bar{H}(I) = \Omega I.$$

For this case $\omega(I) = \mathrm{d}\bar{H}/\mathrm{d}I = \Omega$ is independent of I, and (7.24) becomes

$$\theta(t) = \theta(0) + \Omega t.$$

From (7.27) we have for the harmonic oscillator

$$\partial S/\partial q = [2m(\Omega I - \tfrac{1}{2}m\Omega^2 q^2)]^{1/2}.$$

which on integration and application of (7.22) gives

$$q = (2I/m\Omega)^{1/2}\cos\theta,$$
$$p = -(2mI\Omega)^{1/2}\sin\theta.$$

The trajectory in p, q phase space is an ellipse on which the orbit circulates one time every period of oscillation $2\pi/\Omega$ (Figure 7.6(a)). (Since $N = 1$ we have a 'one-dimensional torus', namely a closed curve. For $N > 1$ we typically have N frequencies and quasiperiodic motion.) The harmonic oscillator is exceptional in that $\omega(I)$ is independent of I. As an example of the more typical situation where $\omega(I)$ depends on I consider the case of a hard-wall potential: $V(q) = 0$ for $|q| < \delta$ and $V(q) = +\infty$ for $|q| > \delta$. In this case the trajectory in phase space is as shown in Figure 7.6(b). The integral for I, Eq. (7.28), is just $(2\pi)^{-1}$ times the phase space area in the rectangle; $I = 4(2\pi)^{-1}(2mE\delta)^{1/2}$. Thus $\bar{H}(I) = (\pi I)^2/8m\delta^2$ and $\omega(I) = \pi^2 I/4m\delta^2$ which increases linearly with I.

7.2 Perturbation of integrable systems

7.2.1 The KAM theorem

We next address a very fundamental question concerning Hamiltonian systems; namely, how prevalent is integrability? One extreme conjecture is that integrability generally applies, and whatever difficulty we might encounter in obtaining the solution to any given problem only arises because we are not clever enough to determine the N independent

Figure 7.5 Particle of energy E in a potential well $V(q)$.

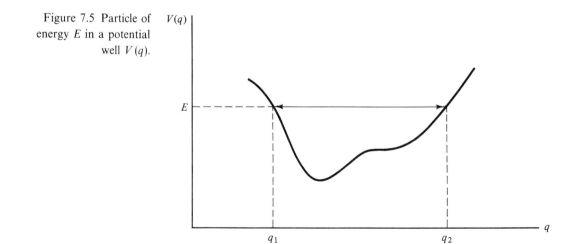

constants of the motion which must surely exist. Another conjecture, which is essentially the opposite of this, is that, given any integrable Hamiltonian $H_0(\mathbf{p},\mathbf{q})$, if we alter it slightly by the addition of a perturbation

$$H(\mathbf{p},\mathbf{q}) = H_0(\mathbf{p},\mathbf{q}) + \varepsilon H_1(\mathbf{p},\mathbf{q}), \qquad (7.29)$$

then we should expect that for a typical form of the perturbation, $H_1(\mathbf{p},\mathbf{q})$, all the constants of the motion for the integrable system, $H_0(\mathbf{p},\mathbf{q})$, except for the energy constant, $E = H(\mathbf{p},\mathbf{q})$, are immediately destroyed as soon as $\varepsilon \neq 0$. Presumably, if this second conjecture were to hold, then, for small ε, orbits would initially approximate the orbits of the integrable system, staying close to the unperturbed N-tori that exist for $\varepsilon = 0$ for some time. Eventually, however, the orbit, if followed for a long enough time, could ergodically wander anywhere on the energy surface. These two opposing views both have some support in experimental observation. On the one hand, the solar system appears to have been fairly stable. In particular, ever since its formation the Earth has been in a position relative to the position of the Sun such that its climate has been conducive to life. Thus,

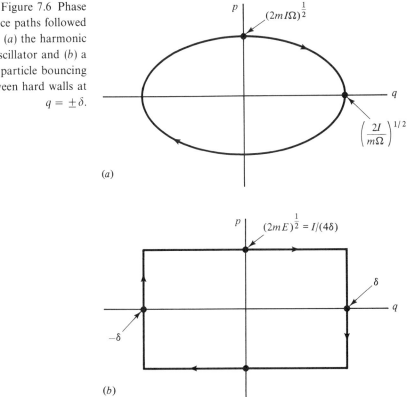

Figure 7.6 Phase space paths followed by (a) the harmonic oscillator and (b) a particle bouncing between hard walls at $q = \pm\delta$.

in spite of the perturbation caused by the gravitational pull of other planets (particularly that of Jupiter), the Earth's orbit has behaved as we would have expected had we neglected all the other planets. (In that case the system is integrable, and we obtain the elliptical Kepler orbit of the Earth around the Sun.) On the other hand, in support of the second conjecture, we have the amazing success of the predictions of statistical mechanics. In statistical mechanics one considers a Hamiltonian system with a large number of degrees of freedom ($N \gg 1$), and then makes the fundamental ansatz that at any given time the system is equally likely to be located at any point on the energy surface (the motion is ergodic on the energy surface). This would not be possible if there were additional constants of the motion constraining the orbit of the system. The success of statistical mechanics in virtually every case to which it may reasonably be applied can be interpreted as evidence supporting the validity of its fundamental ansatz in a wide variety of systems with $N \gg 1$.

Given the discussion above, it should not be too surprising to find out that the true situation lies somewhere between the two extremes that we have discussed. The resolution of the basic question of how prevalent integrability is has come only with the rigorous mathematical work of Kolmogorov, Arnold and Moser (KAM) and with the subsequent extensive computer studies of chaos and integrability in Hamiltonian systems. The basic question considered by Kolmogorov, Arnold and Moser was what happens when an integrable Hamiltonian is perturbed, Eq. (7.29). The research was initiated by Kolmogorov (1954) who conjectured what would happen with the addition of the perturbation. He also outlined an ingenious method which he felt could be used to prove his conjecture. The actual carrying out of this program, accomplished by Arnold and Moser (see Arnold (1963) and Moser (1973)), was quite difficult. The result they obtained is called the KAM theorem. We shall only briefly indicate some of the sources of the difficulty, and then state the main result.

We express (7.29) in the action-angle variables $(\mathbf{I}, \boldsymbol{\theta})$ of the unperturbed Hamiltonian H_0,

$$H(\mathbf{I}, \boldsymbol{\theta}) = H_0(\mathbf{I}) + \varepsilon H_1(\mathbf{I}, \boldsymbol{\theta}). \tag{7.30}$$

We are interested in determining whether this perturbed Hamiltonian has N-dimensional tori to which its orbits are restricted. If there are tori, there is a new set of action-angle variables $(\mathbf{I}', \boldsymbol{\theta}')$ such that

$$H(\mathbf{I}, \boldsymbol{\theta}) = H'(\mathbf{I}'),$$

where, in terms of the generating function S, we have using (7.10)

$$\mathbf{I} = \frac{\partial S(\mathbf{I}', \boldsymbol{\theta})}{\partial \boldsymbol{\theta}}, \quad \boldsymbol{\theta}' = \frac{\partial S(\mathbf{I}', \boldsymbol{\theta})}{\partial \mathbf{I}'}. \tag{7.31}$$

The Hamilton–Jacobi equation for S is

$$H\left(\frac{\partial S}{\partial \boldsymbol{\theta}}, \boldsymbol{\theta}\right) = H'(\mathbf{I}'). \tag{7.32}$$

One approach to solving (7.32) for S might be to look for a solution in the form of a power series in ε,

$$S = S_0 + \varepsilon S_1 + \varepsilon^2 S_2 + \dots . \tag{7.33}$$

For S_0 we use $S_0 = \mathbf{I}' \cdot \boldsymbol{\theta}$ which when substituted in (7.31) gives $\mathbf{I} = \mathbf{I}'$, $\boldsymbol{\theta} = \boldsymbol{\theta}'$, corresponding to the original action-angle variables applicable for $\varepsilon = 0$. Substituting the series (7.33) for S in (7.32) gives,

$$H_0(\mathbf{I}' + \varepsilon\, \partial S_1/\partial \boldsymbol{\theta} + \varepsilon^2\, \partial S_2/\partial \boldsymbol{\theta} + \dots) + \varepsilon H_1(\mathbf{I}' + \varepsilon\, \partial S_1/\partial \boldsymbol{\theta} + \dots, \boldsymbol{\theta})$$
$$= H'(\mathbf{I}'). \tag{7.34}$$

Expanding (7.34) for small ε and only retaining first-order terms, we have

$$H_0(\mathbf{I}') + \varepsilon \frac{\partial H_0}{\partial \mathbf{I}'} \cdot \frac{\partial S_1}{\partial \boldsymbol{\theta}} + \varepsilon H_1(\mathbf{I}', \boldsymbol{\theta}) = H'(\mathbf{I}'). \tag{7.35}$$

We next express $H_1(\mathbf{I}', \boldsymbol{\theta})$ and $S_1(\mathbf{I}', \theta)$ as Fourier series in the angle vector $\boldsymbol{\theta}$,

$$H_1 = \sum_{\mathbf{m}} H_{1,\mathbf{m}}(\mathbf{I}') \exp(\mathrm{i}\mathbf{m} \cdot \boldsymbol{\theta}),$$

$$S_1 = \sum_{\mathbf{m}} S_{1,\mathbf{m}}(\mathbf{I}') \exp(\mathrm{i}\mathbf{m} \cdot \boldsymbol{\theta}),$$

where $\mathbf{m}$ is an N-component vector of integers. Substituting these Fourier series in (7.35), we obtain

$$S_1 = \mathrm{i} \sum_{\mathbf{m}} \frac{H_{1,\mathbf{m}}(\mathbf{I}')}{\mathbf{m} \cdot \boldsymbol{\omega}_0(\mathbf{I}')} \exp(\mathrm{i}\mathbf{m} \cdot \boldsymbol{\omega}) \tag{7.36}$$

where $\boldsymbol{\omega}_0(\mathbf{I}) \equiv \partial H_0(\mathbf{I})/\partial \mathbf{I}$ is the unperturbed N-dimensional frequency vector for the torus corresponding to action $\mathbf{I}$. One question is that of whether the infinite sum (7.36) converges. This same question also arises in taking (7.34) to higher order in ε to determine successively the other terms, S_2, S_3, etc., appearing in the series (7.33).

This problem is precisely the 'problem of small denominators' encountered in Section 6.2 where we treated frequency locking of quasiperiodic orbits for dissipative systems. In particular, clearly (7.36) does not work for values of $\mathbf{I}$ for which $\mathbf{m} \cdot \boldsymbol{\omega}_0(\mathbf{I}) = 0$ for some value of $\mathbf{m}$. These $\mathbf{I}$ define *resonant tori* of the unperturbed system. (These resonant tori are typically destroyed by the perturbation for any small $\varepsilon > 0$.) We emphasize that the resonant tori are dense in the phase space of the unperturbed Hamiltonian. On the other hand, there is still a large set of 'very nonresonant' tori. These are tori for which $\boldsymbol{\omega}$ satisfies the condition

$$|\mathbf{m} \cdot \boldsymbol{\omega}| > K(\boldsymbol{\omega})|\mathbf{m}|^{-(N+1)}, \tag{7.37}$$

for *all* integer vectors $\mathbf{m}$ (except the zero vector). Here $|\mathbf{m}| \equiv |m_1| + |m_2| + \cdots + |m_N|$, and $K(\boldsymbol{\omega}) > 0$ is a number independent of $\mathbf{m}$. The set of N-dimensional vectors $\boldsymbol{\omega}$ which do not satisfy (7.37) has zero Lebesgue measure in $\boldsymbol{\omega}$-space, and thus the 'very nonresonant' tori are, in this sense, very common. For $\boldsymbol{\omega}$ satisfying (7.37), the series (7.36), and others of similar form giving $S_2, S_3, \ldots$, converges. This follows if we assume that H_1 is analytic in $\boldsymbol{\theta}$ which implies that $H_{1,\mathbf{m}}$ decreases exponentially with m; i.e., $|H_{1,\mathbf{m}}| < (\text{constant})\exp(-\alpha|\mathbf{m}|)$ for some constant $\alpha > 0$. (Refer to the discussion in Section 6.2.)

Even given that all the terms $S_1, S_2, \ldots$ exist and can be found, we would still be faced with the problem of whether there is convergence of the successive approximations to S obtained by taking more and more terms in the series (7.33). Actually, the scheme we have outlined (wherein S is expanded in a straightforward series in ε, Eq. (7.33)) is too crude, and the proof of the KAM theorem relies on a more sophisticated method of successive approximations which has much faster convergence properties. We shall not pursue this discussion further. Suffice it to say that the KAM theorem essentially states that under very general conditions for small ε 'most' (in the sense of the Lebesgue measure of the phase space) of the tori of the unperturbed integrable Hamiltonian survive. We say that a torus of the unperturbed system with frequency vector $\boldsymbol{\omega}_0$ 'survives' perturbation if there exists a torus of the perturbed ($\varepsilon \neq 0$) system which has a frequency vector $\boldsymbol{\omega}(\varepsilon) = k(\varepsilon)\boldsymbol{\omega}_0$, where $k(\varepsilon)$ goes continuously to 1 as $\varepsilon \to 0$, and such that the perturbed toroidal surface with frequency $\boldsymbol{\omega}(\varepsilon)$ goes continuously to the unperturbed torus as $\varepsilon \to 0$. Thus, writing $\boldsymbol{\omega} = (\omega_1, \omega_2, \ldots, \omega_N)$, the unperturbed and perturbed frequency vectors $\boldsymbol{\omega}_0$ and $\boldsymbol{\omega}(\varepsilon)$ have the same frequency ratios of their components, $\omega_{0j}/\omega_{01} = \omega_j(\varepsilon)/\omega_1(\varepsilon)$ for $j = 2, 3, \ldots, N$. According to the KAM theorem, for small ε, the perturbed system's phase space volume (Lebesgue measure) not occupied by surviving tori is small and approaches zero as ε approaches zero.

Note, however, that, since the resonant tori on which $\mathbf{m} \cdot \boldsymbol{\omega}_0(\mathbf{I}) = 0$ are dense, we expect that, arbitrarily near surviving tori of the perturbed system, there are regions of phase space where the orbits are not on surviving tori. We shall, in fact, see that these regions are occupied by chaotic orbits as well as new tori and elliptic and hyperbolic periodic orbits all created by the perturbation. In the language of Section 3.9, the set in the phase space occupied by surviving perturbed tori is a fat fractal. That is, it is the same kind of set on which values of the parameter r yielding chaos for the logistic map (2.10) exist and on which values of the parameter w in the circle map (6.11) yield two frequency quasiperiodic orbits (for $k < 1$). The Poincaré–Birkhoff theorem discussed in the next subsection sheds light on the exceedingly complex and intricate situation which arises in the vicinity of resonant tori when an integrable system is perturbed.

7.2.2 The fate of resonant tori

We have seen that most tori survive small perturbation. The resonant tori, however, do not. What happens to them? To simplify the discussion of this question we consider the case of a Hamiltonian system described by a two-dimensional area preserving map. We can view this map as arising from a surface of section for a time-independent Hamiltonian with $N = 2$, as illustrated for the integrable case in Figure 7.7. The tori of the integrable system intersect the surface of section in a family of nested closed curves. Without loss of generality we can take these curves to be concentric circles represented by polar coordinates (r, ϕ). In this case we obtain a map $(r_{n+1}, \phi_{n+1}) = \mathbf{M}_0(r_n, \phi_n)$,

$$\left. \begin{array}{l} r_{n+1} = r_n, \\ \phi_{n+1} = [\phi_n + 2\pi R(r_n)] \text{ modulo } 2\pi. \end{array} \right\} \tag{7.38}$$

Here $R(r)$ is the ratio of the frequencies ω_1/ω_2 where $\boldsymbol{\omega}_0 = (\omega_1, \omega_2) = (\partial H_0/\partial I_1, \partial H_0/\partial I_2)$ for the torus which intersects the surface of section in a circle of radius r, and we have taken the surface of section to be $\theta_2 = $ (const.), where $\boldsymbol{\theta} = (\theta_1, \theta_2)$ are the angle variables conjugate to the actions $\mathbf{I} = (I_1, I_2)$. The quantity ϕ_n is the value of θ_1 at the nth piercing of the surface of section by the orbit. On a resonant torus the rotation number $R(r)$ is rational:

$$R = \omega_1/\omega_2 = \tilde{p}/\tilde{q}; \quad \tilde{q}\omega_1 - \tilde{p}\omega_2 = 0,$$

Figure 7.7 Surface of section for an integrable system.

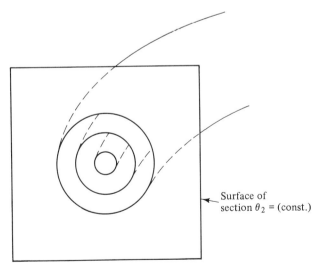

Surface of
section $\theta_2 = $ (const.)

where $\tilde{p}$ and $\tilde{q}$ are integers which do not have a common factor. At the radius $r = \hat{r}(\tilde{p}/\tilde{q})$ corresponding to $R(\hat{r}) = \tilde{p}/\tilde{q}$ we have that application of the map (7.38) $\tilde{q}$ times returns every point on the circle to its original position,

$$\mathbf{M}_0^{\tilde{q}}(r, \phi) = [r, (\phi + 2\pi\tilde{p}) \text{ modulo } 2\pi] = (r, \phi).$$

Now we consider a perturbation of the integrable Hamiltonian H_0, Eq. (7.30). This will perturb the map $\mathbf{M}_0$ to a new area preserving map $\mathbf{M}_\varepsilon$ which differs slightly from $\mathbf{M}_0$,

$$\begin{aligned} r_{n+1} &= r_n + \varepsilon g(r_n, \phi_n), \\ \phi_{n+1} &= [\phi_n + 2\pi R(r_n) + \varepsilon h(r_n, \phi_n)] \text{ modulo } 2\pi. \end{aligned} \tag{7.39}$$

We have seen that, on the intersection $r = \hat{r}(\tilde{p}/\tilde{q})$ of the resonant torus with the surface of section, every point is a fixed point of $\mathbf{M}_0^{\tilde{q}}$. We now inquire, what happens to this circle when we add the terms proportional to ε in (7.39)? Assume that $R(r)$ is a smoothly increasing function of r in the vicinity of $r = \hat{r}(\tilde{p}/\tilde{q})$. (Equation (7.38) is called a 'twist map' if $R(r)$ increases with r.) Then for the unperturbed map we can choose a circle at $r = r_+ > \hat{r}(\tilde{p}/\tilde{q})$ which is rotated by $\mathbf{M}_0^{\tilde{q}}$ in the direction of increasing ϕ (i.e., counterclockwise) and a circle at $r = r_- < \hat{r}(\tilde{p}/\tilde{q})$ which is rotated by $\mathbf{M}_0^{\tilde{q}}$ in the direction of decreasing ϕ (i.e., clockwise). The circle $r = \hat{r}(\tilde{p}/\tilde{q})$ is not rotated at all. See Figure 7.8 (a). If ε is sufficiently small, then $\mathbf{M}_\varepsilon^{\tilde{q}}$ still maps all the points initially on the circle $r = r_-$ to new positions whose ϕ coordinate is clockwise displaced from its initial position (the radial coordinate, after application of the perturbed map, will in general differ from r_-). Similarly, for small enough ε all points on r_+ will be counterclockwise displaced. Given this situation, we have that for any given fixed value of ϕ, as r increases from r_- to r_+, the value of the angle that the point (r, ϕ) maps to increases from below ϕ to above ϕ. Hence, there is a value of r between r_+ and r_- for which the angle is not changed. We conclude that, for the perturbed map, there is a closed curve, $r = \hat{r}_\varepsilon(\phi)$, lying between $r_+ \geq r \geq r_-$ and close to $r = \hat{r}(\tilde{p}, \tilde{q})$, on which points are mapped by $\mathbf{M}_\varepsilon^{\tilde{q}}$ purely in the radial direction. This is illustrated in Figure 7.8 (b). We now apply the map $\mathbf{M}_\varepsilon^{\tilde{q}}$ to this curve obtaining a new curve $r = \hat{r}_\varepsilon'(\phi)$. The result is shown schematically in Figure 7.9. Since $\mathbf{M}_\varepsilon$ is area preserving the areas enclosed by the curve $\hat{r}_\varepsilon(\phi)$ and by the curve $\hat{r}_\varepsilon'(\phi)$ are equal. Hence, these curves must intersect. Generically these curves intersect at an even number of distinct points. (Here by use of the word generic we mean to rule out cases where the two curves are tangent or else (as in the integrable case) coincide. These nongeneric cases can be destroyed by small changes in ε or in the form of the perturbing functions g and h in Eq. (7.39).) The intersections of $\hat{r}_\varepsilon$ and $\hat{r}_\varepsilon'$ correspond to fixed points of $\mathbf{M}_\varepsilon^{\tilde{q}}$. Thus, the circle of fixed points $r = \hat{r}(\tilde{p}/\tilde{q})$ for the unperturbed map $\mathbf{M}_0^{\tilde{q}}$ is replaced by a finite number of fixed points when the map is perturbed.

What is the character of these fixed points of $\mathbf{M}_\varepsilon^{\tilde{q}}$? Recall that for $r > \hat{r}_\varepsilon$ points are rotated counterclockwise by $\mathbf{M}_\varepsilon^{\tilde{q}}$. Also recall that $\mathbf{M}_\varepsilon^{\tilde{q}}$ maps $\hat{r}_\varepsilon$ to $\hat{r}_\varepsilon'$. Thus, we have the picture shown in Figure 7.10, where the arrows indicate the displacements experienced by points as a result of applying the map $\mathbf{M}_\varepsilon^{\tilde{q}}$. We see that elliptic and hyperbolic fixed points of $\mathbf{M}_\varepsilon^{\tilde{q}}$

Figure 7.8(*a*) Three invariant circles of the unperturbed map. (*b*) The curve $r = \hat{r}_\varepsilon(\phi)$.

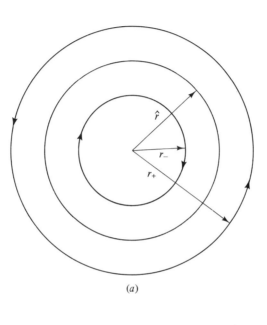

(*a*)

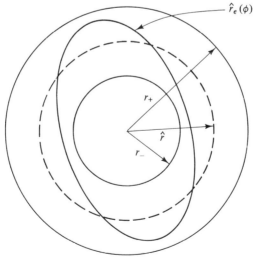

(*b*)

alternate. Hence, perturbation of the resonant torus with rational rotation number $\tilde{p}/\tilde{q}$ results in an equal number of elliptic and hyperbolic fixed points of $\mathbf{M}_\varepsilon^{\tilde{q}}$. This result is known as the Poincaré–Birkhoff theorem (Birkhoff, 1927). Since fixed points of $\mathbf{M}_\varepsilon^{\tilde{q}}$ necessarily are on period $\tilde{q}$ orbits of $\mathbf{M}_\varepsilon$, we see that there are $\tilde{q}$ (or a multiple of $\tilde{q}$) elliptic fixed points of $\mathbf{M}_\varepsilon^{\tilde{q}}$ and the same number of hyperbolic fixed points. Thus, for example, the two elliptic fixed points of $\mathbf{M}_\varepsilon^{\tilde{q}}$ shown in Figure 7.10 might be a single

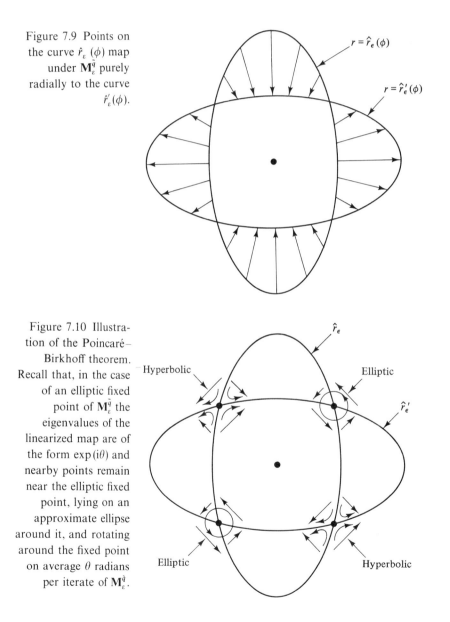

Figure 7.9 Points on the curve $\hat{r}_\varepsilon(\phi)$ map under $\mathbf{M}_\varepsilon^{\tilde{q}}$ purely radially to the curve $\hat{r}'_\varepsilon(\phi)$.

Figure 7.10 Illustration of the Poincaré–Birkhoff theorem. Recall that, in the case of an elliptic fixed point of $\mathbf{M}_\varepsilon^{\tilde{q}}$ the eigenvalues of the linearized map are of the form $\exp(i\theta)$ and nearby points remain near the elliptic fixed point, lying on an approximate ellipse around it, and rotating around the fixed point on average θ radians per iterate of $\mathbf{M}_\varepsilon^{\tilde{q}}$.

periodic orbit of $\mathbf{M}_\varepsilon$ of period two (and similarly for the two hyperbolic fixed points in the figure). Thus, in this case, we have $\tilde{q} = 2$. Near each resonant torus of the unperturbed map we can expect a structure of elliptic and hyperbolic orbits to appear as illustrated schematically in Figures 7.11(a) and (b) where we only include the $\tilde{q} = 3$ and the $\tilde{q} = 4$ resonances.

Points near the elliptic fixed points rotate around them as shown by the

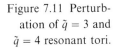

Figure 7.11 Perturbation of $\tilde{q} = 3$ and $\tilde{q} = 4$ resonant tori.

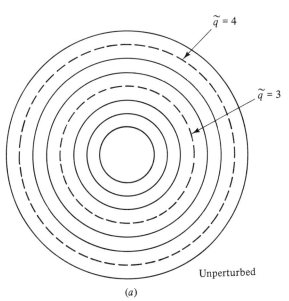

(a)

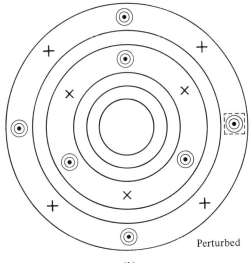

(b)

linear theory (cf. Section 7.1.3). *Very* near an elliptic fixed point the linear approximation is quite good, and in such a small neighborhood the map can again be put in the form of Eq. (7.39). Thus, if we examine the small region around one of the elliptic points of a periodic orbit, such as the area indicated by the dashed box in Figure 7.11(*b*), then what we will see is qualitatively similar to what we see in Figure 7.11(*b*) itself. Thus, surrounding an elliptic point there are encircling KAM curves between which are destroyed resonant KAM curves that have been replaced by elliptic and hyperbolic periodic orbits. Furthermore, this repeats *ad infinitum*, since any elliptic point has surrounding elliptic points of destroyed resonances which themselves have elliptic points of destroyed resonances, and so on.

What influence do the hyperbolic orbits created from the destroyed resonant tori have on the dynamics? If we follow the stable and unstable manifolds emanating from the hyperbolic points, they typically result in heteroclinic intersections, as shown in Figure 7.12. As we have seen in

Figure 7.12 Stable and unstable manifolds of hyperbolic periodic orbits.

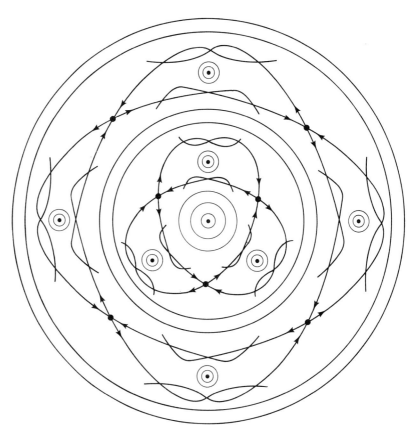

Chapter 4 (see Figure 4.10 (d)), one such heteroclinic intersection between the stable and unstable manifolds of two hyperbolic points implies an infinite number of intersections between them.[5] Furthermore (as we have discussed for the homoclinic case, Figure 4.11), this also implies the presence of horseshoe type dynamics and hence chaos. Thus, not only do we have a dense set of destroyed resonance regions containing elliptic and hyperbolic orbits, but now we find that these regions of destroyed resonances also have embedded within them chaotic orbits. Furthermore, this repeats on all scales as we successively magnify regions around elliptic points. A very fascinating and intricate picture indeed!

7.3 Chaos and KAM tori in systems describable by two-dimensional Hamiltonian maps

Numerical examples clearly show the general phenomenology described for small perturbations of integrable systems in the previous section. In addition, numerical examples give information concerning what occurs when the perturbations are not small. Such information in turn points the way for theories applicable in the far-from-integrable regime. The clearest and easiest numerical experiments are those that result in a two-dimensional map (a two-dimensional Poincaré surface of section).

7.3.1 The standard map

As an example, we consider the standard map, Eq. (7.15), which results from periodic impulsive kicking of the rotor in Figure 7.3. Setting the kicking strength to zero, $K = 0$, the standard map becomes

$$\theta_{n+1} = (\theta_n + p_n) \text{ modulo } 2\pi, \tag{7.40a}$$

$$p_{n+1} = p_n. \tag{7.40b}$$

This represents an integrable case. The intersections of the tori in the (θ, p) surface of section are just the lines of constant p (according to (7.40b) p is a constant of the motion). On each such line the orbit is given by $\theta_n = (\theta_0 + np_0)$ modulo 2π, and, if $p_0/2\pi$ is an irrational number, a single orbit densely fills the line $p = p_0$. If $p_0/2\pi$ is a rational number, then orbits on the line return to themselves after a finite number of iterates (the unperturbed orbit is periodic), and we have a resonant torus.

Increasing K slightly from zero introduces a small nonintegrable perturbation to the integrable case (7.40). Figure 7.13 shows plots of $\bar{p} \equiv p$ modulo 2π versus θ modulo 2π resulting from iterating a number of different initial conditions for a long time and for various values of K. If the initial condition is on an invariant torus it traces out the closed curve corresponding to the torus. If the initial condition yields a chaotic orbit, then it will wander throughout an area densely filling that area. We see that, for the relatively small perturbation, $K = 0.5$, Figure 7.13(a), there are many KAM tori running roughly horizontally from $\theta = 0$ to $\theta = 2\pi$. These tori are those that originate from the nonresonant tori of the unperturbed system ($p = p_0, p_0/2\pi$ irrational) and have survived the perturbation. Also, clearly seen in Figure 7.13(a) are tori, created by the perturbations nested around elliptic periodic orbits originating from resonant tori. In particular, the period one elliptic orbits, $(\theta, p) = (\pi, 0)$ and $(\theta, p) = (\pi, 2\pi)$, and the period two elliptic orbit, $(0, \pi) \rightleftarrows (\pi, \pi)$, are clearly visible. We call the structure surrounding a period $\tilde{q}$ elliptic periodic orbit a *period $\tilde{q}$ island chain.*

An important property of two-dimensional smooth area preserving maps is that the area bounded by two invariant KAM curves is itself invariant. This is illustrated in Figure 7.14 where we show two invariant curves (tori) bounding a shaded annular shaped region. Since the two curves are invariant and areas are preserved, the shaded region must map

Figure 7.13 Plots of p modulo 2π for four values of K: (a) $K = 0.5$; (b) $K = 1.0$; (c) $K = 2.5$; (d) $K = 4.0$. (This figure courtesy of Y. Du.)

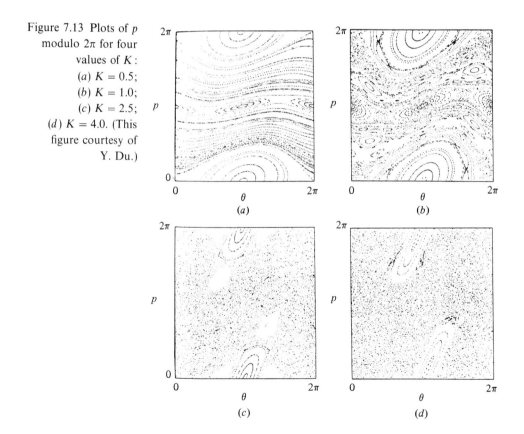

into itself. Thus, while there may be chaotic orbits sandwiched between KAM curves (as, for example, in the island structures surrounding elliptic orbits), these chaotic orbits are necessarily restricted to lie between the bounding KAM curves. (As we shall discuss later, this picture is fundamentally different for systems of higher dimensionality.)

As K is increased, more of the deformed survivors originating from the unperturbed tori are destroyed. At $K = 1$ (Figure 7.13(b)) we see that there are none left; that is, there are no tori running as continuous curves from $\theta = 0$ to $\theta = 2\pi$. In their place we see chaotic regions with interspersed island chains. As K is increased further (Figures 7.13(c) and (d)) many of the KAM surfaces associated with the island chains disappear, and the chaotic region enlarges. At $K = 4.0$, for example, we see (Figure 7.13(d)) that the only discernable islands are those associated with the period one orbits at $(\theta, p) = (\pi, 0), (\pi, 2\pi)$. Increasing K, Chirikov (1979) numerically found values of K (e.g., $K \simeq 8\frac{8}{9}$) for which there are no discernable tori, and the entire square $0 \leq (\theta, p) \leq 2\pi$ is, to within the available numerical resolution, ergodically covered densely by a single orbit. Thus, if any island chains are present, they are very small.

7.3.2 The destruction of KAM surfaces and island chains

Considering the standard map, the absence of a period $\tilde{q}$ island chain at some value $K = K'$, implies that the period $\tilde{q}$ elliptic periodic orbit has become unstable as K increases from $K = 0$ to $K = K'$. How does this occur? The answer is that as K increases, the eigenvalues of the $\tilde{q}$th iterate of the linearized map $\mathbf{DM}^{\tilde{q}}$ evaluated on the period $\tilde{q}$ orbit eventually change from complex and of magnitude 1 (i.e., $\exp(\pm i\theta)$) to real and negative with one eigenvalue with magnitude larger than 1 and one with

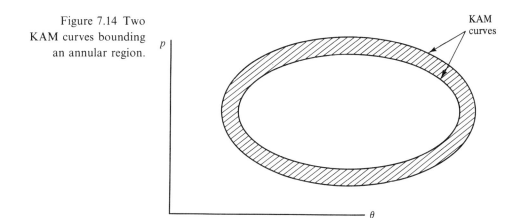

Figure 7.14 Two KAM curves bounding an annular region.

magnitude less than 1 (i.e., λ and $1/\lambda$ with $|\lambda| > 1$). That is, the periodic orbit of period $\tilde{q}$ changes from elliptic to hyperbolic with reflection as K passes through some value $K = K_{\tilde{q}}$. When a periodic orbit becomes hyperbolic with reflection its eigenvalues in the elliptic range, $\exp(\pm i\theta)$, both approach -1 by having θ approach π as K approaches $K_{\tilde{q}}$. The migration of the eigenvalues in the complex plane as K passes through $K_{\tilde{q}}$ is illustrated in Figure 7.15. This leads to a period doubling bifurcation and is typically followed by an infinite period doubling cascade (Bountis, 1981; Greene *et al.*, 1981). In such a cascade, the period $\tilde{q}$ elliptic orbit destabilizes (becomes hyperbolic) simultaneously with the appearance of a period $2\tilde{q}$ elliptic orbit, which then period doubles to produce a period $2^2\tilde{q}$ elliptic orbit, and so on. Eventually, at some finite amount past $K_{\tilde{q}}$, all orbits of period $2^n\tilde{q}$ have been stably created and then rendered unstable (hyperbolic) as they period double. This is a Hamiltonian version of the period doubling cascade phenomena we have discussed for one-dimensional maps in Chapter 2. As in that situation, there are universal numbers that describe the scaling properties of such cascades (cf. Chapter 8), although these numbers differ in the Hamiltonian case from those given in Chapter 2. Note that, in this period doubling cascade, whenever K is in the range where there is an elliptic period $2^n\tilde{q}$ periodic orbit, there is a nested set of invariant tori surrounding that orbit (i.e., there is a period $2^n\tilde{q}$ island chain).

When $K = 0$ the standard map is integrable. As K is increased, chaotic regions occupy increasingly large areas, and the original KAM tori of the integrable system are successively destroyed. Say we identify a particular nonresonant KAM torus by its rotation number

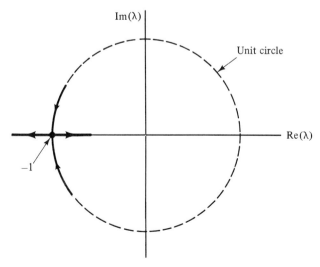

Figure 7.15
Path followed by the eigenvalues as an elliptic orbit changes to hyperbolic.

$$R = \lim_{m \to \infty} \frac{1}{2\pi m} \sum_{n=1}^{m} p_n$$

(p_n is the amount by which θ increases on each iterate; see Eq. (7.15a)). As we increase K, the torus deforms from the straight horizontal line, $p = 2\pi R$, that it occupied for $K = 0$. Past some critical value $K > K_{\text{crit}}(R)$ the torus no longer exists. How does one numerically calculate $K_{\text{crit}}(R)$? To answer this question we note the result from number theory that the irrational number R can be represented as an infinite continued fraction,

$$R = a_1 + \cfrac{1}{a_2 + \cfrac{1}{a_3 + \cfrac{1}{a_4 + \dots}}},$$

where the a_i are integers. As a shorthand we write $R = [a_1, a_2, a_3, \dots]$. If one cuts off the continued fraction at a_n,

$$R_n = [a_1, a_2, \dots, a_n, 0, 0, 0, \dots],$$

then one obtains a rational approximation to R which converges to R as $n \to \infty$,

$$R = \lim_{n \to \infty} R_n.$$

If we examine the $K > 0$ island chain with rotation number R_n, we find that the elliptic Poincaré–Birkhoff periodic orbits (Figures 7.10–7.12) for the island chain approach the nonresonant torus of irrational rotation number R as n increases. This leads one to investigate the stability of these periodic orbits. As illustrated in Figure 7.15, the complex eigenvalues $\exp(i\theta)$ of the Jacobian matrix corresponding to such a periodic orbit change to real negative eigenvalues λ and $1/\lambda$ at some critical K-value (which depends on the particular periodic orbit). It is found numerically that the critical K-values of these Poincaré–Birkhoff periodic orbits of rational rotation number R_n rapidly approach the value $K_{\text{crit}}(R)$ as n increases. Since efficient numerical procedures exist for finding such orbits, this provides an efficient way of accurately determining $K_{\text{crit}}(R)$. Schmidt and Bialek (1982) have used this procedure to investigate the pattern accompanying torus destruction of arbitrary irrational tori. Greene (1979) conjectured that, since the golden mean $R_g = (\sqrt{5} - 1)/2$ is the 'most irrational' number in the sense that it is most slowly approached by cutoffs of its continued fraction expansion,

$$R_g = [1, 1, 1, 1, \dots] = 1 + \cfrac{1}{1 + \cfrac{1}{1 + \cfrac{1}{1 + \cfrac{1}{1 + \dots}}}},$$

the torus with $R = R_g$ will be the last surviving torus as K is increased (i.e., $K_{crit}(R)$ is largest for $R = R_g$). Using the periodic orbit technique described above, Greene finds that $K_{crit}(R_g) = 0.97\ldots$. Figure 7.16 shows the standard map for $K = 0.97$ (Greene, 1979) with the $R = R_g$ tori and some chaotic orbits plotted (there are two such tori in $0 \le p \le 2\pi$). An important result concerning the $R = R_g$ torus is that the phase space structure in its vicinity exhibits intricate scaling properties at and near $K = K_{crit}(R_g)$, and this phenomenon has been investigated by the renormalization group technique (Kadanoff, 1981; Escande, 1982; MacKay, 1983).

   In constructing the figure plotted in Figure 7.13 we have made use of the fact that the standard map is invariant to translation in momentum by an integer multiple of 2π, $p \to p + 2\pi k$. This periodicity of the map allowed us to use p modulo 2π for the vertical coordinate in Figure 7.13 rather than p. Thus, if we were to ask for the structure of the solutions for all p, our answer would be given by pasting together an infinite string of pictures obtained by successively translating the basic unit (as in Figure 7.13) by 2π. For example, for the case of $K = 4.0$ (Figure 7.13(c)), we see that a single chaotic component connects regions of the line $p = 0$ with regions of the line $p = 2\pi$. By the periodicity in p, this implies that this chaotic region actually runs from $p = -\infty$ to $p = +\infty$. Thus, in terms of the rotor model (Figure 8.3), if we start an initial condition in this chaotic component, it can wander with time to arbitrarily large rotor energies, $p^2/2$ (here we have taken the rotor's moment of inertia to be 1). On the other hand, if we were to start an initial condition for $K = 4.0$ inside the period one island surrounding one of the period one fixed points, it would

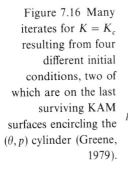

Figure 7.16 Many iterates for $K = K_c$ resulting from four different initial conditions, two of which are on the last surviving KAM surfaces encircling the (θ, p) cylinder (Greene, 1979).

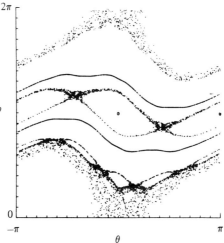

remain there forever; its energy would thus be bounded for all time. Note that if we plot the actual momentum, $+\infty > p > -\infty$, versus θ, then we are treating the phase space of the two-dimensional standard map as a cylinder. On the other hand, our plot where we utilized p modulo 2π reduced the phase space to the surface of a torus. While the toroidal surface representation is convenient for displaying the structure of intermixed chaotic and KAM regions, we emphasize that p and $p + 2\pi k$ ($k =$ an integer) are not physically equivalent, since they generally represent different kinetic energies of the rotor.

The case shown in Figure 7.16 corresponds to the largest value of K for which there are KAM curves running completely around the (θ, p) cylinder in the θ-direction. The presence of a KAM curve running around the (θ, p) cylinder implies an infinite number of such curves by translation of p by multiples of 2π. Furthermore, any orbit lying between two such curves cannot cross them (Figure 7.14) and so is restricted to lie between them forever. Thus, the energy of the rotor cannot increase without bound. When K increases past the critical value $K_c \simeq 0.97$ the last invariant tori encircling the cylinder are destroyed, and a chaotic area connecting $p = -\infty$ and $p = +\infty$ exists. This means that for $K > K_c$ the rotor energy can increase without bound if the initial condition lies in the chaotic component connecting $p = -\infty$ and $p = +\infty$.

7.3.3 Diffusion in momentum

Let us now consider the case of large K such that there are no discernible KAM surfaces present, and the entire region of a plot of p modulo 2π versus θ appears to be densely covered by a single chaotic orbit. Referring to Eq. (7.15b), we see that the change in momentum (not taken modulo 2π), $\Delta p_n \equiv p_{n+1} - p_n = K \sin \theta_{n+1}$, is typically large (i.e., of the order of K). If we assume $K \gg 2\pi$, then p will also typically be large compared to 2π. Thus, by Eq. (7.15a), we expect θ (which is taken modulo 2π) to vary very wildly in $[0, 2\pi]$. We, therefore, treat θ_n as effectively random, uniformly distributed, and uncorrelated for different times (i.e., different n). With these assumptions, the motion in p becomes a random walk with step size $\Delta p_n = K \sin \theta_{n+1}$. Thus, over momentum scales larger than K, the momentum evolves according to a diffusion process with diffusion coefficient,

$$\frac{\langle (\Delta p_n)^2 \rangle}{2} = \frac{K^2}{2} \langle \sin^2 \theta_{n+1} \rangle, \tag{7.41}$$

where the angle brackets denote a time average, and by virtue of the randomness assumption for the θ_n we have $\langle \sin^2 \theta_{n+1} \rangle = \frac{1}{2}$. Inserting the latter in (7.41) gives the so-called *quasilinear* approximation to the diffusion coefficient,

$$D \cong D_{\mathrm{QL}} = K^2/4. \qquad (7.42)$$

If we imagine that we spread a cloud of initial conditions uniformly in θ and p in the cell $-\pi \le p \le \pi$, then the momentum distribution function $f(p, n)$ at time n, coarse grained over intervals in p greater than 2π, is

$$f(p, n) \simeq \frac{1}{(2\pi n D)^{1/2}} \exp\left(-\frac{p^2}{2nD}\right). \qquad (7.43)$$

That is, the distribution is a spreading Gaussian. This result follows from the fact that the process is diffusive. Taking the second moment of the distribution function, $\int p^2 f \, dp$, we see that the average rotor energy increases linearly with time,[6]

$$\langle p^2/2 \rangle \simeq Dn. \qquad (7.44)$$

The quasilinear result (7.42) is valid for very large K. For moderately large, but not very large, values of K, neglected correlation effects can significantly alter the diffusion coefficient from the quasilinear value. These effects have been analytically calculated by Rechester and White (1980) (see also Rechester *et al.* (1981), Karney *et al.* (1981), and Carey *et al.* (1981)). Figure 7.17 shows a plot of the diffusion coefficient D normalized to D_{QL} as a function of K from the paper of Rechester and White. The solid curve is their theory, and the dots are obtained by

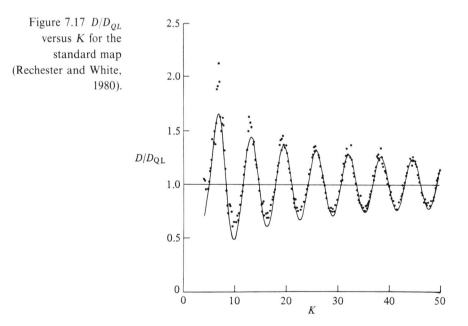

Figure 7.17 D/D_{QL} versus K for the standard map (Rechester and White, 1980).

numerically calculating the spreading of a cloud of points and obtaining D from Eq. (7.44). Note the decaying oscillations about the quasilinear value as K increases.[7]

7.3.4 Other examples

So far in this section we have dealt exclusively with the standard map. We now discuss some other examples, also reducible to two-dimensional maps, where similar phenomena are observed.

We first consider a time-independent two-degree-of-freedom system investigated by Schmidt and Chen (1991). This system, depicted in Figure 7.18, consists of two masses, a large mass M connected to a linearly behaving spring of spring constant k_s and a small mass m which elastically bounces between a fixed wall on the left and the oscillating large mass on the right. The motion in space is purely one-dimensional. This represents a time-independent Hamiltonian system which Schmidt and Chen call the 'autonomous Fermi system'. Since the Hamiltonian is time-independent, the total energy of the system, consisting of the sum of the kinetic energy of the two masses and the potential energy in the spring, is conserved. (This is unlike the rotor system (Figure 7.3) which has external kicking that enables the energy to increase without bound for $K > K_c$.) Schmidt and Chen numerically calculate a Poincaré surface of section and plot the state of the system at the instants of time just after the masses m and M collide. Plots corresponding to three cases are shown in Figure 7.19. In this figure v is the velocity of the small mass and ϕ is the phase of the large mass in the sinusoidal oscillation that it experiences between bounces. The maximum value of v is 1 (for the normalization used) and is attained if all the system energy is in the small mass. The three cases shown correspond to

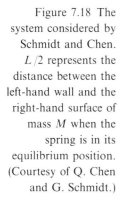

Figure 7.18 The system considered by Schmidt and Chen. $L/2$ represents the distance between the left-hand wall and the right-hand surface of mass M when the spring is in its equilibrium position. (Courtesy of Q. Chen and G. Schmidt.)

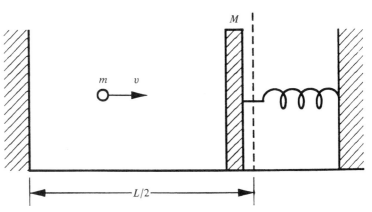

successively larger values of $\omega_0 \bar{T}$, where $\omega_0 = (k_s/M)^{1/2}$ is the natural oscillation frequency of the large mass and $\bar{T}$ is the mean time between bounces. We note that at low $\omega_0 \bar{T}$ (Figure 7.19(a)) we see many KAM surfaces as well as island chains and chaos for lower v ($v \gtrsim 0.25$). At higher $\omega_0 \bar{T}$ (Figure 7.19(b)), the chaotic region enlarges substantially, while at the highest value plotted (Figure 7.19(c)) a single orbit appears to cover the available area of the surface of section ergodically. This latter situation corresponds to ergodic wandering of the orbit over the energy surface in the full four-dimensional phase space. Accordingly, for the case of Figure 7.19(c) Schmidt and Chen numerically confirm that there is a time average energy equipartition between the energies of the two masses and the energy of the spring, each having on average very close to one third of the total energy of the system. The equipartition of time averaged kinetic energy is a familiar result in the statistical mechanics of many degree of freedom systems $(\langle p_1^2 \rangle / 2m_1 = \langle p_2^2 \rangle / 2m_2 = \cdots = \langle p_N^2 \rangle / 2m_N)$. Here equipartition of kinetic energy for a system of only two degrees of freedom

Figure 7.19 Poincaré sections of v versus ϕ. (a), (b) and (c) correspond to three successively larger values of $\omega_0 \bar{T}$. (Courtesy of Q. Chen and G. Schmidt.)

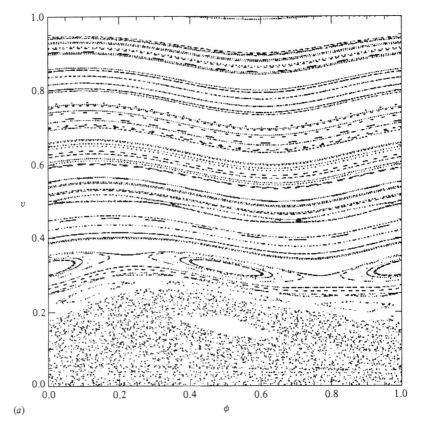

(a)

holds because the system is essentially ergodic on the energy surface. (Indeed it is the most important fundamental assumption of statistical mechanics that typical many-degree-of-freedom systems are egodic on their energy surface. The justification of this assumption, however, is far from obvious, and remains an open problem.)

We could go on to cite many other examples of mechanical systems displaying the type of behaviour seen in the two examples of the kicked rotor (Figure 7.3) and the autonomous Fermi system (Figure 7.19). It is, perhaps, somewhat more surprising that these same phenomena apply to situations in which one is not dealing with straightforward problems of mechanics. The point is that these problems are also described by Hamilton's equations. Three examples of this type are the following.

(1) Nonturbulent mixing in fluids.
(2) The trajectories of magnetic field lines in plasmas.
(3) The ray equations for the propagation of short wavelength waves in inhomogeneous media.

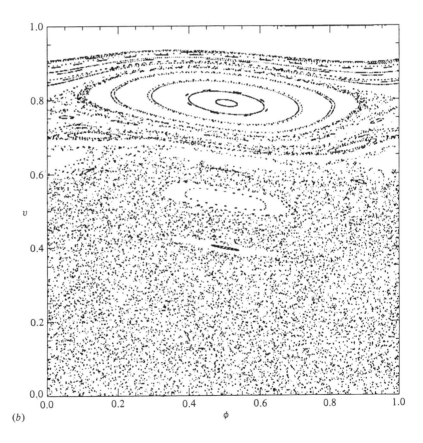

(b)

In the case of mixing in fluids we restrict ourselves to the situation of a two-dimensional incompressible flow: $\mathbf{v}(\mathbf{x}, t) = v_x(x, y, t)\mathbf{x}_0 + v_y(x, y, t)\mathbf{y}_0$ with $\partial v_x/\partial x + \partial v_y/\partial y = \nabla \cdot \mathbf{v} = 0$. The incompressibility condition, $\nabla \cdot \mathbf{v} = 0$, means that we can express $\mathbf{v}$ in terms of a stream function ψ,

$$\mathbf{v} = \mathbf{z}_0 \times \nabla\psi(x, y, t)$$

or

$$v_x = -\partial\psi/\partial y, v_y = \partial\psi/\partial x.$$

Now consider the motion of an impurity particle convected with the fluid. The location of this particle is given by $d\mathbf{x}/dt = \mathbf{v}(\mathbf{x}, t)$, or, using the stream function,

$$dx/dt = -\partial\psi/\partial y, \tag{7.45a}$$

$$dy/dt = \partial\psi/\partial x. \tag{7.45b}$$

Comparing Eqs. (7.45) with Eqs. (7.1), we see that (7.45) are in the form of a one-degree-of-freedom ($N = 1$) time-dependent Hamiltonian system, if

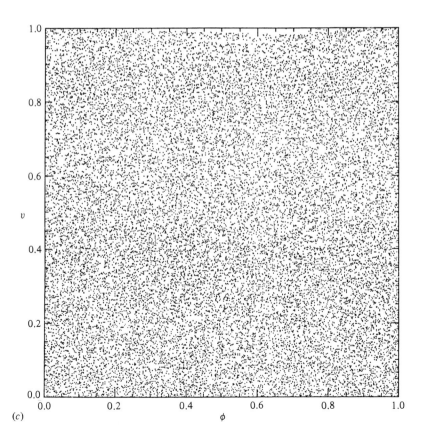

(c)

we identify the stream function ψ with the Hamiltonian H, x with the momentum p, and y with the 'position' q:

$$\psi(x, y, t) \leftrightarrow H(p, q, t),$$

$$x \leftrightarrow p,$$

$$y \leftrightarrow q.$$

Thus, in our fluid problem the canonically conjugate variables are x and y.

As an example, we consider the 'blinking vortex' flow of Aref (1984). In this flow there are two vortices of equal strength, one located at $(x, y) = (a, 0)$ and the other located at $(x, y) = (-a, 0)$. The vortices (which may be thought of as thin rotating stirring rods) are taken to 'blink' on and off with time with period $2T$. That is, for $2kT \leq t \leq (2k + 1)T$ $(k = 0, 1, 2, 3, \ldots)$, the vortex at $(a, 0)$ is on, while the vortex at $(-a, 0)$ is turned off, and, for $(2k + 1)T \leq t \leq 2(k + 1)T$, the vortex at $(a, 0)$ is off while the vortex at $(-a, 0)$ is on. The flow induced by a single vortex of strength Γ can be expressed in (ρ, θ) polar coordinates centered at the vortex as

$$v_\theta = \Gamma/2\pi\rho \text{ and } v_\rho = 0.$$

Thus, the blinking vortex has the effect of alternatively rotating points in concentric circles first about one vortex center and then about the other vortex center, each time by an angle $\Delta\theta = \Gamma T/2\pi\rho^2$. Sampling the position of a particle at times $t = 2kT$ defines a two-dimensional area preserving map which depends on the strength parameter $\mu = \Gamma T/2\pi a^2$. Figure 7.20 from Doherty and Ottino (1988) shows results from iterating several different initial conditions for successively larger values of the strength parameter μ. For very small μ (Figure 7.20(a), $\mu = 0.1$) the result is very close to the completely integrable case where both vortices act simultaneously and steadily in time. As μ is increased we see that the area occupied by chaotic motion increases.

The practical effect of this type of result for fluid mixing can be seen by considering a small dollop of dye in such a flow. For example, say the dye is initially placed in the location indicated by the shaded circle in Figure 7.20(a). In the near integrable case Figure 7.20(a), as time goes on, this dye would always necessarily be located between the two KAM curves that initially bound it. Due to the different rotation rates on the different KAM surfaces, the dye will mix throughout the annular region bounded by these KAM curves, but (in the absence of molecular diffusion) it can never mix with the fluid outside this annular region. In contrast, for the case $\mu = 0.4$ (Figure 7.20(e)), we see that there is a large single connected chaotic region, and an initial dollop of dye in the same location as before would thus mix uniformly throughout this much larger region. Thus, we

Figure 7.20 Blinking
vortex orbits for
(*a*) $\mu = 0.01$,
(*b*) $\mu = 0.15$,
(*c*) $\mu = 0.25$,
(*d*) $\mu = 0.3$, and
(*e*) $\mu = 0.4$ (Doherty
and Ottino, 1988).

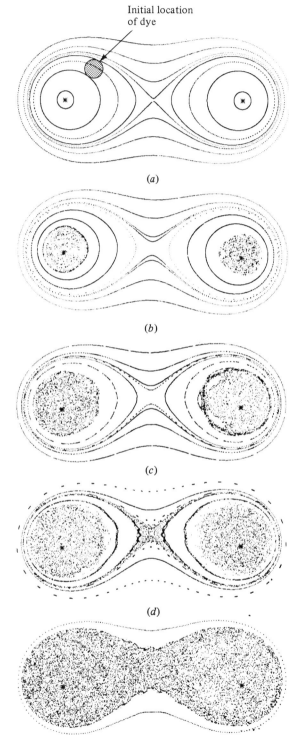

Initial location
of dye

(*a*)

(*b*)

(*c*)

(*d*)

(*e*)

see that, for the purposes of achieving the most uniform mixing in fluids, chaos is a desirable attribute of the flow that one should strive to maximize.

Several representative references on chaotic mixing in fluids are Aref and Balachandar (1986), Chaiken *et al.* (1986), Dombre *et al.* (1986), Feingold *et al.* (1988), Ott and Antonsen (1989), Rom-Kedar *et al.* (1990), and the comprehensive book on the subject by Ottino (1989).

We now discuss the second of the three applications mentioned above, namely, the trajectory of magnetic field lines in plasmas. Let $\mathbf{B}(\mathbf{x})$ denote the magnetic field vector. The field line trajectory equation gives a parametric function $\mathbf{x}(s)$ for the curve on which a magnetic field line lies, where s is a parameter which we can think of as a (distorted) measure of distance along the field line. The equation for $\mathbf{x}(s)$ is

$$d\mathbf{x}(s)/ds = \mathbf{B}(\mathbf{x}). \qquad (7.46)$$

(Alternatively, we can multiply the right-hand side of (7.46) by any positive scalar function of $\mathbf{x}$.) Since $\nabla \cdot \mathbf{B} = 0$, Eq. (7.46) represents a conservative flow, if we make an analogy between s and time. Thus, the magnetic field lines in physical space are mathematically analogous to the trajectory of a dynamical system in its phase space. In Problem 3 you are asked to establish for a simple example that (7.46) can be put in Hamiltonian form.

The Hamiltonian nature of 'magnetic field line flow' means that under many circumstances we can expect that some magnetic field line trajectories fill up toroidal surfaces, while other field lines wander chaotically over a volume which may be bounded by tori. In other words, the situation can be precisely as depicted in Figure 7.12.

These considerations are of great importance in plasma physics and controlled nuclear fusion research. In the latter, the fundamental problem is to confine a hot plasma (gas of electrons and ions) for a long enough time that sufficient energy-releasing nuclear fusion reactions take place. If the magnetic field is strong, then, to a first approximation, the motion of the charged particles constituting the plasma is constrained to follow the magnetic field lines. (This approximation is better for the lighter mass electrons than for the ions.) In this view, the problem of confining the plasma becomes that of creating a magnetic field line configuration such that the magnetic field lines are confined. That is, the magnetic field lines do not connect the plasma interior to the walls of the device. The most simple example of such a configuration is provided by the tokamak device, originally invented in the Soviet Union. (This device is currently the one on which most of the attention of the nuclear fusion community is focused.) Figure 7.21 (*a*) illustrates the basic idea of the tokamak. An external current system (in the figure the wire with current I_0) creates a

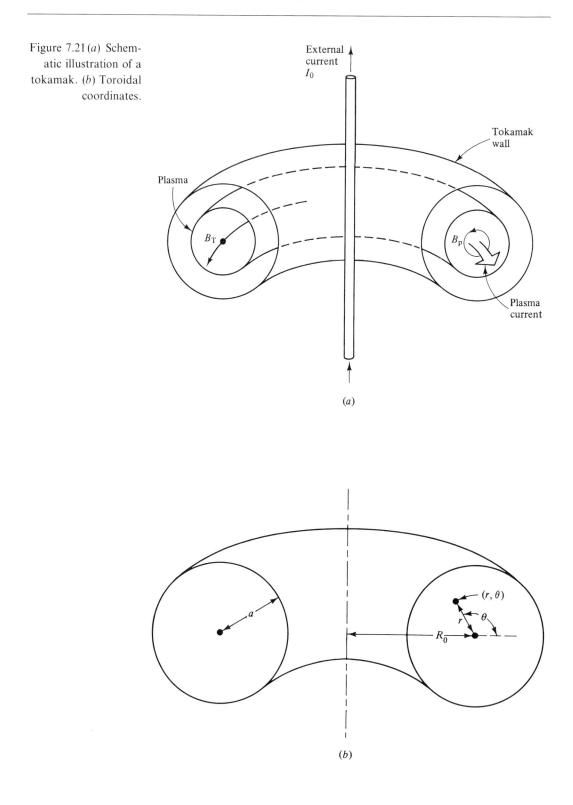

Figure 7.21 (*a*) Schematic illustration of a tokamak. (*b*) Toroidal coordinates.

(*a*)

(*b*)

magnetic field B_T running the long way (called the 'toroidal direction') around a toroid of plasma. At the same time, another current is induced to flow in the plasma in the direction running the long way around the torus. (This toroidal plasma current is typically created by transformer action wherein the plasma loop serves as the secondary coil of a transformer.) The toroidal plasma current then creates a magnetic field component B_p which circles the short way around the torus (the 'poloidal direction'). Assuming that the configuration is perfectly symmetric with respect to rotations around the axis of the system, the superposition of the toroidal and poloidal magnetic fields leads to field lines that typically circle on a toroidal surface, simultaneously in both the toroidal and poloidal directions, filling the surface ergodically. Thus, the field lines are restricted to lie on a nested set of tori and never intersect bounding walls of the device. This is precisely analogous to the case of an integrable Hamiltonian system.

This is the situation if there is perfect toroidal symmetry. Unfortunately, symmetry can be destroyed by errors in the external field coils, by necessary asymmetries in the walls, and, most importantly, by toroidal dependences of the current flowing the plasma. (The latter can arise due to collective motions of the plasma as a result of a variety of instabilities that have been very extensively investigated.) Such symmetry-breaking magnetic field perturbations play a role analogous to nonintegrable perturbations of an integrable Hamiltonian system. Thus, they can destroy some of the nested set of toroidal magnetic surfaces that exists in the symmetric case. If the perturbation is too strong, chaotic field lines can wander from the interior of the plasma to the wall. This leads to rapid heat and particle loss of the plasma. (Refer to Figure 7.13 and think of K as the strength of the asymmetric field perturbation.)

Some representative papers which discuss chaotic magnetic field line trajectories in plasmas and their physical effects are Rosenbluth *et al.* (1966), Sinclair *et al.* (1970), Finn (1975), Rechester and Rosenbluth (1978), Cary and Littlejohn (1983), Hanson and Cary (1984), and Lau and Finn (1991).

As our final example, we consider the ray equations describing the propagation of short wavelength waves in a time-independent spatially inhomogeneous medium. In the absence of inhomogeneity, we assume that the partial differential equations governing the evolution of small amplitude perturbations of the dependent quantities admit plane wave solutions in which the perturbations vary as $\exp(i\mathbf{k} \cdot \mathbf{x} - i\omega t)$, where ω and $\mathbf{k}$ are the frequency and wavenumber of the wave. The quantities ω and $\mathbf{k}$ are constrained by the governing equations (e.g., Maxwell's equations if we are dealing with electromagnetic waves) to satisfy a dispersion relation

$$D(\mathbf{k}, \omega) = 0.$$

Now assume that the medium is inhomogeneous with variations occurring on a scale size L which is much longer than the typical wavelength of the wave, $|\mathbf{k}|L \gg 1$. For propagation distances small compared to L, waves behave approximately as if the medium were homogeneous. For propagation distances of the order of L or longer, the spatial part of the homogeneous medium expression for the phase, namely $\mathbf{k} \cdot \mathbf{x}$, is distorted. We, therefore, assume that the perturbations have a rapid (compared to L) spatial variation of the form

$$\exp[i\tilde{S}(\mathbf{x}) - i\omega t], \tag{7.47}$$

where the function $\tilde{S}(\mathbf{x})$ is called the *eikonal* and replaces the homogeneous medium phase term $\mathbf{k} \cdot \mathbf{x}$. The *local* wavenumber $\mathbf{k}$ is given by

$$\mathbf{k} = \nabla\tilde{S}(\mathbf{x}). \tag{7.48}$$

We wish to find an equation for the propagation of a wave along some path (called the 'ray path'). Along this path we seek parametric equations for $\mathbf{x}$ and $\mathbf{k}$. That is, we seek $(\mathbf{x}(s), \mathbf{k}(s))$, where s measures the distance along the ray. In terms of these ray path functions, we can determine the function $\tilde{S}(\mathbf{x})$ using Eq. (7.48),

$$\tilde{S}(\mathbf{x}) = \tilde{S}(\mathbf{x}_0) + \int_{\mathbf{x}_0}^{\mathbf{x}} \mathbf{k}(s) \cdot d\mathbf{x}(s).$$

(In addition, by a systematic expansion in $|\mathbf{k}|L$, one can also determine the variation of the wave amplitudes along the rays as well as higher order corrections. See, for example, Courant and Hilbert (1966) and references therein.) Using the local wavenumber $\mathbf{k}$, we can write down a local dispersion relation,

$$D(\mathbf{k}, \omega, \mathbf{x}) = 0, \tag{7.49}$$

which is identical with the homogeneous medium dispersion relation if we use the local parameters of the inhomogeneous medium at the point $\mathbf{x}$. Since the medium is assumed time-independent, the frequency ω is constant. Differentiating (7.49) with respect to $\mathbf{x}$, we obtain

$$\frac{\partial D}{\partial \mathbf{x}} + \nabla\mathbf{k} \cdot \frac{\partial D}{\partial \mathbf{k}} = 0. \tag{7.50}$$

The x-component of this equation is

$$\frac{\partial D}{\partial x} + \frac{\partial D}{\partial k_x}\frac{\partial k_x}{\partial x} + \frac{\partial D}{\partial k_y}\frac{\partial k_y}{\partial x} + \frac{\partial D}{\partial k_z}\frac{\partial k_z}{\partial x} = 0$$

However, from (7.48) we have $\nabla \times \mathbf{k} = 0$, so that $\partial k_y/\partial x = \partial k_x/\partial y$ and $\partial k_z/\partial x = \partial k_x/\partial z$. Thus, the above equation becomes

$$\frac{\partial D}{\partial x} + \frac{\partial D}{\partial \mathbf{k}} \cdot \frac{\partial k_x}{\partial \mathbf{x}} = 0.$$

Since similar equations apply for the y- and z-components, (7.50) becomes

$$(\partial D / \partial \mathbf{k}) \cdot \nabla \mathbf{k} = -\partial D / \partial \mathbf{x}. \tag{7.51}$$

If we regard $\partial D / \partial \mathbf{k}$ as being like a velocity, then $(\partial D / \partial \mathbf{k}) \cdot \nabla$ is like the time derivative following a point with the 'velocity' $\partial D / \partial \mathbf{k}$. Thus,

$$\frac{\mathrm{d}\mathbf{k}}{\mathrm{d}s} = -\frac{\partial D}{\partial \mathbf{x}}, \tag{7.52a}$$

$$\frac{\mathrm{d}\mathbf{x}}{\mathrm{d}s} = \frac{\partial D}{\partial \mathbf{k}}, \tag{7.52b}$$

which are the ray equations, the second of which comes from our interpretation of $\partial D / \partial \mathbf{k}$ as being like a velocity with the parametric variable s playing the role of the fictitious time variable. More formally, Eqs. (7.52) represent the solution of the first-order partial differential equation, Eq. (7.51), by the method of characteristics (Courant and Hilbert, 1966). In the special case where the dispersion relation is written in the form, $D(\mathbf{k}, \omega, \mathbf{x}) = \hat{\omega}(\mathbf{k}, \mathbf{x}) - \omega$, we have that $\partial D / \partial \mathbf{k} = \partial \hat{\omega} / \partial \mathbf{k}$ which is just the group velocity of a wavepacket. Thus, letting τ denote time, we have $\mathrm{d}\mathbf{x} / \mathrm{d}\tau = \partial \hat{\omega} / \partial \mathbf{k}$, and hence $s = \tau$ in this case. This yields

$$\mathrm{d}\mathbf{k} / \mathrm{d}\tau = -\partial \hat{\omega} / \partial \mathbf{x}, \tag{7.53a}$$

$$\mathrm{d}\mathbf{x} / \mathrm{d}\tau = \partial \hat{\omega} / \partial \mathbf{k}, \tag{7.53b}$$

which can be interpreted as giving the temporal evolution of the position $\mathbf{x}$ and wavenumber $\mathbf{k}$ of a wavepacket. Both (7.52) and (7.53) are Hamiltonian with D and $\hat{\omega}$ respectively playing the role of the Hamiltonian, and $(\mathbf{k}, \mathbf{x})$ being the canonically conjugate momentum ($\mathbf{k}$) and position ($\mathbf{x}$) variables.

We now discuss a particular example where chaotic solutions of the ray equations play a key role (Bonoli and Ott, 1982; Wersinger et al., 1978). One of the central problems in creating a controlled thermonuclear reactor lies in raising the temperature of the confined plasma sufficiently to permit fusion reactions to take place. One way of doing this is by launching waves from outside the plasma that then propagate to the plasma interior, where they dissipate their energy to heat (similar, in principle, to a kitchen microwave oven). Clearly, conditions must be such that the wave is able to reach the plasma interior. In this case, in the terminology of the field, the wave is said to be 'accessible.' Of the various types of waves that can be used for plasma heating, the so-called 'lower hybrid' wave is one of the most attractive from a technological point of view. The accessibility problem for this wave was originally considered by

Stix (1965) for the case in which there are two symmetry directions. For example, in a straight circular cylinder, k_z and $m = k_\theta r$ are constants of the ray equations due to the translational symmetry along the axis of the cylinder (the z-axis) and to the rotational symmetry around the cylinder (in θ). The accessibility situation for this case is illustrated in Figure 7.22 for a cylinder with an applied magnetic field $\mathbf{B} = B_0\mathbf{z}_0 + B_\theta\boldsymbol{\theta}_0$, and a wave launched from vacuum with $m = 0$. Let $n_\| = k_z c/\omega$ (where c is the speed of light) and $n_\perp = k_r c/\omega$. We assume that the plasma density N_0 increases with decreasing radius r from $N_0 = 0$ at the plasma edge ($r = a$) to its maximum value in the center of the cylinder ($r = 0$). Figure 7.22 shows plots of $n_\perp^2$ (obtained from the dispersion relation) as a function of N_0. For $n_\| < n_a$, $n_\| = n_a$, and $n_\| > n_a$, Figures 7.22(a), (b) and (c) apply, respectively, where n_a is a certain critical value (Stix, 1965). Between $N_0 = 0$ and $N_0 = N_S$, there is a narrow cutoff region through which a slow wave (i.e., lower hybrid wave), launched from the vacuum region, typically has little trouble in tunneling. Figure 7.22(a) shows that for $n_\| < n_a$ an additional, effectively much wider, cutoff region between $N_0 = N_{T1}$ and $N_0 = N_{T2}$ exists. This cutoff region presents a barrier for

Figure 7.22 Plots of $n_\perp^2$ versus density N_0 for (a) $n_\| < n_a$, (b) $n_\| = n_a$ and (c) $n_\| > n_a$.

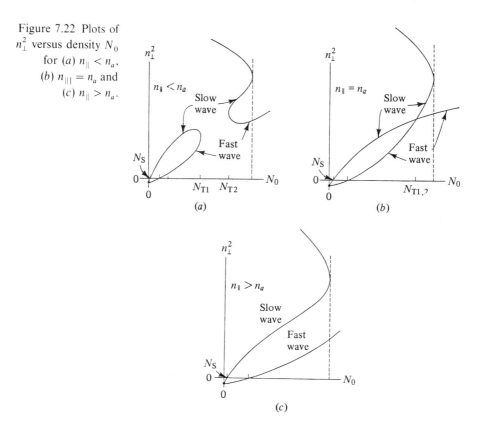

propagation to the plasma center and prevents accessibility. Figure 7.22(c) shows that for $n_\parallel > n_a$ this barrier is absent, and the lower hybrid (slow) waves becomes accessible.

Now we consider heating a circular cross section toroidally symmetric plasma (a tokamak). We use toroidal coordinates wherein r, θ are circular polar coordinates centered in the circular cross section of the tokamak plasma such that the distance of a point from the major axis of the torus is $R = R_0 + r \cos \theta$, where R_0 is the distance from the major axis of the torus to the center of the plasma cross section (Figure 7.21(b)). We refer to θ as the poloidal angle, and we denote by ϕ the toroidal angle (i.e., the angle running the long way around the plasma torus). Let $\varepsilon = a/R_0$, where $r = a$ denotes the plasma boundary. As $\varepsilon \to 0$ with a fixed, the straight cylinder limit is approached. However, for finite ε the plasma equilibrium depends on θ. Thus, it is no longer expected that $m = rk_\theta$ is a constant of the motion, although the toroidal symmetry still guarantees that a constant of the motion analogous to k_z in the cylinder still exists; namely, $n = Rk_\phi$ is a constant. The questions that now arise are what happens to the constant m, and how is the accessibility condition for lower hybrid waves affected? For finite ε there may still be some other constant $\tilde{m} = \tilde{m}(r, \theta, k_r, m)$, which takes the place of m. For small ε, regions where $\tilde{m}$ exist (KAM tori) occupy most of the phase space. As ε increases the regions occupied by chaotic trajectories increase, until almost all regions where KAM tori exist are gone. A ray in the region with no tori may eventually approach the plasma interior and be absorbed even if $n_\parallel$ at launch does not satisfy the straight cylinder accessibility condition. Thus we need to know at what value of ε most of the tori are gone.

Figure 7.23 shows numerical results (Bonoli and Ott, 1982) testing for the existence of tori by the surface of section method with $\theta = 0 \pmod{2\pi}$ as the surface of section. Figure 7.23(a) shows that for $\varepsilon = 0.10$, most tori are not destroyed, and initially inaccessible rays (i.e., $n_\parallel < n_a$ at launch) do not reach the plasma interior. Figures 7.23(b) and (c) show a case for $\varepsilon = 0.15$, illustrating the coexistence of chaotic and integrable orbits including (Figure 7.23(c)) higher-order island structures. For $\varepsilon = 0.25$ all appreciable KAM surfaces are numerically found to be completely destroyed, and even waves launched with $n_\parallel$ substantially below n_a are absorbed in the plasma interior after a few piercings of the surface of section.

7.4 Higher-dimensional systems

There is a very basic topological distinction to be made between the case of time-independent Hamiltonians with $N = 2$ degrees of freedom, on the one hand, and $N \geq 3$ degrees of freedom, on the other. (For a time periodic Hamiltonian the same distinction applies for the cases $N = 1$ and $N \geq 2$.) In particular, since the energy is a constant of the motion for a time-independent system, the motion is restricted to the $(2N - 1)$-dimensional energy surface $H(\mathbf{p}, \mathbf{q}) = E$. Thus, we can regard the dynamics as taking place in an effectively $(2N - 1)$-dimensional space. In general, in order for a closed surface to divide a $(2N - 1)$-dimensional space into two distinct parts, one inside the closed surface and another outside, the closed surface must have a dimension one less than the dimension of the space; i.e., its dimension is $2N - 2$. Thus, KAM surfaces which are N-dimensional tori, only satisfy this condition for $N = 2$. In particular, for $N = 2$, the energy surface has a dimension 3, and a two-dimensional toroidal surface in a three dimension space has an inside and an outside. As an example of a toroidal 'surface' which does not divide the space in which it lies, consider a circle (which can be regarded as a 'one-dimensional torus') in a three-dimensional Cartesian space. For

Figure 7.23 Surface of section ($\theta = 0$) plots for several different initial conditions. $n_a = 2.0$. (a) $1.3 \leq n_{\parallel} \leq 1.4$, $a/R_0 = 0.10$; (b) $1.25 \leq n_{\parallel} \leq 1.4$, $a/R_0 = 0.15$; (c) same as (b) but with a different initial condition (Bonoli and Ott, 1981).

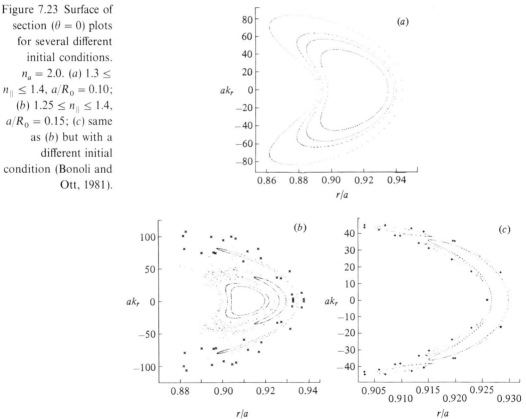

$N > 2$ the situation for KAM tori in the energy surface is similar (e.g., for $N = 3$ we have $2N - 1 = 5$ and $2N - 2 = 4 > 3 = N$).

Now consider the situation where an integrable system is perturbed. In this case tori begin to break up and are replaced by chaotic orbits. For the case $N = 2$ these chaotic regions are necessarily sandwiched between surviving KAM tori (Figure 7.15). In particular, if such an orbit is outside (inside) a particular torus it remains outside (inside) that torus forever. Because of this sandwiching effect, the chaotic orbit of a slightly perturbed integrable two-degree-of-freedom system must lie close to the orbit on a torus of the unperturbed integrable system for all time. Hence, two-degree-of-freedom integrable systems are relatively stable to perturbations. The situation for $N \geq 3$ is different because chaotic orbits are not enclosed by tori, and hence their motions are not restricted as in the case $N = 2$. In fact, it is natural to assume that all the chaos created by destroyed tori can form a single connected ergodic chaotic region which is dense in the phase space. Under this assumption a chaotic orbit can, in principle, come arbitrarily close to *any* point in phase space, if we wait long enough. This phenomenon was first demonstrated for a particular example by Arnold (1964) and is known as 'Arnold diffusion'. For further discussion of Arnold diffusion and other aspects of chaos in Hamiltonian systems with more than two degrees of freedom we refer the reader to the exposition of these topics in the book by Lichtenberg and Lieberman (1983).

7.5 Strongly chaotic systems

We have seen in Sections 7.2 and 7.3 that when elliptic periodic orbits are present there is typically an exceedingly intricate mixture of chaotic regions and KAM tori: surrounding each elliptic orbit are KAM tori, between which are chaotic regions and other elliptic orbits, which are themselves similarly surrounded, and so on *ad infinitum*. It would seem that the situation would be much simpler if there were no elliptic periodic orbits (i.e., all were hyperbolic). In such a case one would expect that the whole phase space would be chaotic and no KAM tori would be present at all.

As a model of such a situation, one can consider two-dimensional area preserving maps which are hyperbolic. One example is the cat map, Eq. (4.29), discussed in Chapter 4. In that case we saw that the map took the picture of the cat and stretched it out (chaos) and reassembled it in the square (Figure 4.13). (Recall that the square is an unwrapping of a two-dimensional toroidal surface.) More iterations mix the striations of the unfortunate cat more and more finely within the square. Given any small fixed region $\mathscr{R}$ within the square, as we iterate more and more times, the fraction of the area of the region $\mathscr{R}$ occupied by black striations that were originally part of the face of the cat approaches the fraction of area of the entire square that was originally occupied by the cat's face (the black region in Figure 4.13). We say that the cat map is *mixing* on the unit square with essentially the same meaning that we use when we describe the mixing of cream as a cup of coffee is stirred. More formally, an area preserving map $\mathbf{M}$ of a compact region S is mixing on S if given any two subsets σ and σ' of S where σ and σ' have positive Lebesgue measure $(\mu_L(\sigma) > 0, \mu_L(\sigma') > 0)$, then

$$\frac{\mu_L(\sigma)}{\mu_L(S)} = \lim_{m \to \infty} \frac{\mu_L[\sigma' \cap \mathbf{M}^m(\sigma)]}{\mu_L(\sigma')} \tag{7.54}$$

As another example of an area preserving mixing two-dimensional hyperbolic map, we mention the generalized baker's map (Figure 3.4) in the area preserving case, $\lambda_a = \alpha$, $\lambda_b = \beta$. (It is easy to check that the Jacobian determinant, Eq. (4.28), is one in this case.)

Another group of strongly chaotic systems can be constructed from certain classes of 'billiard' problems. A billiard is a two-dimensional planar domain in which a point particle moves with constant velocity along straight line orbits between specular bounces ((angle of incidence) = (angle of reflection)) from the boundary of the domain, Figure 7.24(a). Figures 7.24(b)–(g) show several shapes of billiards. The circle (b) and the rectangle (c) are completely integrable. The two constants of the motion for the circle are the particle energy and the angular momentum about the center of the circle. The two angular frequencies associated with the action-angle variables are the inverses of the time between bounces and the time to make a complete rotation around the center of the circle. For the rectangle the two constants are the vertical and horizontal kinetic energies, and the two frequencies are the inverses of twice the time between successive bounces off the vertical walls and twice the time between successive bounces off the horizontal walls. The billiard shapes shown in Figures 7.24(d)–(g) are strongly chaotic (Bunimovich, 1979; Sinai, 1970) in the same sense as the cat map and the area preserving generalized baker's map: almost every initial condition yields a chaotic orbit which eventually comes arbitrarily close to every point in the phase space. (In Figures 7.24(d)–(g) the curved line segments are arcs of circles.) In particular, for these chaotic billiard problems the orbit generates a density which is uniform in the angle of the velocity vector and is also uniform in the accessible area of the billiard. Note that, unlike the cat map and the area preserving generalized baker's map, the billiard systems shown in Figures 7.24(d), (f), and (g) are not hyperbolic. This is because, in

Figure 7.24 Billiard systems.

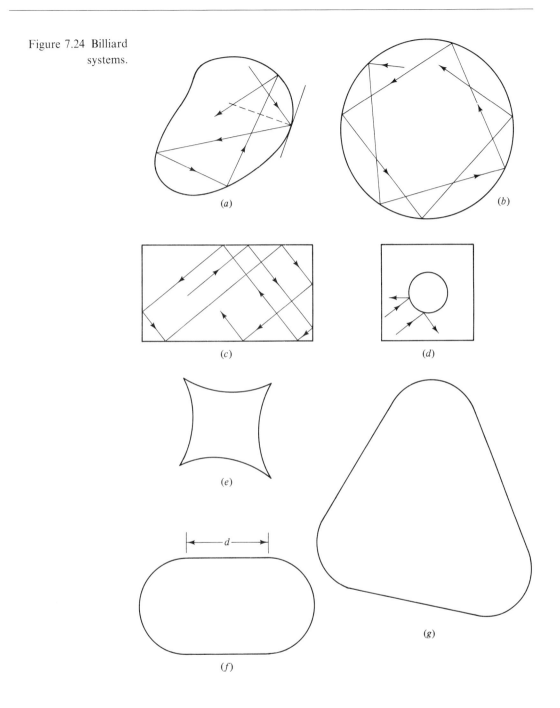

violation of the definition of hyperbolicity in Section 4.3, they possess periodic orbits with zero Lyapunov exponent. (For (d) and (f) there are simple zero exponent periodic orbits bouncing vertically between the straight parallel walls (and, for (d), not intersecting the interior circle). For (g) there is a zero exponent periodic orbit tracing out a triangle as it bounces between points on the straight line component segments of the boundary.)

For the cases (d) and (e) in Figure 7.24 the origin of the chaos is intuitively fairly clear. The curved boundary segments have curvature vectors pointing out of the billiard area. As shown in Figure 7.24 (d) this has a 'dispersing' effect on two parallel orbits, and one expects this to lead to rapid separation of nearby orbits after a few bounces. Indeed, Sinai (1970) showed that such systems were chaotic. Further work by Bunimovich (1979) showed that a certain class of nondispersing billiards, of which (f) and (g) of Figure 7.24 are members, are also chaotic. The so-called stadium billiard (f) is of particular interest in that it has been the subject of a number of numerical studies, particularly in the context of quantum chaos (Chapter 10). Note that the stadium is chaotic for any value of $d > 0$, where d is the length of the two parallel line segments joining the two semicircular segments at the left and right end of the stadium. Thus, we have the somewhat suprising result that the stadium is chaotic for any $d > 0$, but, as soon as we make $d = 0$, we have the circular billiard, Figure 7.24 (b), which is integrable.

The author is unaware of any example of a smooth two-dimensional potential $V(x, y)$ in Cartesian spatial variables $\mathbf{q} = (x, y)$ such that the Hamiltonian $H = (p_x^2/2m) + (p_y^2/2m) + V(x, y)$ yields chaotic orbits that are ergodic on the energy surface. For a time it was thought that the potential $V(x, y) = x^2 y^2$ would yield chaotic orbits for almost all initial conditions, but very careful numerical analysis (Dahlqvist and Russberg, 1990) has found KAM surfaces surrounding a small region about an elliptic periodic orbit of high period (period 11 in the surface of section).

An example of a continuous time two-dimensional dynamical system which is completely chaotic, in the sense we have been using, is the case of a point mass moving along geodesics on a closed two-dimensional surface of negative Gaussian curvature (the two principal curvature vectors at each point on the surface point to opposite sides of the surface, Figure 7.25) (Hadamard, 1898). Two geodesics on such a surface that are initially close and parallel separate exponentially as they are followed forward in time. Note, however, that a closed surface of negative curvature cannot be embedded in a three-dimensional Cartesian space (four dimensions are required), so this example is somewhat nonphysical, although it has proven to be very fruitful for mathematical study.

7.6 The succession of increasingly random systems

In the previous section we have discussed 'strongly chaotic' systems by which we meant chaotic systems that were mixing throughout the phase space. One often encounters in the literature various terms used to describe the degree of randomness of a Hamiltonian system. In particular, one can make the following list in order of 'increasing randomness':

ergodic systems,
mixing systems,
K-systems,
C-systems,
Bernoulli systems.

We now discuss and contrast these terms, giving some examples of each.

Ergodicity is defined in Section 2.3.3 for maps. For an ergodic invariant measure of a dynamical system, phase space averages are the same as time averages. That is, for the case of a continuous time system, ergodicity implies

$$\lim_{T \to \infty} \frac{1}{T} \int_0^T f(\tilde{\mathbf{x}}(t)) \, dt = \langle f(\tilde{\mathbf{x}}) \rangle, \tag{7.55}$$

where $f(\tilde{\mathbf{x}})$ is any smooth function of the phase space variable $\tilde{\mathbf{x}}$, $\tilde{\mathbf{x}}(t)$ represents a trajectory in phase space, $\langle f(\tilde{\mathbf{x}}) \rangle$ represents the average of $f(\tilde{\mathbf{x}})$ over the phase space weighted by the invariant measure under consideration, and (7.55) holds for almost every initial condition with respect to the invariant measure. As an example, consider the standard map, Eqs. (7.15). In the case $K = 0$, we have (Eq. (7.40)) $\theta_{n+1} = (\theta_n + p_n)$ modulo 2π, $p_{n+1} = p_n$. Thus the lines $p = $ const. are invariant. Any region $p_a > p > p_b$ is also invariant. Orbits are not egodic in $p_a > p > p_b$. Orbits are, however, ergodic on the lines $p = $ const., provided that $p/2\pi$ is irrational. Ergodicity is the weakest form of randomness and does not necessarily imply chaos. This is clear since the example, $\theta_{n+1} = (\theta_n + p)$

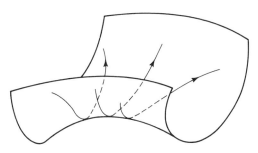

Figure 7.25 Geodesics on a surface of negative curvature.

modulo 2π, $p/2\pi$ irrational, is ergodic on the line $p = $ const. but is nonchaotic (its Lyapunov exponent is zero). In the case of $K > 0$ there are regions of positive Lebesgue measure in the (θ, p)-space over which orbits wander chaotically (see Figure 7.13). In this case, for each such region, ergodicity applies (where the relevant measure of a set A is just the fraction of the area (Lebesgue measure) of the ergodic chaotic region in A).

Mixing is defined by (7.54). An example of a nonmixing system is the map, $\theta_{n+1} = (\theta_n + p)$ modulo 2π, $p/2\pi = $ irrational, which is just a rigid rotation of the circle by the angle increment p. An example, which is chaotic but not mixing, occurs when we have a $\tilde{p}/\tilde{q}$ island chain ($\tilde{q} > 1$) for the standard map. In this case, there are typically chaotic sets consisting of component areas (within the $\tilde{p}/\tilde{q}$ island chain), each of which map successively one to another, returning to themselves after $\tilde{q}$ iterates. To see that these chaotic sets are not mixing according to (7.54), let σ' be the area of one of the $\tilde{q}$ components, and σ be the area of another. Then, as m (time) increases, the quantity $[\mu_L(\sigma' \cap \mathbf{M}^m(\sigma))]/\mu_L(\sigma')$ is equal to one once every $\tilde{q}$ iterates and is equal to zero for the other iterates. Hence, the limit in (7.54) does not exist.[8]

A system is said to be a K-system if there exists invariant sets with positive metric entropy. Basically, in terms of our past terminology, this is the same as saying that the system is chaotic (possesses a positive Lyapunov exponent for typical initial conditions).

A C-system is one which is chaotic and is hyperbolic at every point in the phase space (not just on the invariant set). Examples of C-systems are the cat map, a compact surface of negative geodesic curvature, and the billiard of Figure 7.24(e). An example, which is a K-system, but not a C-system, is the stadium billiard, Figure 7.24(f).

A Bernoulli system is a system which can be represented as a symbolic dynamics consisting of a full shift on a finite number of symbols (see Section 4.1). An example, of such a system, for the case of an area preserving map, is the generalized baker's map with $\lambda_a = \alpha$ and $\lambda_b = \beta$.

Problems

1. (a) Show that the change of variables specified by
$$\mathbf{q} = \frac{\partial \hat{S}(\mathbf{p}, \bar{\mathbf{q}}, t)}{\partial \mathbf{p}}, \quad \bar{\mathbf{p}} = \frac{\partial \hat{S}(\mathbf{p}, \bar{\mathbf{q}}, t)}{\partial \bar{\mathbf{q}}}$$
is symplectic.
   (b) Find a function $\hat{S}(p_n, \theta_{n+1})$ in terms of which the map (7.15) is given by $\theta_n = \partial \hat{S}/\partial p_n$, $p_{n+1} = \partial \hat{S}/\partial \theta_{n+1}$.

2. Consider the following four-dimensional map (Ding *et al.*, 1990a),

$$x_{n+1} = 2\alpha x_n - p_{x,n} - \rho x_n^2 + y_n^2,$$

$$p_{x,n+1} = x_n,$$

$$y_{n+1} = 2\beta y_n - p_{y,n} + 2x_n y_n,$$

$$p_{y,n+1} = y_n.$$

Is it volume preserving? Using Eq. (7.13) test to see whether the map is symplectic.

3. Consider a magnetic field in a plasma given by

$$\mathbf{B}(x, y, z) = B_0 \mathbf{z}_0 + \nabla \times \mathbf{A},$$

where B_0 is a constant and the vector potential $\mathbf{A}$ is purely in the z-direction, $\mathbf{A} = A(x, y, z)\mathbf{z}_0$. Denote the path followed by a field line as $\mathbf{x}(z) = x(z)\mathbf{x}_0 + y(z)\mathbf{y}_0 + z\mathbf{z}_0$. Show that the equations for $x(z)$ and $y(z)$ are in the form of Hamilton's equations where z plays the role of time and $A(x, y, z)/B_0$ plays the role of the Hamiltonian.

4. Consider the motion of a charged particle in an electrostatic wave field in which the electric field is given by $\mathbf{E}(x, t) = E_x(x, t)\mathbf{x}_0$ with

$$E_x(x, t) = \sum_{\kappa, \omega} E_{\kappa, \omega} \exp(i\kappa x - i\omega t).$$

(This situation arises in plasma physics where the wave field E_x is due to collective oscillations of the plasma.) In the special case where there is only one wavenumber, $\kappa = \pm k_0$, the frequencies ω form a discrete set, $\omega = 2\pi n/T$ (where T is the fundamental period and n is an integer; $n = \ldots, -2, -1, 0, 1, 2, \ldots$), and the amplitudes $E_{\kappa, \omega}$ are real and independent of ω and κ, $E_{\kappa, \omega} = E_0/2$, the above expression for E_x reduces to

$$E_x(x, t) = E_0 \cos(k_0 x) \sum_n \exp(2\pi i n t/T)$$

$$= E_0 \cos(k_0 x) \sum_m \delta(t - mT).$$

Show that the motion of a charged particle is described by a map which is of the same form as the standard map Eq. (7.15).

5. Find the fixed points of the standard map (7.15) that lie in the strip $\pi > p > -\pi$. Determine their stability as a function of K. In what range of K is there an elliptic fixed point (assume $K \geq 0$)?

6. Write a computer program to iterate the standard map, Eqs. (7.15).
   (a) Plot p modulo 2π versus θ for orbits with $K = 1$ and the following five initial conditions, $(\theta_0, p_0) = (\pi, \pi/5), (\pi, 4\pi/5), (\pi, 6\pi/5), (\pi, 8\pi/5), (\pi, 2\pi)$.
   (b) For $K = 21$ plot versus iterate number the average value of p^2 averaged over 100 different initial conditions, $(\theta_0, p_0) = (2n\pi/11, 2m\pi/11)$ for $n = 1, 2, \ldots, 10$ and $m = 1, 2, \ldots, 10$, and hence estimate the diffusion coefficient D. How well does your numerical result agree with the quasilinear value Eq. (7.42)?

7. The 'sawtooth map' is obtained from the standard map, Eqs. (7.15), by replacing the function, $\sin \theta_{n+1}$ in (7.15b) by the sawtooth function, saw θ_{n+1}, where

$$\text{saw } \theta \equiv \begin{cases} \theta, & \text{for } 0 \le \theta < \pi, \\ \theta - \pi, & \text{for } \pi < \theta \le 2\pi, \end{cases}$$

and saw $\theta \equiv$ saw $(\theta + 2\pi)$. Show that the sawtooth map is an example of a C-system if $K > 0$ or $K < -2$ and calculate the Lyapunov exponents.

Notes

1. Additional useful material on chaos in Hamiltonian systems can be found in the texts by Sagdeev *et al.* (1990), by Ozorio de Almeida (1988), by Lichtenberg and Lieberman (1983), and by Arnold and Avez (1968), in the review articles by Berry (1978), by Chirikov (1979) and by Helleman (1980), and in the reprint selection edited by MacKay and Meiss (1987).

2. See books which cover the basic formulation and analysis of Hamiltonian mechanics, such as Goldstein (1980), Ozorio de Almeida (1988), and Arnold (1978, 1982).

3. Our review in Section 7.1 is meant to refresh the memory, rather than to be a self-contained first-principles exposition. Thus, the reader who wishes more detail or clarification should refer to one of the texts cited above.[2]

4. If only k-independent relations of the form $\mathbf{m} \cdot \boldsymbol{\omega} = 0$ hold with $1 < k < N - 1$, then orbits on the N-torus are $(N - k)$-frequency quasiperiodic and do not fill the N-torus. Rather individual orbits fill $(N - k)$-tori which lie in the N-torus.

5. In the area preserving case, the areas of lobes bounded by stable and unstable manifold segments must be the same if these lobes map to each other under iteration of the map. For example, consider one of the finger shaped areas bounded by stable and unstable manifold segments in Figure 4.10(c). This area must be the same as the areas of the regions shown in the figure to which it successively maps.

6. Long-time power law correlations of orbits have been observed numerically in two-dimensional maps (Karney, 1983; Chirikov and Shepelyanski, 1984) and in higher-dimensional systems (Ding *et al.*, 1990a). This comes about due to the 'stickiness' of KAM surfaces: an orbit in a chaotic component which comes near a KAM surface bounding that component tends to spend a longer time there, and this time is typically longer the nearer the orbit comes. This behaviour has been examined theoretically using self-similar random walk models (Hanson *et al.*, 1985; Meiss and Ott, 1985). This type of behavior has also been shown to result in anomalous diffusion wherein the average of the square of a map variable increases as n^α with $\alpha > 1$ (in contrast to ordinary diffusive behavior where $\alpha = 1$ as in Eq. (7.44)). See Geisel *et al.* (1990), Zaslavski *et al.* (1989), and Ishizaki *et al.* (1991).

7. Note that near the peaks of the graph in Figure 7.17 it appears that the numerically computed D values can be much larger than the analytical

estimate. In fact, it was subsequently found that D diverges to infinity in these regions, and the actual behavior of $\langle p^2/2 \rangle$ is anomalous[6] in that $\langle p^2/2 \rangle \sim n^\alpha$ with $\alpha > 1$. This behavior is due to the presence of 'accelerator modes' in the range of K-values near the peaks of the graph. Accelerator modes are small KAM island chains such that, when an orbit originates in an island, it returns periodically to that island but is displaced in p by an integer multiple of 2π. Hence, the orbit experiences a free acceleration, $p \sim n$. Orbits in the large chaotic region can stick close to the outer bounding KAM surfaces of these accelerator islands, thus leading to the above mentioned anomalous behavior[6] (Ishizaki *et al.*, 1991).

8. We note, however, that, if we consider the $\tilde{q}$ times iterated map, then there may be mixing regions in σ for the map $\mathbf{M}^{\tilde{q}}$, since $\mathbf{M}^{\tilde{q}}(\sigma) = \sigma$.

CHAPTER EIGHT

Chaotic transitions

A central problem in nonlinear dynamics is that of discovering how the qualitative dynamical properties of orbits change and evolve as a dynamical system is continuously changed. More specifically, consider a dynamical system which depends on a single scalar parameter. We ask, what happens to the orbits of the system if we examine them at different values of the parameter? We have already met this question and substantially answered it for the case of the logistic map, $x_{n+1} = rx_n(1 - x_n)$. In particular, we found in Chapter 2 that as the parameter r is increased there is a period doubling cascade, terminating in an accumulation of an infinite number of period doublings, followed by a parameter domain in which chaos and periodic 'windows' are finely intermixed. Another example of a context in which we have addressed this question is our discussion in Chapter 6 of Arnold tongues and the transition from quasiperiodicity to chaos. Still another aspect of this question is the types of generic bifurcations of periodic orbits which can occur as a parameter is varied. In this regard recall our discussions of the generic bifurcations of periodic orbits of one-dimensional maps (Section 2.3) and of the Hopf bifurcation (Chapter 6).

In this chapter we shall be interested in transitions of the system behavior with variation of a parameter such that the transitions involve chaotic orbits. Some changes of this type are the following:

As the system parameter is changed, a chaotic attractor appears.

As the system parameter is changed, a chaotic transient is created from a situation where there were only nonchaotic orbits.

As the system parameter is changed, a formerly nonfractal basin boundary becomes fractal.

As the system parameter is changed, a scattering problem changes from being nonchaotic to chaotic.

As the system parameter is changed, a chaotic set experiences a sudden jump in its size in phase space (the set may or may not be an attractor, e.g., it could be a fractal basin boundary).

Of the above types of chaotic transitions, the one which initially received the most interest was the first, namely, the question of characterizing the various 'scenarios' by which chaotic attractors can appear with variation of a system parameter. One such scenario is the period doubling cascade to chaos, which is so graphically illustrated by the logistic map (see Chapter 2). Furthermore, we have seen in Chapter 2 that the period doubling cascade route to a chaotic attractor occurs in many other dynamical systems (Section 2.4).

8.1 The period doubling cascade route to chaotic attractors

Perhaps, the most notable aspect of the period doubling cascade is the existence of the 'universal' numbers $\hat{\delta}$ and $\hat{\alpha}$ (cf. Chapter 2). These numbers apply not only for the logistic map, but typically give a *quantitative* characterization near the accumulation of period doublings for any dissipative system undergoing a period doubling cascade *independent of the details of the system*. A similar universality applies in statistical mechanics in the study of critical phenomena near phase transitions. In that case 'critical exponents' governing the behavior near the phase transition point have been derived which are universal in the same sense; *viz.*, their numerical values are precisely the same for a large class of physical systems. The same general mathematical technique which has allowed the calculation of critical exponents in statistical mechanics has been used by Feigenbaum (1978, 1980a) to derive the values of $\hat{\delta}$ and $\hat{\alpha}$ for period doubling cascades of dissipative systems. This general technique is called the *renormalization group* and has also been used for other problems in nonlinear dynamics. These include period doubling cascades in conservative systems (Section 7.3.2), the study of the scaling properties of the last surviving KAM surface (Section 7.3.2), the destruction of two frequency quasiperiodic attractors (Section 6.2), and the transition from Hamiltonian to dissipative systems (Chen *et al.*, 1987). In this section we supplement our previous discussion in Chapter 2 of period doubling cascades by summarizing part of Feigenbaum's use of the renormalization group technique for studying period doubling cascades. Since we only wish to indicate briefly the spirit of the renormalization group technique, we shall limit our discussion to the derivation of the universal constant $\hat{\alpha}$. (The reader is referred to Feigenbaum (1980a) and Schuster (1988, Section 3.2) for nice expositions of the further treatment giving the constant $\hat{\delta}$.)

To focus the discussion we consider the logistic map,

$$M(x,r) = rx(1-x).$$

Let $\bar{r}_n$ denote the value of r at which the periodic orbit of period 2^n is superstable; i.e., the period 2^n orbit passes through the critical point where $M'(x,r) = 0$ (namely, $x = \frac{1}{2}$). Figures 8.1–8.3 show $M(x,r)$ and iterates of M for $r = \bar{r}_0, \bar{r}_1$ and $\bar{r}_2$. In Figure 8.1 we show M at $r = \bar{r}_0$. Figure 8.2(a) shows $M(x,\bar{r}_1)$ and the superstable period two orbit. Figure 8.2(b) shows $M^2(x,\bar{r}_1)$. Note that the elements of the superstable period two of Figure 8.2(b) fall on critical points of M^2 (i.e., points where $(M^2)' = 0$). Also note the square box in Figure 8.2(b) whose upper right-hand corner is located at the (now unstable) period one fixed point of $M(x,\bar{r}_1)$. The important point is that, if we reflect this square about the point $(\frac{1}{2},\frac{1}{2})$, then the appearance of the resulting graph in the square is nearly the same as that for Figure 8.1, although on smaller scale. Figure 8.3(a) shows $M^2(x,\bar{r}_2)$. Now the period two orbit shown in Figure 8.2 is unstable. In its place the attractor is now the period four orbit given by the four critical points on the dashed line in Figure 8.3(b). Note that we have again blocked out a small square. This square has at its center the element of the period four orbit which falls on $(\frac{1}{2},\frac{1}{2})$, the critical point of $M(x,\bar{r}_2)$. The lower left-hand corner of this square is the element of the unstable period two orbit which was at the critical point $(\frac{1}{2},\frac{1}{2})$ when r was $\bar{r}_1$. Again note that the graph of $M^4(x,\bar{r}_2)$ in the small square is similar to the reflection about the point $(\frac{1}{2},\frac{1}{2})$ of the graph in the blocked out square of Figure 8.2(b) and is similar to the entire graph of $M(x,\bar{r}_0)$ shown in Figure 8.1.

Figure 8.1 $M(x,\bar{r}_0)$ versus x.

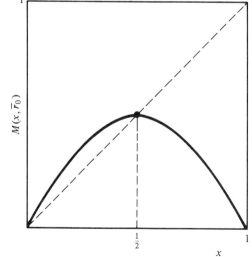

The situation with respect to the small squares shown in Figures 8.2(b) and 8.3(b) repeats as we consider higher period 2^n superstable cycles. Furthermore, for large n, the edge length of the square reduces geometrically at some limiting factor, and this factor is the universal constant $\hat{\alpha}$ defined in Chapter 2. In fact, not only the size of the square scales, but also the functional form of $M^{2^n}(x, \bar{r}_n)$ within the square. Thus shifting the coordinate x so that the critical point of M now falls on $x = 0$ (rather than $x = \frac{1}{2}$), we have that near $x = 0$

$$M^{2^n}(x, \bar{r}_n) \simeq (-\hat{\alpha}) M^{2^{n+1}}(-x/\hat{\alpha}, \bar{r}_{n+1}) \tag{8.1}$$

for large n. The minus signs accounts for the reflection of the square about the point $(\frac{1}{2}, \frac{1}{2})$ line on each increase of n by 1. This implies that the limit

Figure 8.2 (a) $M(x, \bar{r}_1)$ versus x. (b) $M^2(x, \bar{r}_1)$ versus x.

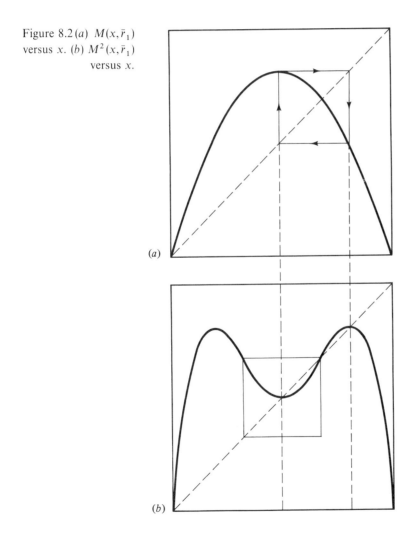

(a)

(b)

$$\lim_{n \to \infty} (-\hat{\alpha})^n M^{2^n}(x/(-\hat{\alpha})^n, \bar{r}_n) = g_0(x) \tag{8.2}$$

exists and gives the large n behavior of $M^{2^n}(x, \bar{r}_n)$ near the critical point. In analogy with (8.2), we define

$$g_m(x) \equiv \lim_{n \to \infty} (-\hat{\alpha})^n M^{2^n}(x/(-\hat{\alpha})^n, \bar{r}_{n+m}). \tag{8.3}$$

The functions $g_m(x)$ are related by the 'doubling transformation' $\hat{T}$

$$g_{m-1}(x) = (-\hat{\alpha})g_m[g_m(-x/\hat{\alpha})] \equiv \hat{T}[g_m(x)], \tag{8.4}$$

which can be verified from (8.3). This transformation has a fixed point in function space which we denote $g(x)$,

$$g(x) = \hat{T}[g(x)]. \tag{8.5}$$

Figure 8.3(a)
$M^2(x, \bar{r}_2)$ versus x.
(b) $M^4(x, \bar{r}_2)$ versus x.

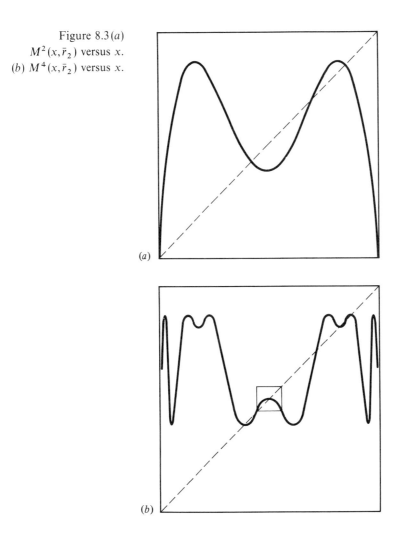

(a)

(b)

Here $g(x)$ is just the $m \to \infty$ limit of g_m

$$g(x) = \lim_{m \to \infty} g_m(x), \tag{8.6}$$

and (8.5) follows by letting $m \to \infty$ in (8.4). The objective is now to solve (8.5) for the function $g(x)$ and the constant $\hat{\alpha}$. (Equation (8.5), although nonlinear, may be regarded as similar to an eigenvalue equation, with $g(x)$ playing the role of the eigen-function, and $\hat{\alpha}$ playing the role of the eigenvalue.) If $g(x)$ is a solution of (8.5), then, by the definition of $\hat{T}$ given in (8.4), for any constant k, the function $k^{-1}g(kx)$ is also a solution. Thus, $g(x)$ is arbitrary to within changes of the scale in x. To remove this ambiguity we arbitrarily specify that

$$g(0) = 1. \tag{8.7}$$

Since the original $M(x,r)$ was quadratic about its maximum we have that $g(x)$ must also be quadratic. Indeed it follows that it is even. One way of solving the fixed point equation, Eq. (8.5), is to expand $g(x)$ and $\hat{T}[g(x)]$ in a Taylor series in powers of x^{2n} and equate the coefficients of x^{2n} on each side of the equation $g = \hat{T}[g]$. As an illustration, we follow this procedure retaining only the first two terms,

$$g(x) = 1 - ax^2 + O(x^4). \tag{8.8}$$

Substituting in $\hat{T}[g] \equiv -\hat{\alpha}g[g(-x/\hat{\alpha})]$, we obtain for (8.5),

$$1 - ax^2 \simeq -\hat{\alpha}(1 - a) - (2b^2/\hat{\alpha})x^2.$$

Thus, at this (rather crude) level of approximation, we have $-\hat{\alpha}(1 - a) = 1$ and $a = 2b^2/\hat{\alpha}$. Solution of these equations yields

$$\hat{\alpha} \simeq 1 + \sqrt{3} = 2.73\ldots,$$

which is reasonably close to the exact value found by Feigenbaum,

$$\hat{\alpha} = 2.50280787\ldots.$$

Although our discussion and the plots in Figures 8.1–8.3 were for the logistic map, the reader should note that at no point in obtaining the fixed point equation, Eq. (8.5), did we make use of the functional form of the map. The only property of the original map that we made use of was that it was quadratic about its maximum. (Note that that property was used in the solution of (8.5) and not in its derivation.) Thus, the value of $\hat{\alpha}$ obtained by Feigenbaum from (8.5) must apply for period doubling cascades of all one-dimensional maps which have quadratic maxima. That is, the value of $\hat{\alpha}$ obtained is universal within this class of dynamical systems. In fact subsequent work has extended this universality class to higher-dimensional systems.

Renormalization group theory has also been utilized to calculate the effect of noise on the period doubling cascade (Shraiman *et al.*, 1981;

Crutchfield *et al.*, 1981; Feigenbaum and Hasslacher, 1982). A principal
result is that noise terminates the period doubling cascade after a finite
number of period doublings past which chaos ensues. If σ denotes the
noise level, and $n(\sigma)$ denotes the number of doublings in the noisy cascade,
then $n(\sigma)$ scales with σ according to

$$\sigma \sim \hat{\mu}^{-n(\sigma)},$$

where $\hat{\mu}$ is a universal constant. Another result of the period doubling
universality is that the frequency power spectra of orbits near the
accumulation of period doublings have universal features (Feigenbaum,
1980b; Nauenberg and Rudnick, 1981).

8.2   The intermittency transition to a chaotic attractor

In the intermittency transition to a chaotic attractor (Pomeau and
Manneville, 1980) the phenomenology is as follows. For values of the
parameter (call it p) less than a critical transition value p_T the attractor is a
periodic orbit. For p slightly larger than p_T there are long stretches of time
('laminar phases') during which the orbit appears to be periodic and
closely resembles the orbit for $p < p_T$, but this regular (approximately
periodic) behavior is intermittently interrupted by a finite duration 'burst'
in which the orbit behaves in a decidedly different manner. These bursts
occur at seemingly random times, but one can define a mean time $\bar{T}(p)$
between the bursts. As p approaches p_T from above the mean time between
bursts approaches infinity,

$$\lim_{p \to p_T^+} \bar{T}(p) = +\infty,$$

and the attractor orbit thus becomes always 'laminar' so that the motion
is periodic. As p increases substantially above p_T, the bursts become so

Figure 8.4 Solutions of the Lorenz system of equations, Eqs. (2.30), with $\tilde{\sigma} = 10$, $\tilde{b} = 8/3$ and increasing values of $\tilde{r}$ (Pomeau and Manneville, 1980).

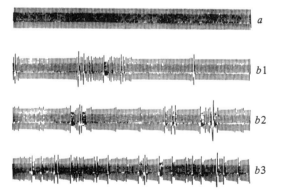

frequent that the regular oscillation (laminar phase) can no longer be distinguished. The above phenomenology is nicely illustrated in Figure 8.4 by the numerical solution from the paper of Pomeau and Manneville (1980) of the Lorenz system for four successively larger values of the parameter $\tilde{r}$ in Eq. (2.30b). The smaller value, $\tilde{r} = 166$, labeled a in the figure, corresponds to stable periodic motion. As $\tilde{r}$ is increased through $\tilde{r}_T \cong 166.06$, the plots labeled $b1, b2$ and $b3$ are obtained. We see that the bursts become more frequent as r increases.

In the intermittency transition one has a simple periodic orbit which is replaced by chaos as p passes through p_T. This necessarily implies that the stable attracting periodic orbit either becomes unstable or is destroyed as p increases through p_T. Furthermore, when this happens, the orbit is not replaced by another nearby stable periodic orbit, as occurs, for example, in the forward period doubling bifurcation (Figure 2.15); this is implied by the fact that during the bursts the orbit goes far from the vicinity of the original periodic orbit. Three kinds of generic bifurcations which meet these requirements are the saddle-node bifurcation (in which stable and unstable orbits coalesce and obliterate each other as illustrated in Figure 8.5 (in the context of one-dimensional maps this is also called a tangent bifurcation; see the diagram labeled backward tangent bifurcation in Figure 2.15)), the inverse period doubling bifurcation (in which an unstable periodic orbit collapses onto a stable periodic orbit of one half its period and the two are replaced by an unstable periodic orbit of the lower period, Figure 2.15), and the subcritical Hopf bifurcation of a periodic orbit. In the Hopf bifurcation of a periodic orbit the orbit goes unstable by having a complex conjugate pair of eigenvalues of its linearized surface of section map pass through the unit circle; in the saddle-node bifurcation, a stable eigenvalue (inside the unit circle) and an unstable eigenvalue (from outside the unit circle) come together and coalesce at the point $+1$; in the inverse period doubling bifurcation an eigenvalue goes from inside to outside the unit circle by passing through -1. The word subcritical

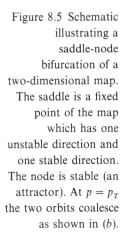

Figure 8.5 Schematic illustrating a saddle-node bifurcation of a two-dimensional map. The saddle is a fixed point of the map which has one unstable direction and one stable direction. The node is stable (an attractor). At $p = p_T$ the two orbits coalesce as shown in (b).

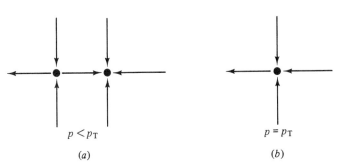

applied to the Hopf bifurcation of a periodic orbit signifies that, as the parameter is increased, an unstable two frequency quasiperiodic orbit (a closed curve in the surface of section) collapses on to the stable periodic orbit, and the latter is rendered unstable as r passes through the bifurcation point.

Pomeau and Manneville distinguish three types of intermittency transitions corresponding to the three types of generic bifurcations mentioned above:

Type I: saddle-node,
Type II: Hopf,
Type III: inverse period doubling.

These are illustrated in Figure 8.6 where we schematically show the orbits of the surface of section map as a fucntion of p using heavy solid lines for stable orbits and dashed lines for unstable orbits.

Each of the three types of intermittency transitions displays distinct characteristic behavior near p_T. For example, the scaling of the average time between bursts is given by

Figure 8.6 Schematic illustration of the three types of intermittency transitions to chaos.

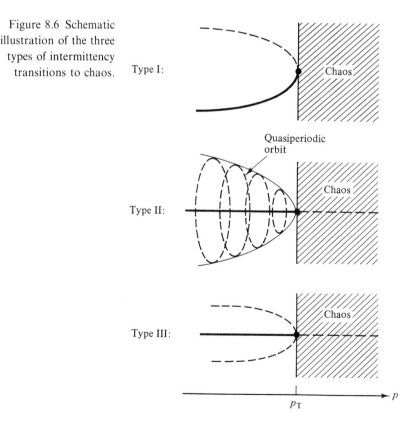

$$\overline{T}(p) \sim \begin{cases} (p - p_T)^{-1/2} & \text{for Type I,} \\ (p - p_T)^{-1} & \text{for Type II,} \\ (p - p_T)^{-1} & \text{for Type III.} \end{cases} \tag{8.9}$$

We note that, while the scaling behavior of the average interburst time $\overline{T}(p)$ is the same for Type II and Type III, the characteristic probability distributions of the interburst times are quite different in the two cases.

As in the case of the period doubling route to chaos, intermittency has been studied extensively in experiments. An example is the experiment of Bergé *et al.* (1980) on Rayleigh–Bénard convection, data from which are shown in Figure 8.7.

In order to clarify the nature of the intermittency transition to chaos, we now give a derivation of Eq. (8.9) for the case of the Type I transition. To simplify matters we assume that the dynamics is well described by a one-dimensional map just before (Figure 8.8(*a*)) and just after (Figure 8.8(*b*)) the saddle-node (or 'tangent') bifurcation to Type I intermittency. We see from Figure 8.8(*b*) that, for p just slightly greater than p_T, the orbit takes many iterates to traverse the narrow tunnel between the map function and the 45° line. While in the tunnel, the orbit is close to the value of x that applies for the stable fixed point for p slightly less than p_T. Thus, we identify the average time to traverse the tunnel with $\overline{T}(p)$. After traversing the tunnel, the orbit undergoes chaotic motion determined by the specific form of the map (not shown in Figure 8.8) away from the vicinity of the tunnel, and is then reinjected into the tunnel when 'by chance' the chaotic orbit lands in the tunnel. To calculate the typical time to traverse the tunnel we approximate the map as being quadratic to lowest order,

$$x_{n+1} = x_n + \varepsilon + x_n^2, \tag{8.10}$$

where $\varepsilon \sim (p - p_T)$. Note that (8.10) undergoes a saddle-node bifurcation as ε increases through zero. Considering ε as small and positive, we utilize the fact that the steps in x with successive iterates in the tunnel are small.

Figure 8.7 The *z*-component of the fluid velocity at a point in the interior of the container in the experiment of Bergé *et al.* for three different values of the Rayleigh number (normalized temperature difference).

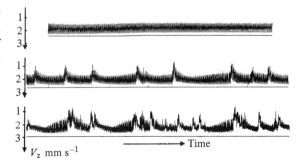

This allows us to approximate (8.10) as a differential equation. Replacing x_n by $x(n)$ and considering n as a continuous variable, we have $x_{n+1} - x_n \simeq dx(n)/dn$. Equation (8.10) thus becomes

$$dx/dn = x^2 + \varepsilon,$$

For an orbit reinjected into the tunnel by landing at a point $-x_0 \gg -\varepsilon$, this yields an approximate time to traverse the tunnel given by

$$\int_{x_0}^{+\infty} \frac{dx}{x^2 + \varepsilon} \cong \int_{-\infty}^{+\infty} \frac{dx}{x^2 + \varepsilon} = \pi/\varepsilon^{1/2} \sim \varepsilon^{-1/2}$$

in agreement with Eq. (8.9).

It is interesting to note that Type I intermittency transitions to chaos are already present in the simple quadratic logistic map example that we have extensively examined in Chapter 2. In particular, we noted there that each periodic window is initiated by a tangent bifurcation (e.g., see Figure

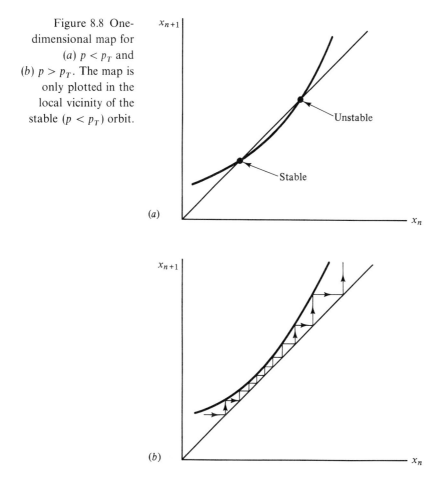

Figure 8.8 One-dimensional map for (a) $p < p_T$ and (b) $p > p_T$. The map is only plotted in the local vicinity of the stable $(p < p_T)$ orbit.

2.13 for the case of the period three). For example, referring to Figure 2.12(a) which shows the period three window, we see that just above r_{*3} we have a stable period three orbit, while just below r_{*3} there is chaos. Thus, as the parameter r is *decreased* through r_{*3} we have an intermittency transition from a periodic to a chaotic attractor. Examination of an orbit for r just below r_{*3} shows that it has long stretches where it approximately follows the period three that exists above r_{*3}.

Some examples of further works on the theory of intermittency transitions are the following: Hu and Rudnick (1982), who use the renormalization group; Hirsch *et al.* (1982), who treat the effect of noise; and Ben-Mizrachi *et al.* (1985) who discuss the low-frequency power specter of chaotic orbits near an intermittency transition.

8.3 Crises

Sudden changes in chaotic attractors with parameter variation are seen very commonly (two early examples are Simó (1979) and Ueda (1980)). Such changes, caused by the collision of the chaotic attractor with an unstable periodic orbit or, equivalently,[1] its stable manifold, have been called crises and were first extensively studied by Grebogi *et al.* (1982, 1983c). Three types of crisis can be distinguished according to the nature of the discontinuous change that the crisis induces in the chaotic attractor: in the first type a chaotic attractor is suddenly destroyed as the parameter passes through its critical value; in the second type the size of the attractor in phase space suddenly increases; in the third type (which can occur in systems with symmetries) two or more chaotic attractors merge to form one chaotic attractor. The inverse of these processes (i.e., sudden creation, shrinking or splitting of a chaotic attractor) occurs as the parameter is varied in the other direction. The sudden destruction of a chaotic attractor occurs when the attractor collides with a periodic orbit on its basin boundary and is called a *boundary crisis*. The sudden increase in the size of a chaotic attractor occurs when the periodic orbit with which the chaotic attractor collides is in the interior of its basin and is called an *interior crisis*. In an *attractor merging crisis* two (or more) chaotic attractors *simultaneously* collide with a periodic orbit (or orbits) on the basin boundary which separates them.

8.3.1 Boundary crises

In a boundary crisis, as a parameter p is raised, the distance between the chaotic attractor and its basin boundary decreases until at a critical value $p = p_c$ they touch (the crisis). At this point the attractor also touches an

unstable periodic orbit that was on the basin boundary before the crisis.[1]
For $p > p_c$ the chaotic attractor no longer exists but is replaced by a
chaotic transient. In particular, for p just slightly greater than p_c, consider
an initial condition placed in the phase space region occupied by the basin
of attraction of the chaotic attractor that existed for $p < p_c$. This initial
condition will typically move toward the region of the $p < p_c$ attractor,
bounce around in an orbit that looks like an orbit on the $p < p_c$ chaotic
attractor, and then, after what could be a relatively long time (for p
sufficiently close to p_c), the orbit rather suddenly starts to move off toward
some other distant attractor. (Note that our assumption that the attractor
exists for $p < p_c$ and the chaotic transient exists for $p > p_c$ is merely an
arbitrary convention that might equally well be reversed.)

As an example, Figure 8.9 shows an orbit for the Ikeda map (Hammel et
$al.$, 1985)

$$z_{n+1} = a - bz_n \exp\left(i\kappa - \frac{i\eta}{1 + |z_n|^2}\right), \qquad (8.11)$$

where $z = x + iy$ is a complex number. (Taking real and imaginary parts
(8.11) may be regarded as a two-dimensional real map.) The physical
origin of this map is illustrated in Figure 8.10.

The orbit shown in Figure 8.9 corresponds to a parameter value p just
slightly past that yielding a boundary crisis. The numerals in the figure
label the number of iterates to reach the corresponding point, with 1
denoting the initial condition. We see that the orbit bounces around,
appearing to fill out a chaotic attractor for over 86 000 iterates. Then at
iterate 86 431 the orbit point 'by chance' lands near (and just to the right
of) a stable manifold segment of an unstable period one (fixed point)
saddle on the basin boundary of the $p < p_c$ attractor. The orbit then
moves toward this fixed point (points 86 432; 86 433; ...), following the

Figure 8.9 Orbit for
the Ikeda map Eq.
(8.11) for parameter
values $a = 1.0027$,
$b = 0.9$, $\kappa = 0.4$ and
$\eta = 6.0$ (Grebogi et
$al.$, 1986a).

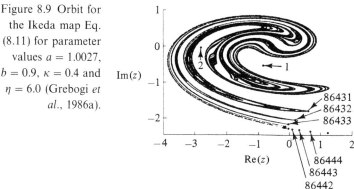

direction of its stable manifold and then gets ejected to the right along its unstable manifold (points ...; 86 442; 86 443; 86 444; ...) moving off to some other attractor.

Basically, the attractor becomes 'leaky' (hence no longer an attractor), developing a region from which orbits can escape. The length of a chaotic transient depends sensitively on the initial condition. However, if one looks at many randomly chosen initial conditions in the $p < p_c$ basin region, then one typically sees that the chaotic transient lengths, τ, have an exponential distribution for large τ,

$$P(\tau) \sim \exp(-\tau/\langle\tau\rangle) \qquad (8.12)$$

(see Chapter 5), where $\langle\tau\rangle$ is the characteristic transient lifetime.

Figure 8.10 The Ikeda map can be viewed as arising from a string of light pulses of amplitude a entering at the partially transmitting mirror M_1. The time interval between the pulses is adjusted to the round-trip travel time in the systems. Let $|z_n|$ be the amplitude and angle (z_n) be the phase of the nth pulse just to the right of mirror M_1. Then the terms in (8.11) have the following meaning: $1 - b$ is the fraction of energy in a pulse transmitted or absorbed in the four reflections from M_1, M_2, M_3, and M_4; κ is the round-trip phase shift experienced by the pulse in the vacuum region; and $-\eta/(1 + |z_n|^2)$ is the phase shift in the nonlinear medium.

Note that while the above discussion was in the context of a crisis destroying a chaotic attractor, one may equally well consider that, as the parameter is varied in the other direction, the crisis creates a chaotic attractor. Thus, along with period doubling and Pomeau–Manneville intermittency, crises represent one of several routes to the creation of chaotic attractors.

The scaling of the characteristic transient lifetime $\langle\tau\rangle$ with $p - p_c$ is of great interest. For cases like that shown in Figure 8.9, it is found that

$$\langle\tau\rangle \sim (p - p_c)^{-\gamma} \qquad (8.13)$$

for p just past p_c. The quantity γ is called the critical exponent of the crisis.

As a simple, almost trivial, example of the scaling Eq. (8.13), consider the logistic map, $M(x,r) = rx(1 - x)$. The basin of attraction of the attractor for finite x is the interval $[0, 1]$, and the basin for the attractor $x = -\infty$ is its complement. The point $x = 0$ is an unstable $(r > 1)$ fixed point on the basin boundary. Looking at the bifurcation diagram, Figure 2.11(a), we see that the chaotic attractor is destroyed as r increases through $r = 4$ (for $r > 4$ the only attractor is $x = -\infty$). Furthermore, we see that the size of the chaotic attractors increases with increasing r

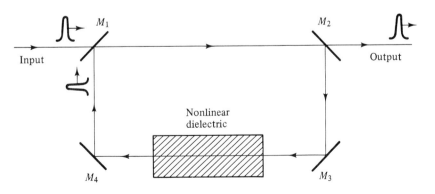

colliding with the fixed point, $x = 0$, at $r = 4$. Thus, this is a simple example of a boundary crisis. For $r > 4$, the chaotic attractor is replaced by a chaotic transient. To estimate $\langle \tau \rangle$ for r just slightly greater than 4, we note that $M(x, r) > 1$ for a small range of x-values about the maximum point $x = \frac{1}{2}$ and that this range has a width of order $(r - 4)^{1/2}$ (this square root dependence of the width is a general consequence of the maximum of the map function being quadratic). An orbit which falls in this narrow range about $x = \frac{1}{2}$ is mapped on one iterate to $x > 1$, on the next to $x < 0$, and on subsequent iterates to increasing negative x-values. Thus, this interval represents a loss region through which orbits initially in $[0, 1]$ can 'leak' out. For r just slightly greater than 4 the orbits in $[0, 1]$ will be similar to those for $r = 4$ (until they fall in the loss region). For $r = 4$ the probability per iterate of an orbit falling in the small region of length ε about $x = \frac{1}{2}$, namely $[(1 - \varepsilon)/2, (1 + \varepsilon)/2]$, is simply proportional to ε, the length of the interval. (This is because the natural measure varies smoothly through $x = \frac{1}{2}$; see Figure 2.7.) Identifying the loss rate $1/\langle \tau \rangle$ with the probability per iterate of falling in the loss region whose length is of the order of $(r - 4)^{1/2}$, we obtain $\langle \tau \rangle \sim (r - 4)^{-1/2}$. That is, the critical exponent γ in Eq. (8.13) is

$$\gamma = \tfrac{1}{2}.$$

This result is general for crises of one-dimensional maps with quadratic maxima and minima. We emphasize, however, that γ is typically greater than $\frac{1}{2}$ for higher-dimensional systems.

A theory yielding the critical exponents for crises in a large class of two-dimensional map systems has been given by Grebogi *et al.* (1986a, 1987b). The class of systems considered is two-dimensional maps in which the crisis is due to a tangency of the stable manifold of a periodic orbit on the basin boundary with the unstable manifold of an unstable periodic orbit on the attractor. These types of crisis appear to be the only kinds that can occur for two-dimensional maps that are strictly dissipative (i.e., the magnitude of their Jacobian determinant is less than 1 everywhere[2]) and is a very common feature of many systems such as the forced damped pendulum, the Duffing equation and the Hénon map. For these systems, crises occur in either one of the following two typical ways:

(1) *Heteroclinic tangency crisis.* In this case, the stable manifold of an unstable periodic orbit (B) becomes tangent with the unstable manifold of an unstable periodic orbit (A) (Figure 8.11(a)). Before (and also at) the crisis A was on the attractor and B was on the boundary.

(2) *Homoclinic tangency crisis.* In this case the stable and unstable manifolds of an unstable periodic orbit B are tangent (Figure 8.11(b)).

In both cases, both at and before the crisis, the basin boundary is the closure of the stable manifold of B. At $p = p_c$, again in both cases, the chaotic attractor is the closure of the unstable manifold of B. For $p = p_c$ (and also $p < p_c$) in the heteroclinic case, the attractor is also the closure of the unstable manifold of A. (Note that in both cases B is on the attractor for $p = p_c$ but not for $p < p_c$).

Grebogi *et al.* (1986a, 1987b) obtain the following formulae for the critical exponent γ,

$$\gamma = \tfrac{1}{2} + (\ln |\alpha_1|)/|\ln |\alpha_2||, \tag{8.14a}$$

for the heteroclinic case, and

$$\gamma = (\ln |\beta_2|)/(\ln |\beta_1 \beta_2|^2), \tag{8.14b}$$

for the homoclinic case. Here α_1 and α_2 are the expanding and contracting eigenvalues of the Jacobian matrix of the map evaluated at A, and β_1 and β_2 are the same quantities for the matrix evaluated at B. (If A and B have a period $P > 1$, then we use the Jacobian of the Pth iterate of the map.) The one-dimensional map result, $\gamma = \tfrac{1}{2}$, is recovered from (8.14) by going to the limit of infinitely strong contraction, $\alpha_2 \to 0$, $\beta_2 \to 0$.

We now give a derivation of the formula (8.14a) for the heteroclinic tangency case. (The interested reader is referred to Grebogi *et al.* (1986a, 1987b) for the derivation of (8.14b).) As p passes p_c the unstable manifold of A crosses the stable manifold of B as shown in Figure 8.12. If an orbit lands in the shaded region ab of the figure it is rapidly lost from the region of the old attractor by being attracted to B along the stable manifold of B and then repelled along the unstable manifold segment of B that goes

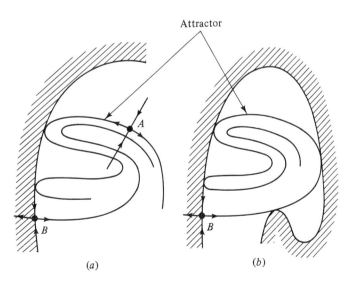

Figure 8.11 Illustrations of heteroclinic (a) and homoclinic (b) tangency crises at $p = p_c$. For simplicity the periodic orbits A and B are shown as fixed points. The crosshatching labels the basin of another attractor.

away from the attractor. For p near p_c the dimensions of the shaded region ab are of order $(p - p_c)$ by $(p - p_c)^{1/2}$ as shown in the figure. Iterating the region ab backward n steps, we have the shaded region $a'b'$ whose dimensions are of order $(p - p_c)^{1/2}/\alpha_1^n$ by $(p - p_c)/\alpha_2^n$. These estimates follow from the fact that for large enough n, except for the first few backwards iterates, the iterated region is near A and its evolution is thus governed by the linearized map at A. The region $a'b'$ is to be regarded as similar to the loss region of our previous logistic map example. As in that case we estimate the loss rate $1/\langle\tau\rangle$ as the attractor measure in the region $a'b'$ with the attractor measure evaluated at $p = p_c$. We denote this measure $m(p - p_c)$ and, in accord with the above assumption, we assume that $m(p - p_c) \sim (p - p_c)^\gamma$. Now, say that we reduce $p - p_c$ by the factor α_2 (i.e., $(p - p_c) \to \alpha_2(p - p_c)$). Iterating the reduced ab region backward by $n + 1$ steps (instead of n steps), the long dimension of the preiterated region is again of order $(p - p_c)/\alpha_2^n$, but the short dimension is of order $[\alpha_2(p - p_c)^{1/2}]/\alpha_1^{n+1}$. Presuming the attractor measure to be smoothly varying along the direction of the unstable manifold of A we then obtain

$$\frac{m(p - p_c)}{m[\alpha_2(p - p_c)]} \sim \frac{(p - p_c)^{1/2}/\alpha_1^n}{[\alpha_2(p - p_c)]^{1/2}/\alpha_1^{n+1}} = \frac{\alpha_1}{\alpha_2^{1/2}}.$$

Now assuming $m(p - p_c) \sim (p - p_c)^\gamma$, the above gives $\alpha_2^{-\gamma} = \alpha_1\alpha_2^{-1/2}$, which, upon taking logs, is the desired result (Eq. 8.14(a)).

An important aspect of the exponent γ is that larger γ makes the chaotic

Figure 8.12 Illustration for the derivation of γ in the heteroclinic case. The long dimension of the shaded region ab is of order $(p - p_c)^{1/2}$ because we asume that the unstable manifold of A is smooth and that the original tangency (Figure 8.11(a)) is quadratic.

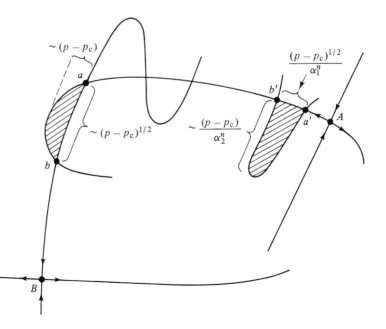

transient phenomena somewhat easier to observe. As an example, say $\langle\tau\rangle = [(p - p_{\mathrm{c}})/p_{\mathrm{c}}]^{\gamma}$, and consider two values of the exponent, $\gamma = \frac{1}{2}$ (as in one-dimensional maps) and $\gamma = 2$. For these hypothetical situations, we see that the range of p yielding transients of length $\langle\tau\rangle > 100$ is only $0 < (p - p_{\mathrm{c}})/p_{\mathrm{c}} < 0.0001$ for $\gamma = \frac{1}{2}$ but is $0 < (p - p_{\mathrm{c}})/p_{\mathrm{c}} < 0.1$ for $\gamma = 2$.

As an example where the power law dependence, Eq. (8.13), does not apply, Grebogi *et al.* (1983b, 1985a) consider a crisis in which the touching of the attractor and its basin boundary occurs as a result of the coalescence and annihilation of an unstable repelling orbit B on the basin boundary and an unstable saddle orbit A on the attractor. The bifurcation of these two orbits at the crisis is illustrated in Figure 8.13. (Note that the map is necessarily expanding in two directions at B so that this situation falls outside the class for which the tangency crises in Figures 8.11(*a*) and (*b*) are claimed to be the only possible cases.) The crisis mediated by this type of bifurcation has been called an 'unstable–unstable pair bifurcation crisis' and results in a characteristic scaling of $\langle\tau\rangle$ with p given by

$$\langle\tau\rangle \sim \exp[k/(p - p_{\mathrm{c}})^{1/2}], \qquad (8.15)$$

where k is a system dependent constant. Comparing (8.15) with (8.13), one sees that $\langle\tau\rangle$ from (8.15) increases faster than any power of $1/(p - p_{\mathrm{c}})$ as $(p - p_{\mathrm{c}}) \to 0$. Thus, in a sense, $\gamma = \infty$ for (8.15).

8.3.2 Crisis induced intermittency

Following a boundary crisis we have chaotic transients. In contrast, following an interior crisis or an attractor merging crisis, we have characteristic temporal behaviors that may be characterized as 'crisis-induced intermittency.'

In particular, for an interior crisis, as the parameter p increases through p_{c} the chaotic attractor suddenly widens. For p slightly larger than p_{c}, the

Figure 8.13 Schematic illustration of the unstable–unstable pair bifurcation. The two orbits A and B move towards each other as p increases, coalesce at $p = p_{\mathrm{c}}$, and cease to exist for $p > p_{\mathrm{c}}$. Note the similarity with the saddle-node bifurcation (Figure 8.5).

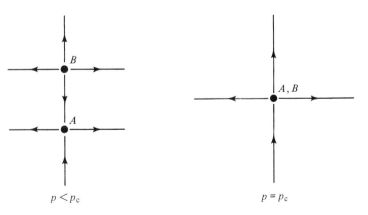

$p < p_{\mathrm{c}}$ $\qquad\qquad\qquad\qquad$ $p = p_{\mathrm{c}}$

orbit on the attractor spends long stretches of time in the region to which the attractor was confined before the crisis. At the end of one of these long stretches the orbit bursts out of the old region and bounces around chaotically in the new enlarged region made available to it by the crisis. It then returns to the old region for another stretch of time, followed by a burst, and so on. The times τ between bursts appear to be random (i.e., 'intermittent') and again have a long-time exponential distribution, Eq. (8.12), with average value which we again denote $\langle \tau \rangle$. As before $\langle \tau \rangle \to \infty$ as p approaches p_c from above. We further note that the critical exponent theory and formulae (Eqs. (8.14)) discussed for boundary crises also apply for interior (and attractor merging) crises, with the difference that, in the interior crisis case, the periodic orbit B of Figure 8.11 is not on the basin boundary.

In an attractor merging crisis of, say, two attractors, each of the two exist for $p < p_c$, and each has its own basin with a basin boundary separating the two. At $p = p_c$ the two attractors both *simultaneously* collide with this boundary. For p slightly greater than p_c, an orbit typically spends long stretches of time moving chaotically in the region of one of the old attractors, after which it abruptly switches to the region of the other old attractor, intermittently switching between the two. Again the times between switches have a long-time exponential distribution with an average $\langle \tau \rangle$ which approaches infinity as p approaches p_c from above.

One may think of the term intermittency as signifying an episodic switching between different types of behavior. Thus, we can schematically contrast crisis induced intermittency with the Pomeau–Manneville intermittency of Section 8.2, as follows:

Poneau–Manneville intermittency:

$$(\text{chaos}) \to (\text{approximately periodic}) \to (\text{chaos})$$

$$\to (\text{approximately periodic}) \to \dots .$$

Crisis-induced intermittency:

$$(\text{chaos})_1 \to (\text{chaos})_2 \to (\text{chaos})_1$$

$$\to (\text{chaos})_2 \to \dots .$$

For the case of intermittent bursting (interior crises), $(\text{chaos})_1$ might denote orbit segments during the bursts, and $(\text{chaos})_2$ might denote orbit segments between the bursts. For the case of intermittent switching (attractor merging crises), $(\text{chaos})_1$ and $(\text{chaos})_2$ would denote chaotic behaviors in the regions of each of the two attractors that exist before the crisis.

As an example, we again consider the Ikeda map, Eq. (8.11), but now for a different parameter set such that there is an interior crisis (Grebogi *et al.*, 1987b). We take $a = 0.85, b = 0.9, \kappa = 0.4$ and vary the parameter η in a small range about its crisis value $\eta_c = 7.26884894\ldots$. Figure 8.14 shows $y_n = \text{Im}(z_n)$ versus n for a value $\eta < \eta_c$ and for three successively larger values of η with $\eta > \eta_c$. The first plot shows the precrisis chaos to be well confined in a range $0.6 \gtrsim y \gtrsim -0.2$. As η is increased above η_c we see that there are occasional bursts of the orbit outside this range and these bursts become more frequent with increasing η. The cover picture on this book shows a computer plot of this attractor for η just past η_c. In this picture the plotted pixels are color coded so that the brighter coloring signifies that the pixel is more frequently visited by the orbit. We see that the attractor consists of a bright inner core region surrounded by a halo representing the region explored during bursting. Comparison of the core region with the precrisis attractor (not shown) reveals that they are essentially identical. Figure 8.15 shows results from numerical experiments (Grebogi *et al.*, 1987b) determining $\langle \tau \rangle$ as a function of $\eta - \eta_c$. We see that a log–log plot of these data is fairly well fit by a straight line in accord with a power law dependence as given by Eq. (8.13). The slope of this straight line gives $\gamma \simeq 1.24$ which has been demonstrated to be in agreement with the theory Eq. (8.14a). This is done in Grebogi *et al.* (1987b) where it is shown that this crisis is a heteroclinic tangency crisis

Figure 8.14 $y_n = \text{Im}(z_n)$ versus n for different values of η: (*a*) $\eta = 7.26 < \eta_c$; (*b*) $\eta = 7.33 > \eta_c$; (*c*) $\eta = 7.35$; (*d*) $\eta = 7.38$ (Grebogi *et al.*, 1987b).

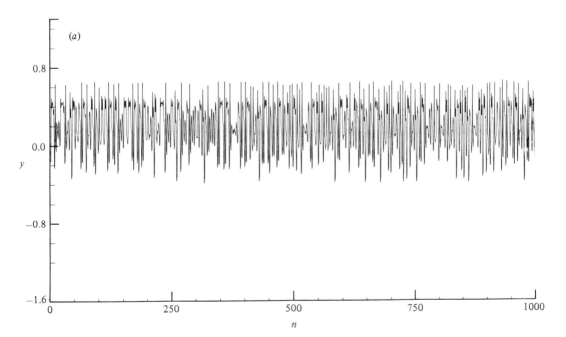

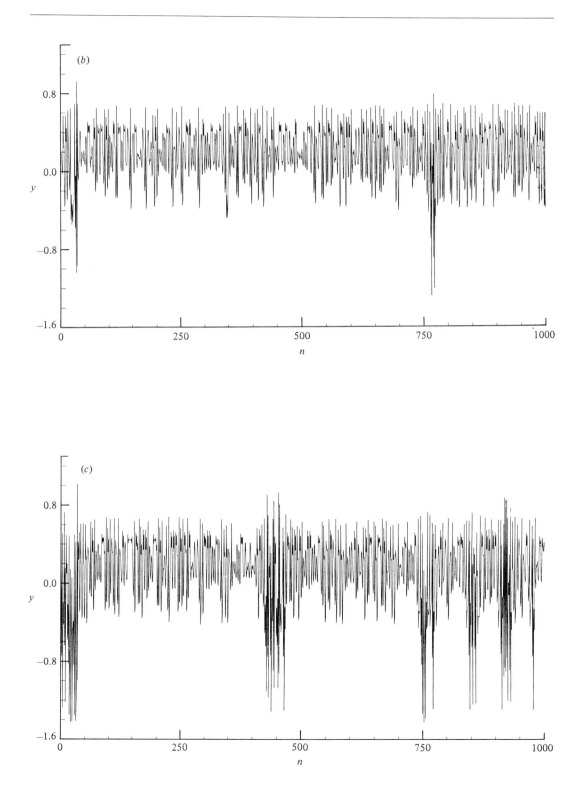

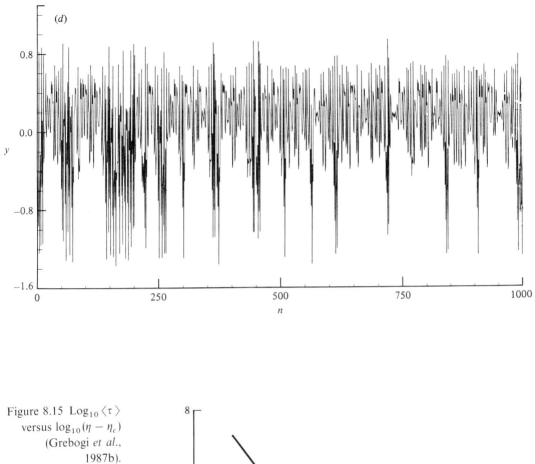

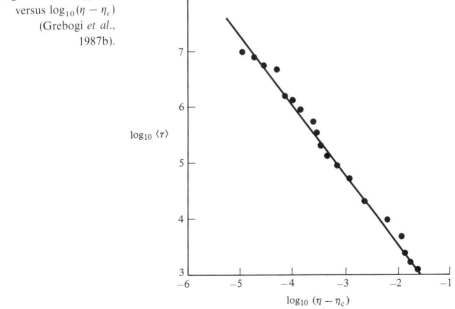

Figure 8.15 $\text{Log}_{10}\langle\tau\rangle$ versus $\log_{10}(\eta - \eta_c)$ (Grebogi *et al.*, 1987b).

and the relevant periodic orbits (*A* and *B* of Figure 8.11(*a*)) and their eigenvalues are determined.

As another example of an attractor widening crisis, refer to the bifurcation diagram for the logistic map in the range of the period three window, Figure 2.12(*a*). We see that, as *r* is increased through the value denoted r_{c3} in the figure, the attractor undergoes a sudden change. Namely, for *r* slightly less than r_{c3}, the attractor is chaotic and consists of three narrow intervals through which the orbit successively cycles; for *r* slightly greater than r_{c3}, the attractor is chaotic, but now consists of one much larger interval which includes the three small intervals of the attractor for $r < r_{c3}$. The event causing this change is an interior crisis. To see how this occurs, we note that at the beginning of the window (i.e., at the value $r = r_{*3}$ in Figure 2.12(a)) there is a tangent bifurcation creating both a period three attractor *and* an unstable period three orbit. As *r* increases from r_{*3} to r_{c3}, the unstable period three created at r_{*3} continues to exist and lies outside the attractor. At the crisis point $r = r_{c3}$, the unstable period three collides with the three piece chaotic attractor (each of the three points in the period three orbit collides with one of the three pieces of the chaotic attractor). Examination of a typical orbit on the attractor for *r* slightly greater than r_{c3} shows crisis induced intermittency in which the orbit stays in and cycles through the three precrisis intervals for long stretches of time, followed by intermittent bursts in which the orbit explores the much wider region in the full interval that the attractor

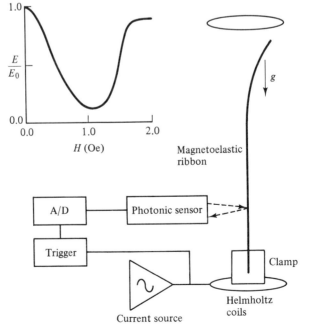

Figure 8.16 Schematic of the magnetoelastic ribbon experiment. The inset shows the Young's modulus *E* versus magnetic field *H*. (Sommerer *et al.*, 1991a).

now occupies. Thus, we see that the period three window is initiated by a Pomeau–Manneville intermittency transition (at $r = r_{*3}$) and terminates by an interior crisis (at $r = r_{c3}$) causing crisis induced intermittency. (The same is true for all windows of logistic map.)

We now discuss an experimental study of crisis-induced intermittency (Sommerer *et al.*, 1991a,b). The physical system studied, Figure 8.16, is a vertically oriented thin magnetoelastic ribbon clamped at its bottom. Due to gravity, the ribbon will tend to buckle to the left or the right. For the particular material that the ribbon is made of the Young's modulus is a strong function of magnetic field. Dynamical motion of the ribbon is induced by applying a time-dependent vertical magnetic field $H(t)$ with a sinusoidally varying component,

$$H(t) = H_{dc} + H_{ac} \sin(2\pi f t).$$

The frequency f is taken as the control parameter that is varied, and it is found that $f \equiv f_c \simeq 0.97\,\text{Hz}$ corresponds to an interior crisis. Using the voltage on the photonic sensor (Figure 8.16) sampled at the period $1/f$ of the driver, Figure 8.17 shows experimental delay-coordinate Poincaré surfaces of section for the precrisis ($f = 0.9760\,\text{Hz}$) attractor (Figure

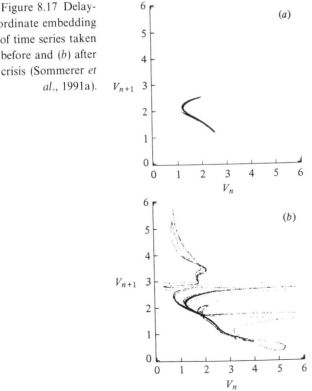

Figure 8.17 Delay-coordinate embedding of time series taken (a) before and (b) after the crisis (Sommerer *et al.*, 1991a).

8.17(*a*)) and for the postcrisis ($f = 0.9630\,\text{Hz}$) attractor (Figure 8.17(*b*)).
We see that the crisis induces a significant expansion of the phase space
extent of the attractor. Figure 8.18 shows a log–log plot of the
experimentally determined average time between bursts (denoted $\hat{\tau}$ in the
figure) versus distance $f - f_c$ from the crisis. Again, a straight line in
accord with Eq. (8.13) is obtained; the measured slope of this line gives a
critical exponent of $\gamma \simeq 1.1$. Furthermore, the authors were able to
determine experimentally the unstable orbit mediating the crisis and its
eigenvalues. Using these they accurately verified that the theoretical
formula for the critical exponent Eq. (8.14a) agreed well with the
experimental value. Another aspect of this work was that they experimentally studied the effect of noise on crises. Using a noise generator, they
added a noise component of controlled strength denoted σ to the magnetic
field $H(t)$. According to a recent theory (Sommerer *et al.*, 1991c), the
scaling of the average time between bursts in the presence of noise should
be (instead of (8.13))

$$\langle \tau \rangle \sim \sigma^{-\gamma} g(|f - f_c|/\sigma), \tag{8.16}$$

where γ denotes the noiseless critical exponent, and the function g depends

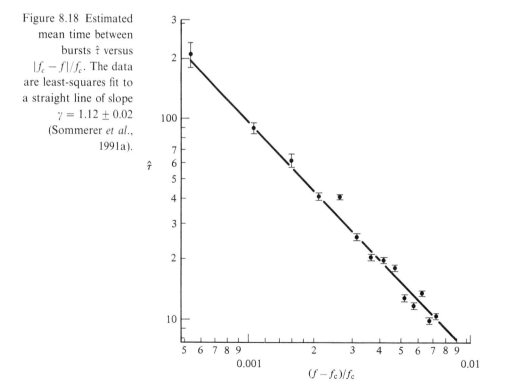

Figure 8.18 Estimated
mean time between
bursts $\hat{\tau}$ versus
$|f_c - f|/f_c$. The data
are least-squares fit to
a straight line of slope
$\gamma = 1.12 \pm 0.02$
(Sommerer *et al.*,
1991a).

on the system and the particular form of the noise. Figure 8.19 (*a*) shows plots of raw data for four different noise levels. Note that, due to the noise, $\hat{\tau}$ can differ by several orders of magnitude at the same value of the control parameter f. Using the noiseless exponent γ to scale the variables as prescribed by Eq. (8.16), Figure 8.19 (*b*) results. We see that the four sets of data collapse onto a single curve (the function g), thus verifying (8.16).

8.4 The Lorenz system: An example of the creation of a chaotic transient

A particularly interesting example is provided by the Lorenz equations (Chapter 2),

$$\mathrm{d}X/\mathrm{d}t = -\tilde{\sigma}X + \tilde{\sigma}Y,$$

$$\mathrm{d}Y/\mathrm{d}t = -XY + \tilde{r}X - Y,$$

$$\mathrm{d}Z/\mathrm{d}t = XY - \tilde{b}Z.$$

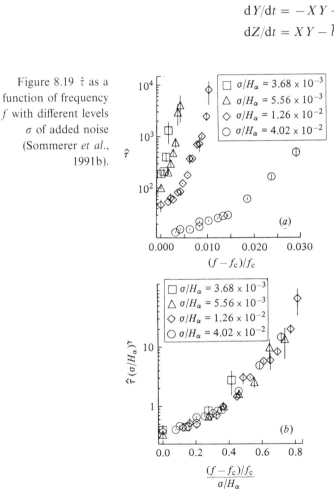

Figure 8.19 $\hat{\tau}$ as a function of frequency f with different levels σ of added noise (Sommerer *et al.*, 1991b).

This system has three possible steady states (i.e., solutions with $dX/dt = dY/dt = dZ/dt = 0$). One is $X = Y = Z = 0$ which we denote O. There are two others, which we denote C and C', which exist for $\tilde{r} > 1$ and are given by

$$C = (X, Y, Z) = ([\tilde{b}(\tilde{r} - 1)]^{1/2}, [\tilde{b}(\tilde{r} - 1)]^{1/2}, \tilde{r} - 1),$$
$$C' = (X, Y, Z) = (-[\tilde{b}(\tilde{r} - 1)]^{1/2}, -[\tilde{b}(\tilde{r} - 1)]^{1/2}, \tilde{r} - 1).$$

Say we fix $\tilde{\sigma}$ and $\tilde{b}$ at the values used by Lorenz ($\tilde{\sigma} = 10$, $\tilde{b} = \frac{8}{3}$) and examine the behavior of the Lorenz system as $\tilde{r}$ increases from zero. The stability of the steady state O is given by the eigenvalues of the Jacobian matrix

$$\begin{bmatrix} -\tilde{\sigma} & \tilde{\sigma} & 0 \\ \tilde{r} & -1 & 0 \\ 0 & 0 & -\tilde{b} \end{bmatrix}$$

which are all negative for $0 < \tilde{r} < 1$ indicating stability (Figure 8.20(a)). In this case O is the only attractor of the system. As $\tilde{r}$ passes through 1, one of the eigenvalues of O becomes positive with the other two remaining negative. This indicates that O has a two-dimensional stable manifold and a one-dimensional unstable manifold. Simultaneous with the loss of stability of O, the two fixed points C and C' are born. C and C' are stable at birth with three real eigenvalues. Thus as $\tilde{r}$ increases through 1 C and C' become the attractors of the system. The basin boundary separating the basins of attraction for the attractors C and C' is the two-dimensional stable manifold of O. Figure 8.20(b) shows the situation for $\tilde{r}$ slightly past 1. Following the unstable manifold of O it goes to the steady states C and C'. As $\tilde{r}$ increases further, a point $\tilde{r}_s$ is reached past which two of the real stable (negative) eigenvalues of C coalesce and become complex conjugate eigenvalues with negative real parts (by symmetry the same happens for C'). In this regime, orbits approach C and C' by spiraling around them. This situation is shown in Figure 8.20(c), while Figure 8.20(d) shows the unstable manifold of O projected onto the (XZ)-plane for the same situation.

For the pictures shown in Figures 8.20(a)–(d) there is no chaotic dynamics at all, neither transient chaos nor a chaotic attractor. To see how the chaos observed by Lorenz at $\tilde{r} = 28$ is formed, we continue to increase $\tilde{r}$. As this is done, the initial spiral of the unstable manifold of O about C and C' increases in size, until at some critical value $\tilde{r} = \tilde{r}_0 \simeq 13.96$ (Kaplan and Yorke, 1979a), we obtain the situation shown in Figure 8.20(e). In this case the orbit leaving O along its unstable manifold comes back to O (i.e., the unstable manifold of O is on the stable manifold of O). The orbit in Figure 8.20(e) is called a homoclinic orbit for O (points on this orbit approach O for both $t \to +\infty$ and $t \to -\infty$). Kaplan and Yorke

(1979a) and Afraimovich *et al.* (1977) prove that as $\tilde{r}$ increases past $\tilde{r}_0$ chaos must be present. However, as emphasized by Kaplan and Yorke (1979a), and by Yorke and Yorke (1979), this chaos is not attracting (it is a chaotic transient), and, just past $\tilde{r}_0$, C and C' continue to be the only attractors. As $\tilde{r}$ increases, however, the transient chaos is converted to a chaotic attractor by a crisis at $\tilde{r} = \tilde{r}_1 \simeq 24.06$. Increasing $\tilde{r}$ still further, the steady states C and C' become unstable at $\tilde{r} = \tilde{r}_2 \simeq 24.74$ as the real parts of their complex conjugate eigenvalues pass from negative to positive (a Hopf bifurcation). Thus, for the relatively narrow range $\tilde{r}_1 < \tilde{r} < \tilde{r}_2$, there are three possible attractors (each with its respective basin), namely C, C' and a chaotic attractor, while for $\tilde{r} > \tilde{r}_2$ there is only a chaotic attractor (i.e., that found by Lorenz). The situation is illustrated schematically in Figure 8.21.

Figure 8.20 Evolution of the fixed points and the unstable manifold of O.

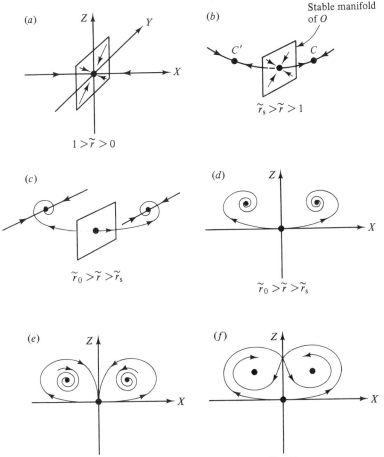

The main interest of this example is the creation of transient chaos by the formation of the homoclinic orbit at $\tilde{r} = \tilde{r}_0$. For the topological arguments leading to this result the reader is referred to the original papers of Kaplan and Yorke (1979a) and Afraimovich *et al.* (1977) and to the book by Sparrow (1982).

In the Lorenz case the homoclinic orbit results for a situation where all the eigenvalues of the fixed point are real. Another important situation occurs when two of the eigenvalues of an unstable fixed point are complex conjugates with the other being real. In this case the fixed point is said to be a saddle-focus. The formation of a homoclinic orbit of a saddle focus is illustrated in Figure 8.22 which shows the stable and unstable manifolds for values of the parameter p below, at, and above the critical homoclinic value $p = p_{\mathrm{h}}$. Shilnikov (1970) proves that chaos results for p-values in the neighborhood of p_{h}, and Arnéodo *et al.* (1982) demonstrate an example of a chaotic attractor with spiral shape based on such a homoclinic structure.

In the Lorenz and Shilnikov cases chaos arises as a result of orbits homoclinic to a single steady state. Another related situation arises when there are heteroclinic orbits connecting two steady states. See Lau and Finn(1992) for a discussion of how chaos is created in this case.

8.5   Basin boundary metamorphoses

As a system parameter is varied the character of a basin boundary can change. For example, as the parameter passes through a critical value, a nonfractal boundary can become fractal and simultaneously experience a jump in its extent in phase space. In this section we discuss these changes, called basin boundary 'metamorphoses' (Grebogi *et al.*, 1986b, 1987c),

Figure 8.21 Major bifurcations of the Lorenz attractor as a function of $\tilde{r}$ (not to scale). In the regime $0 < \tilde{r} < 1$ the only attractor is O. In terms of the physical Rayleigh–Bénard system that the Lorenz equations are meant (very crudely) to model, this represents a situation in which the fluid is at rest and thermal conduction is the only mechanism which transports heat from the bottom to the top plate (Figure 1.4). The steady states C and C' both represent time-independent convective flow patterns as shown in Figure 1.4. (The direction of flow is reversed for C' as compared to C).

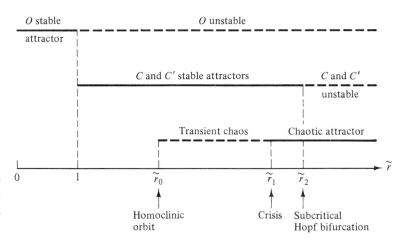

specializing to the case of two-dimensional maps. It is found that metamorphoses occur at the formation of homoclinic crossings of stable and unstable manifolds (Chapter 4) (Grebogi *et al.*, 1986b, 1987c; Moon and Li, 1985; Guckenheimer and Holmes, 1983, p. 114). Furthermore, as we shall see, a key role is played by certain unstable periodic orbits in the basin boundary that are 'accessible' from one basin or the other (Grebogi *et al.*, 1986b, 1987c).

Definition: A point on the boundary of a region is *accessible* from that region if one can construct a finite length curve connecting the point on the boundary to a point in the interior of the region such that no point in the curve, other than the accessible boundary point, lies in the boundary.

For the case of a smooth nonfractal boundary all points are accessible. However, if the basin boundary is fractal, then there are typically inaccessible points[3] in the boundary.

As a specific illustrative example we consider the basin boundary for the

Figure 8.22 Formation of a homoclinic orbit (*b*) of an unstable fixed point with one unstable direction and two complex conjuate stable eigenvalues.

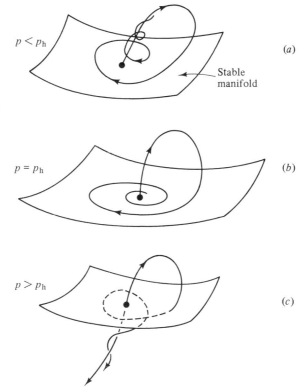

$p < p_\mathrm{h}$

Stable manifold

(*a*)

$p = p_\mathrm{h}$

(*b*)

$p > p_\mathrm{h}$

(*c*)

Hénon map, Eq. (1.14), with $B = -0.3$, and we examine the basin structure as the remaining parameter A is varied (Grebogi *et al.*, 1986b, 1987c). As A increases, there is a saddle-node bifurcation at $A_1 = -(1 - B)/4$. This bifurcation creates a saddle fixed point and an attracting node fixed point. For A just past A_1 there are two attractors, one is the attracting node and the other is an attractor at infinity, $(|x|, |y|) = \infty$. (For $A < A_1$ the only attractor is the attractor at infinity.)

Figure 8.23 (*a*) shows the basins of the node attractor (blank) and of the attractor at infinity (black) for $A = 1.150$. We see that the period one (fixed point) saddle created at $A = A_1$ is on the basin boundary of the period one attractor. In fact, the boundary is the stable manifold of the period one saddle (cf. Chapter 5). This boundary is apparently a smooth curve. Figure 8.23 (*b*) shows the situation for a larger value of A, namely $A = 1.395$ (the period two attractor in this figure results from a period doubling of the attractor in Figure 8.23 (*a*)). The boundary in Figure 8.23 (*b*) is fractal as demonstrated by a numerical application of the uncertainty exponent technique (Chapter 5) which yields a fractal dimension of about 1.53. Evidentally, as A increases from 1.150 to 1.395, the boundary changes from smooth to fractal. We denote the critical value of A where this occurs A_{sf}, and it is found that $A_{sf} \simeq 1.315$. The basin boundary not only becomes fractal as A passes through A_{sf}, but it also experiences a discontinuous jump, in a sense to be discussed below.

Figure 8.23 (*c*) shows the basin boundary at $A = 1.405$ which is just slightly larger than the value $A = 1.395$ used for Figure 8.23 (*b*). Comparing the two figures, it is evident that the basin of infinity (black region) has enlarged by the addition of a set of thin filaments that are well within the interior of the blank region of Figure 8.23 (*b*); note in particular the region $-1.0 \le x \le 0.3$, $2.0 \le y \le 5.0$. This jump in the basin boundary occurs at a value $A \equiv A_{ff} \simeq 1.396$ (the subscripts ff stand for fractal–fractal in accord with the fact that the boundary is fractal both before and after the metamorphosis at A_{ff}). It is important to note that, as A is decreased toward A_{ff}, the set of thin black filaments sent into the old basin become ever thinner, their area going to zero, but they remain essentially fixed in position, *not* contracting to the position of the basin boundary shown in Figure 8.23 (*b*).

Both the metamorphosis at A_{sf} and that at A_{ff} are accompanied by a change in the periodic orbit on the boundary that is 'accessible' from the basin of the finite attractor. In particular, for $A_1 < A < A_{sf}$, the periodic orbit accessible from the basin of the finite attractor is the period one saddle created at A_1 and labeled in Figure 8.23 (*a*). For $A_{sf} < A < A_{ff}$, the periodic orbit accessible from the basin of the finite attractor is the period four saddle orbit labeled in Figure 8.23 (*b*) $(1 \to 2 \to 3 \to 4 \to 1 \to \ldots)$.

As A passes through A_{sf} the period one saddle becomes inaccessible from the basin of the finite attractor. (It is still accesible from the basin of the attractor at infinity.) This inaccessibility of the period one saddle is a result of the fact that the passing of A through A_{sf} corresponds with the formation of a homoclinic intersection of the stable and unstable manifolds of the period one saddle. This is illustrated in Figure 8.24. As a result of the homoclinic intersection a series of progressively longer and thinner tongues of the basin of the attractor at infinity accumulate on the

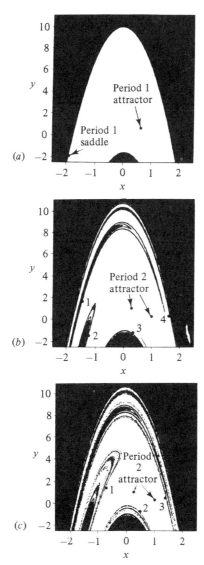

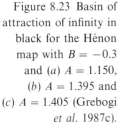

Figure 8.23 Basin of attraction of infinity in black for the Hénon map with $B = -0.3$ and (a) $A = 1.150$, (b) $A = 1.395$ and (c) $A = 1.405$ (Grebogi *et al.* 1987c).

right-hand side of the stable manifold segment through the period one saddle. A finite length curve connecting a point in the interior of the blank region with the period one saddle cannot now be drawn since it would have to circumvent all these tongues, and the length of the nth tongue approaches infinity as $n \to \infty$. Thus, the period one is not accesible from the blank region. We also note that, as discussed in Chapter 5, the fractal boundary contains an invariant chaotic set, embedded in which there is an infinite number of unstable periodic orbits. All of these, except for the period one and the period four shown in the figure, are, however, inaccessible from either basin. They are each essentially 'buried' under an infinite number of thin alternating striations of the two basins accumulating on them from both sides.

Thus, as A increases through A_{sf} the boundary saddle accessible from the blank region suddenly changes from a period one to a period four. Both orbits exist before and after the metamorphosis. We emphasize, however, that the period four remains well within the interior of the blank region as we let A approach A_{sf} from below. Thus, as A increases through A_{sf}, the basin boundary suddenly jumps inward to the period four saddle.

The situation with respect to the metamorphosis at $A = A_{ff}$ is similar: the metamorphosis is due to a homoclinic crossing of stable and unstable manifolds of the boundary saddle accessible from the basin of the finite attractor (now the period four saddle) which then becomes inaccessible and is replaced by the period three saddle. To sum up, we see that basin boundary metamorphoses of two-dimensional maps typically result in changes of the accessible boundary saddles and jumps of the basin boundary location with a possible transition of the boundary character

Figure 8.24 Schematic of the accumulation of tongues on the stable manifold segment through the period one saddle (Grebogi *et al.*, 1987c).

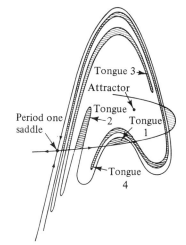

from smooth to fractal. Furthermore, these metamorphoses are induced by the formation of homoclinic intersections of manifolds of certain special (i.e., accessible) periodic saddles in the basin boundary.

8.6 Bifurcations to chaotic scattering

Referring to Figure 5.16, which shows the scattering angle ϕ as a function of the impact parameter b for a particular two-dimensional potential scattering problem, we see that the ϕ versus b curve is smooth when the particle energy is large (Figure 5.16(a)), but becomes complicated on arbitrarily fine scale at lower energy (Figures 5.16(b) and 5.17). As discussed in Chapter 5, the latter indicates the presence of chaotic dynamics. The transition from regular (Figure 5.16(a)) to chaotic (Figure 5.16(b)) scattering with decrease of the particle energy is a typical feature of scattering from smooth finite potential scatterers. Let $V(\mathbf{x})$ denote the potential; we assume that $V(\mathbf{x})$ is smooth, bounded, and rapidly approaches zero with increasing $|\mathbf{x}|$ outside the scattering region.

The problem we wish to address is how does the scattering go from being regular at large incident particle energies to being chaotic at smaller energies. For the case of two-dimensional ($\mathbf{x} = (x, y)$) potential scattering it has been found that this can occur in one of two possible ways (Bleher *et al.*, 1989, 1990; Ding *et al.*, 1990b). We shall illustrate these two routes to chaotic scattering by reference to a simple model situation.

As background for the analysis, we now review some relevant facts concerning scattering from a single circularly symmetric monotonic potential hill. Figure 8.25 shows trajectories incident on such a hill for $E > E_m$, where E is the particle energy and E_m is the potential maximum at the hilltop. As illustrated in the figure, we see that, as the impact parameter decreases from large values, the scattering angle increases,

Figure 8.25 Trajectories incident on a circularly symmetric monotonic potential hill for $E > E_m$.

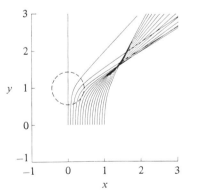

reaches a maximum value $\phi_m(E) < 90°$, and then decreases (ϕ is zero at both $b = \infty$ and $b = 0$). Furthermore, it can be shown that $\phi_m(E)$ increases as E is decreased, approaching $90°$ as E approaches E_m from above,

$$\lim_{E \to E_m^+} \phi_m(E) = 90°.$$

As soon as E drops below E_m, ϕ_m jumps to $\phi_m = 180°$, since now the orbit with $b = 0$ is backscattered. For $E < E_m$, the scattering angle increases monotonically from $\phi = 0$ to $\phi = 180°$ as b decreases from $b = \infty$ to $b = 0$.

We now consider the following situation. We take the potential to consist of three monotonic hills whose separation is large compared to their widths. Further, we assume that the potential is locally circularly symmetric about each of the three hill tops, Figure 8.26. (The case of noncircularity is discussed in Bleher *et al.* (1990).) We label these hills 1, 2 and 3 and denote the potential maxima at the hilltops by E_{m1}, E_{m2} and E_{m3} where by convention, $E_{m3} > E_{m2} > E_{m1}$. We distinguish two cases as shown in Figure 8.27. Case 1 is shown in Figure 8.27(*a*), and case 2 is shown in Figure 8.27(*b*). In case 1, the hill of lowest maximum potential energy (hill 1 with maximum potential energy E_{m1}) is outside the circle whose diameter is the line joining the two hills of larger maximum potential energy. In case 2, hill 1 is inside this circle. (In both cases we presume that hill 1 is far from the circle compared with the width of a hill.) We now consider each case in turn.

Case I.   Say $E < E_{m2} \leq E_{m3}$ and assume that our orbit is deflected from hill 2 (or hill 3) and travels toward hill 1. In order for this orbit to remain

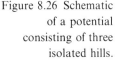
Figure 8.26 Schematic of a potential consisting of three isolated hills.

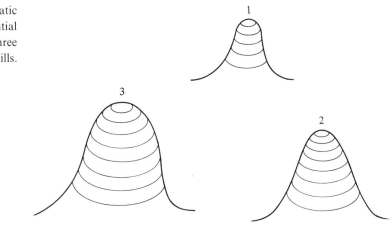

trapped it must be deflected back toward hill 2 or hill 3. Since hill 1 lies outside the circle, the minimum required deflection angle ϕ_{m*} is greater than 90°. Thus, recalling the result for a single hill, we see that for $E > E_{m1}$ there are no bounded orbits reflecting from hill 1. Consequently, the only periodic orbit that can exist is the one bouncing back and forth between hills 2 and 3. Recalling that there is an infinite number of periodic orbits embedded in the chaotic invariant set, we see that there is no chaos for case 1 when $E > E_{m1}$. When E drops below E_{m1}, chaos is immediately created, since now the number of unstable periodic orbits increases exponentially with period: we can represent the periodic orbits as a sequence of symbols representing the order in which each hill is visited, and any sequence is possible. Furthermore, when chaos is created the bounded invariant set in phase space may be shown to be hyperbolic (e.g., there are no KAM tori present). Bleher *et al.* (1989, 1990) have called this transition an *abrupt* bifurcation to chaotic scattering. They also investigate how the fractal dimension characterizing the set of singularities of the scattering function behaves near the bifurcation point. They find that it increases from $d = 0$ at $E = E_{m1}$ roughly as

$$d \sim \frac{1}{\ln[1/(E_{m1} - E)]}$$

for $E < E_{m1}$. See also Tél (1991). Note that the abrupt creation of chaos as E decreases through the critical value $E_c = E_{m1}$ is accompanied by a change in the topology of the energy surface: For $E < E_{m1}$ a forbidden region ($V(\mathbf{x}) > E$), where orbits cannot penetrate, is created about the maximum of hill 1.

Figure 8.27(a) Case 1 and (b) case 2.

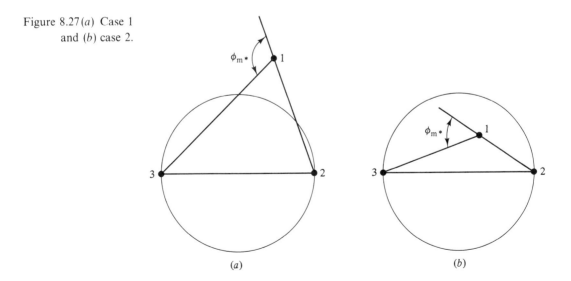

(a)   (b)

Case 2. We take the energy to be in the range $E_{m1} < E < E_{m2} \le E_{m3}$. As E is decreased from E_{m2} to E_{m1} the maximum deflection hill 1 is capable of producing, $\phi_{m1}(E)$, increases monotonically from some value $\phi_{m1}(E_{m2}) < 90°$ to $\phi_{m1}(E_{m1}) = 90°$. Let ϕ_{m*} illustrated in Figure 8.27(b) denote the deflection required by an orbit incident on hill 1 from hill 2 (respectively, hill 3) to be reflected toward hill 3 (respectively, hill 2). Now, however, since hill 1 is inside the circle, $\phi_{m*} < 90°$. We can now distinguish two subcases within case 2: E_{m1} is small enough that $\phi_{m*} > \phi_{m1}(E_{m2})$, or $\phi_{m*} < \phi_{m1}(E_{m2})$. In the latter case, as soon as E drops below E_{m2} it can be shown that there is a transition to hyperbolic chaotic scattering; i.e., this is an abrupt bifurcation to chaotic scattering as in case 1. On the other hand, for $\phi_{m*} > \phi_{m1}(E_{m2})$, there is a qualitatively different kind of bifurcation to chaotic scattering. We therefore restrict our consideration of case 2 in what follows to $\phi_{m*} > \phi_{m1}(E_{m2})$. In this case as E decreases from E_{m2}, $\phi_{m1}(E)$ will increase until at some particle energy $E \equiv E_{m*} > E_{m1}$, we have $\phi_{m*} = \phi_{m1}(E_{m*})$. For $E_{m1} < E \le E_{m*}$, we can have orbits traveling back and forth between hills 2 and 3 in two possible ways: either the path between hills 2 and 3 can pass through the region of hill 1 or it can bypass hill 1 going directly between hills 2 and 3. This is illustrated schematically in Figure 8.28. Thus, we expect that for E below (but not too close) to E_{m*} there will be unstable periodic orbits made up of all possible combinations of the two types of paths between hills 2 and 3 shown in Figure 8.28. Thus, there are an infinite number of periodic orbits and hence chaotic scattering. The way in which this situation arises in this case is very different from what we have for case 1 and is *not* an abrupt bifurcation. In particular, as E decreases, producing chaos, there is no change in the topology of the energy surface. How can the infinite number of unstable periodic orbits necessary for chaos be created in this case? In the abrupt bifurcation it is the change in the energy surface topology that occurs when E passes through one of the E_{mi} which creates the infinity of periodic orbits. In the absence of such a change in topology, the only mechanisms

Figure 8.28 Two possible orbits connecting hills 2 and 3 for case 2.

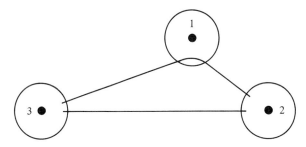

available for the creation of unstable periodic orbits are the standard generic bifurcations of smooth Hamiltonian systems with two degrees of freedom. In this case chaotic scattering first appears when a saddle-node bifurcation occurs in the scattering region (Ding *et al.*, 1990b). This bifurcation produces two periodic orbits bouncing between hills 2 and 3 by way of hill 1, one of these periodic orbits is stable (the node) and the other is unstable (the saddle). Following the bifurcation there is a set of nested KAM tori surrounding the node periodic orbit. Moving outward from the node, we eventually encounter the last KAM torus enclosing the node orbit. Past this last KAM torus there is a region of phase space where chaotic trajectories occur, and which can be reached by scattering particles (the region enclosed by the last KAM surface is inaccessible to scattering particles since they must originate from outside it). It is the chaos in the region surrounding the last KAM torus around the node orbit which is responsible for the appearance of chaotic scattering in this case. Note that in this case the chaotic invariant set is nonhyperbolic since it is bounded by a KAM curve (for which the Lyapunov exponents are necessarily zero). Decreasing E still further, for the situation of narrow hills, all the KAM surfaces will be destroyed, and we obtain a situation where the chaotic set is hyperbolic. In this case all bounded orbits are composed of legs as shown in Figure 8.28.

Problems

1. Consider a one-dimensional map with a nonquadratic maximum, $x_{n+1} = a - |x|^z$, and use the renormalization group analysis of Section 8.1 to obtain an approximate value for the Feigenbaum number $\hat{\alpha}$ for this case.

2. Obtain Eq. (8.4) from Eq. (8.3).

3. The second iterate of a map with an inverse period doubling bifurcation at $\varepsilon = 0$ may be put in the normal form $x_{n+1} = x_n^3 + (1 + \varepsilon)x_n$ (why is this reasonable?). Using this normal form, verify the result $\bar{T} \sim \varepsilon^{-1}$ for Type III intermittency.

4. Using a computer demonstrate Pomeau–Manneville Type I intermittency by plotting orbits for the logistic map $x_{n+1} = rx_n(1 - x_n)$ for a value of r just below the value at the bottom of the period three window. Also plot x_{3n} versus n (i.e., plot every third iterate), and comment on the result.

5. Repeat Problem 4 but for a value of r just above the value at the top of the period three window thus demonstrating crisis induced intermittency.

6. Consider the map $M(x, r)$ shown in Figure 8.29. As the parameter r changes continuously from r_1 to r_2, the map function $M(x, r)$ changes continuously from the shape shown in Figure 8.29(a) to that in Figure 8.29(b). Show that there is a basin boundary metamorphosis for some

r-value between r_1 and r_2 and verify that the basin boundary experiences a discontinuous jump. Also discuss how the metamorphosis affects the accessible periodic boundary orbits for this case. Show that the fixed point of $M(x, r_2)$ at $x = 0$ is inaccessible.

Notes

1. Since any point on its stable manifold maps forward to the periodic boundary orbit as time tends to $+\infty$, and since the attractor is an invariant closed set, the attractor must touch the unstable periodic boundary orbit if it touches its stable manifold.

2. Subsequently we shall see a two-dimensional map example where the map is not strictly dissipative (it has a local region in which there is expansion in two dimensions). In this case, a power law dependence, Eq. (8.13), does not hold, and the crisis is not due to a tangency of stable and unstable manifolds.

3. This seems to be true for many typical nonlinear systems. It is not, however, true for Julia set basin boundaries of complex analytic polynomial maps (discussed, for example, in Devaney (1986)). In the latter case, even though the boundary is fractal, all its points can be accessible. By taking real and imaginary parts, complex analytic maps may be regarded as two-dimensional real maps, and we adopt the point of view that, within the space of two-dimensional maps, these are nontypical, since the complex analyticity condition imposes a special restriction on these maps that endows them with (fascinating) special properties. See McDonald *et al.* (1985) for further discussion.

Figure 8.29 The map $M(x, r)$ for Problem 5.

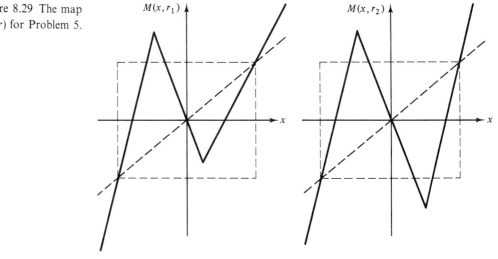

CHAPTER NINE

Multifractals

In Chapters 3 and 5 we have discussed the fractal dimension of strange attractors, as well as the fractal dimension of nonattracting chaotic sets. We have found that, not only are these sets fractal, but the measures associated with them can also have fractal-like properties. To characterize such fractal measures we have discussed the dimension spectrum D_q. A measure for which D_q varies with q is called a *multifractal* measure.[1] In this chapter we shall extend our discussion, begun in Chapter 3, of multifractals. In particular, we shall treat several more advanced developments concerning this topic.

9.1 The singularity spectrum $f(\alpha)$

Imagine that we cover an invariant set A with cubes from a grid of unit size ε. Let $\mu_i = \mu(C_i)$, where C_i denotes the ith cube, and μ is a probability measure on A, $\mu(A) \equiv 1$. We assume that ε is small so that the number of cubes is very large. To each cube we associate a *singularity index* α_i via

$$\mu_i \equiv \varepsilon^{\alpha_i}. \tag{9.1}$$

We then count the number of cubes for which α_i is in a small range between α and $\alpha + \Delta\alpha$. For very small ε, and, hence, large numbers of cubes, we can, as an idealization, pass to the continuum limit and replace $\Delta\alpha$ by a differential $d\alpha$. We then assume that the number of cubes with singularity index in the range α to $\alpha + d\alpha$ is of the form

$$\rho(\alpha)\varepsilon^{-f(\alpha)} \, d\alpha. \tag{9.2}$$

This form can be motivated by the following (see Benzi *et al.* (1984) and Frisch and Parisi (1985) whose considerations included the fractal properties of the spatial distribution of viscous energy dissipation of a turbulent fluid in the limit of large Reynolds number (i.e., small viscosity)). For every point $\mathbf{x}$ on the invariant set A we calculate its pointwise dimension $D_p(\mathbf{x})$ (refer back to Section 3.6 for the definition of the pointwise dimension). We then consider the set of all points with a particular value α of the pointwise dimension, $D_p(\mathbf{x}) = \alpha$, and we calculate the box-counting dimension of this set. We denote this dimension $\hat{f}(\alpha)$. If we interpret (9.1) as resulting from the point in the center of the box having a pointwise dimension α_i, then we might be tempted to identify the quanity $f(\alpha)$ defined by (9.2) with $\hat{f}(\alpha)$, the dimension of the set $D_p(\mathbf{x}) = \alpha$. This is a justification for the assumed form in (9.2). Work by Bohr and Rand (1987) and by Collet *et al.* (1987) shows that $f(\alpha)$ is the dimension of the set $D_p(\mathbf{x}) = \alpha$ (i.e., $f(\alpha) = \hat{f}(\alpha)$) in the case of hyperbolic invariant sets. We caution, however, that this interpretation is not always correct. For a nonhyperbolic attractor (e.g., the Hénon attractor), it has been found that $f(\alpha)$ can indeed be the fractal dimension of the set of $\mathbf{x}$ for which $D_p(\mathbf{x}) = \alpha$, but only when α is in a certain range, while when α is outside that range, $f(\alpha)$ is not the fractal dimension of the set $D_p(\mathbf{x}) = \alpha$ (see, for example, Ott *et al.* (1989a)).

Following Halsey *et al.* (1986), we now relate the quantity $f(\alpha)$ defined by (9.2) to the dimension spectrum D_q. Again using a covering from a grid of edge length ε we have (Eq. (3.14))

$$D_q = \frac{1}{1-q} \lim_{\varepsilon \to 0} \frac{\ln I(q, \varepsilon)}{\ln(1/\varepsilon)}, \tag{9.3}$$

where

$$I(q, \varepsilon) = \sum_i \mu_i^q, \tag{9.4}$$

and $\mu_i = \mu(C_i)$. Making use of (9.2) and (9.1), we have

$$I(q, \varepsilon) = \int d\alpha' \rho(\alpha') \varepsilon^{-f(\alpha')} \varepsilon^{q\alpha'}$$

$$= \int d\alpha' \rho(\alpha') \exp\{[f(\alpha') - q\alpha'] \ln(1/\varepsilon)\}. \tag{9.5}$$

Since we are interested in the limit as ε goes to zero, we can regard $\ln(1/\varepsilon)$ as large. In this case, the main contribution to the integral over α' comes from the neighborhood of the maximum value of the function $f(\alpha') - \alpha' q$. Assuming $f(\alpha)$ is smooth, the maximum is located at $\alpha' = \alpha(q)$ given by

$$\frac{d}{d\alpha'}[f(\alpha') - \alpha' q]|_{\alpha' = \alpha(q)} = 0,$$

provided that

$$\frac{d^2}{d(\alpha')^2}[f(\alpha') - \alpha' q]|_{\alpha' = \alpha(q)} < 0,$$

or

$$f'(\alpha(q)) = q, \tag{9.6a}$$

$$f''(\alpha(q)) < 0. \tag{9.6b}$$

We then have from (9.5) that

$$I(q,\varepsilon) \simeq \exp\{[f(\alpha(q)) - q]\ln(1/\varepsilon)\} \int d\alpha' \rho(\alpha')\varepsilon^{1/2 f''[\alpha' - \alpha(q)]^2}$$

$$\sim \exp\{[f(\alpha(q)) - q\alpha(q)]\ln(1/\varepsilon)\},$$

which, when inserted in (9.3), yields

$$D_q = \frac{1}{q-1}[q\alpha(q) - f(\alpha(q))]. \tag{9.7}$$

Thus, if we know $f(\alpha)$, then we can determine D_q from Eqs. (9.6a) and (9.7). In particular, for each value of α, Eq. (9.6a) determines the corresponding q. Substituting these q- and α-values in (9.7) gives the value of D_q corresponding to the determined value of q. Varying α, we therefore obtain a parametric specification of D_q. To proceed in the other direction, we multiply (9.7) by $q - 1$, differentiate with respect to q, and use (9.6a), to obtain

$$\alpha(q) = \frac{d}{dq}[(q-1)D_q] = \tau'(q), \tag{9.8a}$$

where we have introduced $\tau'(q) \equiv d\tau/dq$ and

$$\tau(q) = (q-1)D_q. \tag{9.8b}$$

Thus, if D_q is given, $\alpha(q)$ can be determined from (9.8a), and then $f(\alpha)$ can be determined from (9.7),

$$f(\alpha(q)) = q\frac{d}{dq}[(q-1)D_q] - (q-1)D_q$$

$$= q\tau'(q) - \tau(q). \tag{9.9}$$

For each value of q, (9.8) and (9.9) give a value of α and the corresponding $f(\alpha)$, thus parametrically specifying the function $f(\alpha)$.

As an example, consider the attractor of the generalized baker's map (Figure 3.4). In this case D_q is the solution of the transcendental equation,[2]

$$\tilde{\alpha}^q \lambda_a^{(1-q)(D_q-1)} + \tilde{\beta}^q \lambda_b^{(1-q)(D_q-1)} = 1. \tag{3.23}$$

This gives (we assume for definiteness that $\ln\tilde{\beta}/\ln\lambda_b > \ln\tilde{\alpha}/\ln\lambda_a$)

$$D_\infty = 1 + \frac{\ln\tilde{\alpha}}{\ln\lambda_a},$$

$$D_1 = 1 + \frac{\tilde{\alpha}\ln\tilde{\alpha} + \tilde{\beta}\ln\tilde{\beta}}{\tilde{\alpha}\ln\lambda_a + \tilde{\beta}\ln\lambda_b},$$

$$D_{-\infty} = 1 + \frac{\ln\tilde{\beta}}{\ln\lambda_b}.$$

Figure 9.1(a) shows a plot of D_q versus q (recall that D_q is a nonincreasing function of q (Eq. (3.16))). This dependence is typical for a hyperbolic attractor with a multifractal measure. Figure 9.1(b) shows the corresponding $f(\alpha)$.

From (9.8), $\alpha(q)$ increases from $\alpha_{min} = D_\infty$ to $\alpha_{max} = D_{-\infty}$ as q decreases from $+\infty$ to $-\infty$. From (9.1) we associate $\alpha = \alpha_{max}$ with the most rarefied regions of the measure, while we associate $\alpha = \alpha_{min}$ with the most concentrated regions of the measure. By (9.6a) the slope of $f(\alpha)$ versus α is infinite at $\alpha = D_{\pm\infty}$. We also note the concave down shape of the $f(\alpha)$

Figure 9.1(a) D_q versus q for the generalized baker's map, and (b) $f(\alpha)$ versus α.

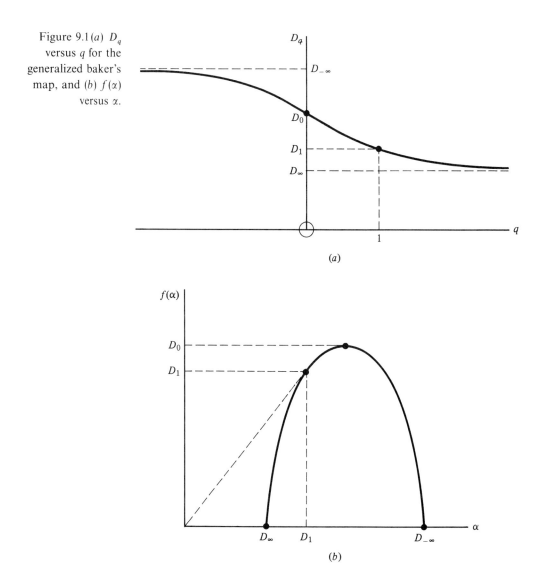

curve in accordance with (9.6b). The maximum value of $f(\alpha)$ occurs at $q = 0$ (Eq. (9.6a)), and at this point $f(\alpha) = D_0$ (Eq. (9.9)). That the maximum value of $f(\alpha)$ is D_0 is very reasonable since this is the dimension of the set A which supports the measure; any of the subsets on which $D_p(\mathbf{x}) = \alpha$ are contained within A and hence cannot have a dimension that exceeds that of A. Finally, we note that at $q = 1$ we have from (9.7) that $f(\alpha) = \alpha$, from (9.6a) that $f'(\alpha) = 1$, and from (9.8) that $\alpha = D_1$. Thus, the straight line $f(\alpha) = \alpha$ (shown dashed in Figure 9.1) is tangent to the $f(\alpha)$ curve at the point $f(\alpha) = \alpha = D_1$.

The fact that α ranges from $D_\infty = 1 + \ln \tilde{\alpha}/\ln \lambda_a$ to $D_{-\infty} = 1 + \ln \tilde{\beta}/\ln \lambda_b$ can be understood by noting that iteration of the unit square n times produces 2^n vertical strips of varying widths $\lambda_a^m \lambda_b^{n-m}$, each of which contain a measure $\tilde{\alpha}^m \tilde{\beta}^{n-m}$ (see Section 3.5). Thus, using (9.1), we obtain a singularity index,

$$\alpha = 1 + \frac{\ln(\tilde{\alpha}^m \tilde{\beta}^{n-m})}{\ln(\lambda_a^m \lambda_b^{n-m})} = 1 + \frac{(m/n)\ln \tilde{\alpha} + (1 - m/n)\ln \tilde{\beta}}{(m/n)\ln \lambda_a + (1 - m/n)\ln \lambda_b}, \quad (9.10)$$

for a square of edge length $\varepsilon = \lambda_a^m \lambda_b^{m-n}$ centered in one of the vertical strips of width $\lambda_a^m \lambda_b^{n-m}$. As m/n increases from 0 to 1, the singularity index α given by (9.10) decreases monotonically from $1 + \ln \tilde{\beta}/\ln \lambda_b = D_{-\infty}$ to $1 + \ln \tilde{\alpha}/\ln \lambda_a = D_\infty$ (recall that we assume $\ln \tilde{\beta}/\ln \lambda_b > \ln \tilde{\alpha}/\ln \lambda_a$).

The $f(\alpha)$ spectrum has proven useful for characterizing multifractal chaotic attractors and has been determined experimentally in a number of studies. For example, Jensen *et al.* (1985) determine $f(\alpha)$ for the attractor of an experimental forced Rayleigh–Bénard system. The experiment is done at parameter values corresponding to quasiperiodic motion at the golden mean rotation number and a critical nonlinearity strength such that the golden mean torus is at the borderline of being destroyed. Jensen *et al.* compare their experimental results to results for the circle map, Eq. (6.11), choosing $k = 1$ (corresponding to the borderline of torus destruction) and adjusting ω to obtain the golden mean rotation number. They find that $f(\alpha)$ is, to within experimental accuracy, the same for the Rayleigh–Bénard experiment and for the circle map. They take this as evidence that there is a type of universal global orbit behavior for systems at the borderline of torus destruction.

The relationships between $f(\alpha)$ and α, on the one hand, and D_q and q, on the other hand, are suggestive of the relationship between the free energy and the entropy in thermodynamics. Indeed, a rather complete formal analogy, based upon the partition function formalism to be discussed in the next section, can be constructed (e.g., see Badii (1987), Fujisaka and Inoue (1987), Bohr and Tél (1988), Tél (1988), and Mori *et al.* (1989) and references therein). Introducing the standard quantities of thermodynamics,

Table 9.1. *Analogy between thermodynamics and the multifractal formalism.*

Thermodynamics	Multifractal formalism
$\hat{\beta}\mathscr{F}$	$\tau(q)$
$\mathscr{U}(\hat{\beta})$	$\alpha(q)$
$\mathscr{S}(\mathscr{U})$	$f(\alpha)$
$\mathscr{U} = \dfrac{\mathrm{d}(\hat{\beta}\mathscr{F})}{\mathrm{d}\hat{\beta}}$	$\alpha(q) = \dfrac{\mathrm{d}\tau(q)}{\mathrm{d}q}$
$\mathscr{S} = \hat{\beta}(\mathscr{U} - \mathscr{F})$	$f(\alpha) = q\alpha - \tau(q)$
$\dfrac{\mathrm{d}\mathscr{S}}{\mathrm{d}\mathscr{U}} = \hat{\beta}$	$\dfrac{\mathrm{d}f(\alpha)}{\mathrm{d}\alpha} = q$

$\mathscr{F} \equiv$ the free energy,
$\mathscr{T} \equiv$ the temperature (in energy units),
$\mathscr{U} \equiv$ the internal energy per unit volume,
$\mathscr{S} \equiv$ the entropy,

we have the analogy given by Table 9.1, where $\hat{\beta} = 1/\mathscr{T}$.

In thermodynamics, phase transitions are manifested as nonanalytic dependences of the thermodynamic quantities. In particular, at a first-order phase transition, the free energy $\mathscr{F}$ has a discontinuous derivative with variation of $\hat{\beta}$ (i.e., with variation of the temperature $\mathscr{T}$). From Table 9.1 we see that the analogous phenomenon in the multifractal formalism[3] would thus be a discontinuity of $\mathrm{d}\tau(q)/\mathrm{d}q$ (equivalently, $\mathrm{d}D_q/\mathrm{d}q$; see Eq. (9.8)) with variation of q. A very simple example of a phase transition in the multifractal formalism comes from considering the logistic map (2.10) at $r = 4$ (Ott *et al.*, 1984b). In this case, we have seen that the natural invariant measure results in a density (Eq. (2.13)),

$$\rho(x) = \frac{1}{\pi[x(1 - x)]^{1/2}}, \qquad (9.11)$$

$0 \leq x \leq 1$. Dividing the interval $[0, 1]$ into k intervals, $(i/k, (i + 1)/k]$, of equal size $\varepsilon_i = 1/k$, we have

$$\mu_i = \int_{i/k}^{(i + 1)/k} \rho(x)\,\mathrm{d}x.$$

Using this in $\Sigma\,\mu_i^q$, it can be shown from the definition of D_q, Eq. (3.14), that (Problem 1)

$$D_q = \begin{cases} 1 & \text{for } q \leq 2, \\ q/[2(q - 1)] & \text{for } q \geq 2. \end{cases} \qquad (9.12)$$

Thus, D_q is continuous but has a discontinuous derivative at $q = q_T \equiv 2$ which we refer to as the phase transition point (Figure 9.2(a)). Utilizing (9.8) and (9.9) we see that for $q < q_T$, Eq. (9.12) yields $\alpha = 1$ and $f = 1$, while for $q > q_T$ we have $\alpha = \frac{1}{2}$ and $f = 0$ (see Figure 9.3(a)). The result $\alpha = 1$, $f = 1$ for $q < q_T$ corresponds to the fact that the singularity index for points in $0 < x < 1$ (a one-dimensional set) is 1 by virtue of the smooth variation of $\rho(x)$ in this range. The result $\alpha = \frac{1}{2}$ and $f = 0$ for $q > q_T$ corresponds to the fact that $\rho(x)$ has a singularity index of $\alpha = \frac{1}{2}$ at the two points $x = 0, 1$ (a set of dimension zero; hence $f = 0$). In Figure 9.3(a) we have joined the points $(\alpha, f) = (\frac{1}{2}, 0)$ and $(1, 1)$ by a straight line. The derivations of Eqs. (9.7)–(9.9) assume smooth $f(\alpha)$ with $f'' < 0$ (Eq. (9.6b)); this is not so for this case (and other phase transition cases as well) for which a more careful treatment of the evaluation of the integral (9.5) yields the dependence shown in Figure 9.3(a).

Figure 9.2 Schematic dependences of D_q on q showing phase transitions (discontinuities in the derivative of D_q) for (a) the logistic map at $r = 4$, and (b) the Hénon map.

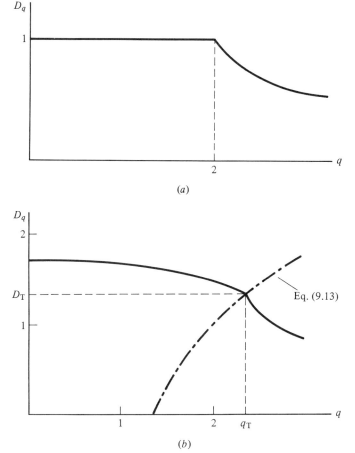

For hyperbolic two-dimensional maps dD_q/dq is expected to be real analytic (cf. Ruelle (1978)) and hence continuous (e.g., Figure 9.1) so that phase transitions do not occur in the hyperbolic case. We do, however, expect phase transitions for nonhyperbolic maps, such as the Hénon map. Indeed, the quadratic maximum of the logistic map (which is responsible for the $\alpha = \frac{1}{2}$ singularities at $x = 0, 1$) can, in some sense, be thought of as corresponding to the points of tangencies of the stable and unstable manifolds of the Hénon map (or similar such maps), and it is these tangencies which are responsible for the nonhyperbolicity. The logistic map at $r = 4$ is not, however, an entirely satisfactory indication of what to expect for typical nonhyperbolic two-dimensional maps.

A detailed consideration of the phase transition behavior of the Hénon map has been carried out by Cvitanović *et al.* (1988), Grassberger *et al.* (1988) and Ott *et al.* (1989a) with the results indicated in Figures 9.2(b) and 9.3(b). The phase transition point q_T occurs at approximately $q_T \simeq 2.3$ at which the following relationship is satisfied

$$(q_T - 1)D_T = 2q_T - 3 \tag{9.13}$$

Figure 9.3 Schematics of the dependences of $f(\alpha)$ on α for the case of (a) the logistic map at $r = 4$ and (b) the Hénon map.

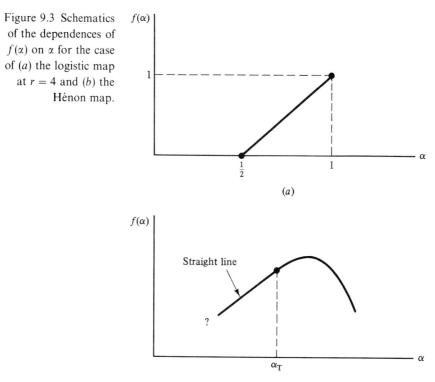

(Grassberger *et al.*, 1988; Ott *et al.*, 1989a). We see (Figure 9.2(*b*)) that D_q decreases smoothly with increasing q in both $q < q_T$ and $q > q_T$ but has a discontinuous decrease in its derivative at $q = q_T$. The singularity spectrum $f(\alpha)$ in Figure 9.3(*b*) has the same characteristic concave down shape ($f''(\alpha) < 0$) as for the generalized baker's map (Figure 9.1(*b*)), but only in the range $\alpha > \alpha_T$. For α just below α_T, there is a range where $f(\alpha)$ is a straight line joining the curve for $\alpha > \alpha_T$ with $df(\alpha)/d\alpha$ continuous. The behavior of $f(\alpha)$ at still smaller α is not at present understood. The range $q < q_T$ in Figure 9.2(*b*) corresponds to the range $\alpha > \alpha_T$ in Figure 9.3(*b*). As in the case of hyperbolic attractors, $f(\alpha)$ in $\alpha > \alpha_T$ (but *not* in $\alpha < \alpha_T$) may be interpreted as the box-counting dimension of the set of points with $D_p(\mathbf{x}) = \alpha$. We, therefore, refer to $\alpha > \alpha_T$ and $q < q_T$ as the 'hyperbolic range.'

9.2 The partition function formalism

For the purposes of this section an understanding of the Hausdorff definition of the dimension of a set will be essential. Therefore, those readers not already familiar with the Hausdorff dimension should refer to the appendix to Chapter 3 where it is described. As in that appendix (and using the same notation), we consider a set A and imagine covering it by a collection of subsets, $S_1, S_2, \ldots, S_N$, such that the diameters of these subsets are all less than or equal to some number δ,

$$|S_i| \leq \delta.$$

We denote this collection of subsets $\{S_i\}$. Let μ be a measure on the set A, and let μ_i be the μ measure of the subset S_i,

$$\mu_i = \mu(S_i).$$

We define the *partition function* (Grassberger, 1985; Halsey *et al.*, 1986),

$$\Gamma_q(\tilde{\tau}, \{S_i\}, \delta) \equiv \sum_{i=1}^{N} \mu_i^q / \varepsilon_i^{\tilde{\tau}}, \qquad (9.14)$$

where ε_i denotes the diameter of S_i or $\varepsilon_i = |S_i|$. Note that (9.14) reduces to the quantity Γ_H^d in Eq. (3.51) if we set $q = 0$ and $\tilde{\tau} = -d$ in (9.14). (The term 'partition function' is motivated by the thermodynamic analogy.) It will later result that, in the limit of large N, the partition function behaves as in Figure 3.15; i.e., it makes an abrupt transition from $+\infty$ to zero with variation of $\tilde{\tau}$. Furthermore, we use the transition point $\tilde{\tau} = \tilde{\tau}(q)$ to define a dimension quantity $\tilde{D}_q$ via $\tilde{\tau}(q) = (q - 1)\tilde{D}_q$. Thus, the partition function given by (9.14) is a natural generalization of the quantity Γ_H^d used in defining the Hausdorff dimension of a set. In particular, for $q = 0$, our Eq. (9.14) shows that the quantity $\tilde{\tau}$ is $-D_H$ at the transition, where D_H is the Hausdorff dimension of A. Thus, $\tilde{D}_0 = D_H$.

Two regions will be considered,

$$\text{Region I: } q \geq 1, \quad \tilde{\tau} \geq 0, \tag{9.15}$$

$$\text{Region II: } q \leq 1, \quad \tilde{\tau} \leq 0. \tag{9.16}$$

In region I we choose the set of coverings $\{S_i\}$ so as to maximize the partition function (subject to the constraint that $|S_i| = \varepsilon_i < \delta$). In region II we choose the covering to minimize the partition function (again subject to the constraint that $\varepsilon_i < \delta$). Thus, we define

$$\Gamma_q(\tilde{\tau}, \delta) = \sup_{S_i} \Gamma_q(\tilde{\tau}, \{S_i\}, \delta) \quad \text{(in region I)}, \tag{9.17a}$$

$$\Gamma_q(\tilde{\tau}, \delta) = \inf_{S_i} \Gamma_q(\tilde{\tau}, \{S_i\}, \delta) \quad \text{(in region II)}. \tag{9.17b}$$

Next, we take the limit of δ going to zero and define

$$\Gamma_q(\tilde{\tau}) = \lim_{\delta \to 0} \Gamma_q(\tilde{\tau}, \delta). \tag{9.18}$$

As in the definition of the Hausdorff dimension, given in the appendix to Chapter 3, the quantity $\Gamma_q(\tilde{\tau})$ makes a transition between $+\infty$ and zero as $\tilde{\tau}$ varies through some critical value which we denote $\tilde{\tau}(q)$,

$$\Gamma_q(\tilde{\tau}) = \begin{cases} +\infty & \text{for } \tilde{\tau} > \tilde{\tau}(q), \\ 0 & \text{for } \tilde{\tau} < \tilde{\tau}(q). \end{cases} \tag{9.19}$$

(The transition as $\tilde{\tau}$ *increases* through $\tilde{\tau}(q)$ is from zero to $+\infty$ (rather than from $+\infty$ to zero) because, as is evident from (9.14), $\Gamma_q(\tilde{\tau}, \{S_i\}, \delta)$ increases monotonically with $\tilde{\tau}$.)

We define the dimension $\tilde{D}_q$ as

$$(q - 1)\tilde{D}_q = \tilde{\tau}(q). \tag{9.20}$$

Figure 9.4 shows a schematic illustrating Eqs. (9.19). (For comparison with Figure 3.15, recall that the quantity d in Figure 3.5 is analogous to $-\tilde{\tau}$.)

In the appendix to Chapter 3 we have found that D_0 is an upper bound on D_H,

$$D_0 \geq D_H. \tag{3.53}$$

However, we have also found an example (namely, the generalized baker's map) for which (3.53) holds with the equality applying. It is conjectured on this basis and on the basis of other examples that $D_0 = D_H$ for typical invariant sets arising in dynamics. For the quantities $\tilde{D}_q$ an analogous result may be shown (Problem 4). Namely, D_q (the dimension defined by Eq. (3.14) using a covering from a grid of *equal* size cubes) provides an upper bound on $\tilde{D}_q$,

$$D_q \geq \tilde{D}_q. \tag{9.21}$$

As in the case of the Hausdorff dimension and D_0, it is found that (9.21) is satisfied with the equality applying for calculable examples. On the assumption that this holds in general for the invariant sets of interest to us in dynamics, we shall henceforth drop the tildes on $\tilde{D}_q$ and on $\tilde{\tau}$ and $\tilde{\tau}(q)$. Granted the above assumption, we now have two methods of obtaining D_q. This can be a great advantage, since one method or the other may be more useful depending on the situation.

As an example, we now use the partition function formalism to calculate D_q for the natural measure of the chaotic attractor of the generalized baker's map. As before (Eq. (3.18)), we write D_q as

$$D_q = 1 + \hat{D}_q,$$

where $\hat{D}_q$ represents the dimension of the measure projected onto the x-axis. Let $\hat{\Gamma}_q$ represent the partition function for this projected measure, and express $\hat{\Gamma}_q(\tau,\delta)$ as

$$\hat{\Gamma}_q(\tau,\delta) = \hat{\Gamma}_{qa}(\tau,\delta) + \hat{\Gamma}_{qb}(\tau,\delta), \tag{9.22}$$

where $\hat{\Gamma}_{qa}$ is the contribution to $\hat{\Gamma}_q$ from the interval $0 \le x \le \lambda_a$, and $\hat{\Gamma}_{qb}$ is the contribution to $\hat{\Gamma}_q$ from $(1 - \lambda_b) \le x \le 1$. If we magnify the interval $0 \le x \le \lambda_a$ by a factor $1/\lambda_a$ we get a replica of the attractor measure in the entire interval $0 \le x \le 1$. Furthermore, since the attractor measure[2] in $0 \le x \le \lambda_a$ is $\tilde{\alpha}$, using the definition Eq. (3.14), we have the following scaling

$$\hat{\Gamma}_{qa}(\tau,\delta) = \tilde{\alpha}^q \lambda_a^{-\tau} \hat{\Gamma}_q(\tau,\delta/\lambda_a). \tag{9.23a}$$

Similarly, we have for the interval $(1 - \lambda_b) \le x \le 1$,

$$\hat{\Gamma}_{qb}(\tau,\delta) = \tilde{\beta}^q \lambda_b^{-\tau} \hat{\Gamma}_q(\tau,\delta/\lambda_b). \tag{9.23b}$$

To proceed, we adopt the ansatz that for small δ

$$\hat{\Gamma}_q(\tau,\delta) \simeq K\delta^{\hat{\tau}(q)-\tau}, \tag{9.24}$$

where $\hat{\tau}(q) = (q-1)\hat{D}_q = (q-1)(D_q - 1)$. (Equation (9.24) is motivated by an analysis of the case of coverings by a grid of cubes, and is analogous to the argument leading to Eq. (3.52). Note, in particular, that, in the limit

Figure 9.4 Schematic of $\Gamma_q(\tilde{\tau})$ versus τ.

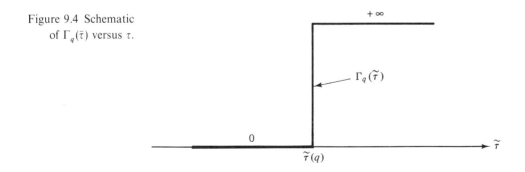

$\delta \to 0$, Eq. (9.24) yields the dependence shown in Figure 9.4.) Combining (9.22)–(9.24), we obtain the following transcental equation for $\hat{\tau}(q)$

$$1 = \tilde{\alpha}^q \lambda_a^{-\hat{\tau}(q)} + \tilde{\beta}^q \lambda_b^{-\hat{\tau}(q)}, \qquad (9.25)$$

which, using $\hat{\tau}(q) = (q - 1)\hat{D}_q$, is identical with Eq. (3.23). Thus, for this example, we have confirmed that the partition function Eq. (9.14) yields the same D_q as the box-counting definition of D_q given by Eq. (3.14).

9.3   Lyapunov partition functions

We have previously seen that it is possible to derive relations between the Lyapunov exponents and the information dimension D_1 of both chaotic attractors (Eq. (4.36)–(4.38)) and nonattracting chaotic sets (Eqs. (5.14) and (5.16)). It is natural to ask whether a similar (possibly not as simple) relationship can be obtained giving the spectrum of dimensions D_q. In this section we shall show that, using the partition function formalism of Section 9.2, this can indeed be done.

We consider a hyperbolic *nonattracting* chaotic invariant set of an invertible two-dimensional map $\mathbf{M}(\mathbf{x})$. We regard this nonattracting chaotic set as the intersection of a Cantor set of one-dimensional stable manifold lines with a similar Cantor set of unstable manifold lines. We call such an invariant set a 'strange saddle' (see the schematic illustration in Figure 5.23). The results for a chaotic *attractor* (Badii and Politi, 1987; Morita et al., 1987) can be obtained as a special case of the strange saddle results (Ott et al., 1989b) by letting $\langle \tau \rangle$, the average decay time[4] for orbits on the strange saddle (defined by (5.12)), approach infinity, $\langle \tau \rangle \to \infty$. (See also Kovács and Tél (1990) for the case of chaotic scattering.)

We begin by recalling the situation studied in Section 5.5 where we considered a strange saddle and defined natural measures, μ_s and μ_u, on the stable and unstable manifolds of the strange saddle[5] (Eqs. (5.13) and (5.15)). We shall be interested in developing results for the dimension spectra D_{qs} of the measure μ_s and D_{qu} of the measure μ_u.

We first derive a formula for the unstable measure μ_u and then use it to estimate the partition function (9.14). As in Section 5.5, we imagine that the strange saddle is contained in some box B (see Figure 5.23). Let $\lambda_1(\mathbf{x}, n) > 1 > \lambda_2(\mathbf{x}, n)$ denote the magnitudes of the eigenvalues of $\mathbf{DM}^n(\mathbf{x})$. Note that the n dependence of $\lambda_1(\mathbf{x}, n)$ is an approximate exponential increase with n, while that of $\lambda_2(\mathbf{x}, n)$ is an exponential decrease with n. An approximation to the unstable manifold in B is obtained by iterating points in B forward in time n steps and seeing which of these forward iterates have not yet left B. This approximation to the unstable manifold is (assuming that points which leave B never return)

$$B \cap \mathbf{M}^n(\mathbf{B})$$

and is illustrated in Figure 9.5(a) where we take B to be a rectangle of dimensions l_1 by l_2 and assume the stable manifold to run vertically and the unstable manifold to run horizontally. Iteration of B forward in time many iterates ($n \gg 1$) results in many long thin horizontal strips which contain the unstable manifold. These rectangles are the set $B \cap \mathbf{M}^n(B)$. Those initial points $\mathbf{x}_i$ which remain in B for n iterates, are contained in the nth preimage of the set $B \cap \mathbf{M}^n(B)$,

$$\mathbf{M}^{-n}(B \cap \mathbf{M}^n(B)) = \mathbf{M}^{-n}(B) \cap B. \tag{9.26}$$

Recall that, from the definition of the decay time $\langle \tau \rangle$, the Lebesgue measure (area) of $\mathbf{M}^{-n}(B) \cap B$ decays as

$$\mu_{\mathrm{L}}(\mathbf{M}^{-n}(B) \cap B) \sim \exp(-n/\langle \tau \rangle), \tag{9.27}$$

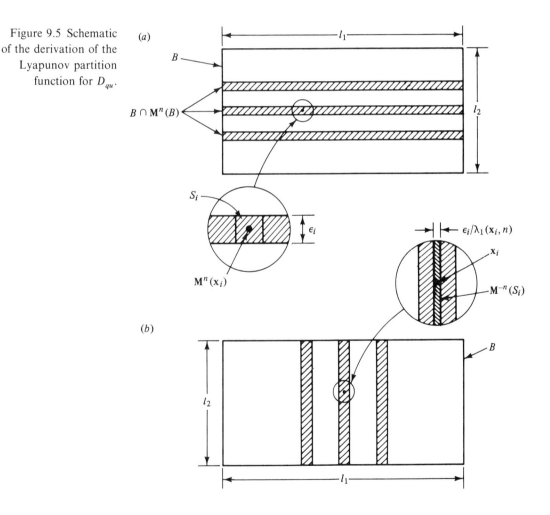

Figure 9.5 Schematic of the derivation of the Lyapunov partition function for D_{qu}.

where μ_L denotes Lebesgue measure. We cover the set $B \cap \mathbf{M}^n(B)$ by a covering $\{S_i\}$ of squares of side $\varepsilon_i \simeq l_2 \lambda_2(\mathbf{x}_i, n)$ where $\mathbf{x}_i$ denotes the initial point whose nth iterate $\mathbf{M}^n(\mathbf{x}_i)$ is in the center of S_i. See Figure 9.5(a). Here we have chosen ε_i to be equal to the local width of the horizontal strip containing $\mathbf{x}_i$ (this is indicated in the magnification in Figure 9.5(a)). Now taking the small square S_i and iterating it backward n iterates (to see where it came from), it goes to a very narrow vertical strip $\mathbf{M}^{-n}(S_i)$ contained within one of the many narrow strips constituting $\mathbf{M}^{-n}(B) \cap B$ (see the magnification shown in Figure 9.5(b)). The width of this very narrow strip $\mathbf{M}^{-n}(S_i)$ is

$$\varepsilon_i / \lambda_1(\mathbf{x}_i, n) \approx l_2 \lambda_2(\mathbf{x}_i, n) / \lambda_1(\mathbf{x}_i, n),$$

and its length is l_1. The unstable measure inside the small square S_i is seen from the definition (5.15) to be

$$\mu_u(S_i) \equiv \frac{\mu_L(\mathbf{M}^{-n}(S_i) \cap B)}{\mu_L(\mathbf{M}^{-n}(B) \cap B)}$$

$$\simeq \exp\left(\frac{n}{\langle \tau \rangle}\right) \lambda_2(\mathbf{x}_i, n)[\lambda_1(\mathbf{x}_i, n)]^{-1},$$

where use has been made of (9.27). Inserting $\varepsilon_i \cong l_2 \lambda_2(\mathbf{x}_i, n) \sim \lambda_2(\mathbf{x}_i, n)$ and the estimate for the measure $\mu_u(S_i)$ in (9.14) we obtain the Lyapunov partition function for the unstable manifold measure μ_u

$$\Gamma_{qL}^u = \sum_i \left[\exp\left(\frac{n}{\langle \tau \rangle}\right) \frac{\lambda_2(\mathbf{x}_i, n)}{\lambda_1(\mathbf{x}_i, n)}\right]^q \lambda_2^{-\tau}(\mathbf{x}_i, n)$$

$$= \exp(qn/\langle \tau \rangle) \sum_i [\lambda_2(\mathbf{x}_i, n)]^{q-\tau}[\lambda_1(\mathbf{x}_i, n]^{-q}.$$

This can be regarded as a Riemann sum (whose area elements are squares of side $\lambda_2(\mathbf{x}_i, n)$) for the integral

$$\exp(qn/\langle \tau \rangle) \int_{B \cap \mathbf{M}^n(B)} [\lambda_2(\mathbf{x}, n)^{q-\tau-2} \lambda_1^{-q}(\mathbf{x}, n) \mathrm{d}^2\mathbf{y},$$

where $\mathbf{y} \equiv \mathbf{M}^n(\mathbf{x})$. Noting that the magnitude of the determinant of the Jacobian matrix of $\mathbf{M}^n(\mathbf{x})$ is $\lambda_1(\mathbf{x}, n)\lambda_2(\mathbf{x}, n)$, we can change the variable of integration in the above integral from $\mathbf{y}$ to $\mathbf{x}$ to obtain

$$\Gamma_{q\Gamma}^u(D, n) = \exp(qn/\langle \tau \rangle) \int_{\mathbf{M}^{-n}(B) \cap B} [\lambda_2^{D-1}(\mathbf{x}, n)\lambda_1(\mathbf{x}, n)]^{1-q} \mathrm{d}^2\mathbf{x}, \quad (9.28)$$

where we have substituted $\tau \equiv (q-1)D$. For the case of a hyperbolic attractor $\langle \tau \rangle = \infty$, and we may replace the domain of integration by any finite area subset $\bar{B}$ of the basin of attraction. Thus, in this case (9.28) can be replaced by

$$\Gamma_{qL}(D, n) = \int_{\bar{B}} [\lambda_2^{D-1}(\mathbf{x}, n)\lambda_1(\mathbf{x}, n)]^{1-q} \mathrm{d}^2\mathbf{x}. \quad (9.29a)$$

Alternatively, we can also write

$$\Gamma_{qL}(D, n) = \langle [\lambda_2^{D-1}(\mathbf{x}, n)\lambda_1(\mathbf{x}, n)]^{1-q} \rangle, \qquad (9.29\text{b})$$

where the angle brackets indicate an average with respect to Lebesgue measure on $\bar{B}$ or, equivalently, an average over the natural measure on the attractor. (In obtaining (9.29b) from (9.29a) we have dropped a factor $[\mu_L(\bar{B})]^{-1}$, since this does not affect the result for D_q.) Also, note that in (9.29), we have dropped the superscript u with the understanding that Γ_{qL} refers to the chaotic attractor (which we assume is the same as its unstable manifold; see Chapter 4).

To find D_{qu}, we recall that $\varepsilon_i \sim \lambda_2(\mathbf{x}, n)$ goes exponentially to zero as $n \to +\infty$. Thus, we let $n \to +\infty$ in the Lyapunov partition function and set D_{qu} equal to that value of D at which the resulting limit transitions from 0 to $+\infty$. Note that our procedure in obtaining the Lyapunov partition function has not involved the optimization with respect to the covering set specified by Eqs. (9.17). Thus, at this stage, we can only conclude that the q-dimension value obtained from the Lyapunov partition function is an upper bound on the true D_{qu}. Nevertheless, as before, we shall assume that the true D_{qu} assumes this upper bound and we shall see that this is supported by application of the Lyapunov partition function to the example of the generalized baker's map.

An important special case occurs when the map in question has a Jacobian determinant that is constant; i.e., det $\mathbf{DM}(\mathbf{x})$ is independent of $\mathbf{x}$ (e.g., for the Hénon map det $\mathbf{DM}(\mathbf{x}) = -B$). Letting J denote this constant, we have

$$\lambda_1(\mathbf{x}, n)\lambda_2(\mathbf{x}, n) = |J|^n,$$

which can be used to eliminate $\lambda_2(\mathbf{x}, n)$ from (9.28) and (9.29),

$$\Gamma_{qL}^u(D, n) = \exp\left\{n\left[\frac{q}{\langle\tau\rangle} + (D-1)(1-q)\ln|J|\right]\right\}$$
$$\int_{\mathbf{M}^{-n}(B)\cap B} [\lambda_1(\mathbf{x}, n)]^\sigma \, d^2\mathbf{x} \qquad (9.28')$$

and

$$\Gamma_{qL}(D, n) = |J|^{n(D-1)(1-q)} \langle [\lambda_1(\mathbf{x}, n)]^\sigma \rangle, \qquad (9.29')$$

respectively. Here

$$\sigma \equiv (2 - D)(1 - q).$$

One of the advantages of the Lyapunov partition function formulation is that it allows the dimension spectrum to be calculated numerically without many of the problems associated with box counting (in particular, the large memory requirements, and the necessity of computing long orbits, when small box sizes are considered). Numerical estimation of

(9.28) proceeds as follows. Take $N(0)$ points uniformly chosen from B, and iterate each n times. Take the $N(n)$ points remaining in B after n iterations, and, for each one, identify its initial condition $\mathbf{x}_i$. Then calculate λ_{1i} and λ_{2i}, the magnitudes of the eigenvalues of $\mathbf{DM}^n(\mathbf{x}_i)$. We then estimate Γ_{qL}^u as

$$\Gamma_{qL}^u(D, n) \simeq \frac{[N(0)]^{q-1}}{[N(n)]^q} \sum_{i=1}^{N(n)} (\lambda_{1i} \lambda_{2i}^{D-1})^{1-q}. \qquad (9.30)$$

This numerical procedure and its obvious modification for the case of attractors have been carried out with good results for D_{qu} in a number of papers.

We now argue heuristically that (9.29) reduces to the Kaplan–Yorke formula in the limit that $q \to 1$. Expanding (9.29) for small $(1 - q)$ we have

$$\Gamma_{qL} \simeq 1 + (1 - q)\{(D - 1)\langle \ln \lambda_2(\mathbf{x}, n)\rangle + \langle \ln \lambda_1(\mathbf{x}, n)\rangle\} + O[(1 - q)^2].$$

Recall the definition of the Lyapunov exponents,

$$h_{1,2}(\mathbf{x}) = \lim_{n \to \infty} \frac{1}{n} \ln \lambda_{1,2}(\mathbf{x}, n),$$

and that $h_{1,2}(\mathbf{x})$ assumes the same value denoted $h_{1,2}$ (i.e., with the argument $\mathbf{x}$ deleted) for almost every $\mathbf{x}$ with respect to the natural measure. Since $h_{1,2}(\mathbf{x}) = h_{1,2}$ for almost every $\mathbf{x}$, its average with respect to the natural measure must also be $h_{1,2}$,

$$h_{1,2} = \langle h_{1,2}(\mathbf{x})\rangle = \lim_{n \to \infty} \frac{1}{n} \langle \ln \lambda_{1,2}(\mathbf{x}, n)\rangle.$$

This yields

$$\Gamma_{qL} \simeq 1 + n(1 - q)\{(D - 1)h_2 + h_1\} + O[(1 - q)^2].$$

For large n the second term becomes large, unless we set the term contained within the curly brackets to zero. Thus, we suspect that the value of D at which this occurs coincides with the value of D at which Γ_{qL} makes a transition from 0 to infinity in the $q \to 1$ limit. Setting the curly bracketed term to zero we obtain

$$D_1 = 1 + (h_1/|h_2|),$$

which is our previously obtained Eq. (4.38). (We assume $h_1 > 0 > h_2$.)

So far we have only been discussing the partition function for the dimension D_{qu} of the measure μ_u on the unstable manifold. For the case of the stable manifold, refer to Figure 5.5(b), and note that the vertical strips converge to the stable manifold as $n \to \infty$. The width of a vertical strip of initial points which do not leave B on n iterates is of the order of $\lambda_1^{-1}(\mathbf{x}_i, n)$ for $\mathbf{x}_i$ in the vertical strip. Thus, taking $\varepsilon_i \sim \lambda_1^{-1}(\mathbf{x}_i, n)$ and $\mu_s(S_i) \sim \varepsilon_i^2 \exp(n/\langle \tau \rangle)$ (with the S_i now covering the vertical strips), and proceeding as in the case of the unstable manifold, we obtain

$$\Gamma^s_{qL}(D, n) = \exp(nq/\langle \tau \rangle) \int_{\mathbf{M}^{-n}(B) \cap B} [\lambda_1(\mathbf{x}, n)]^{(D-2)(q-1)} d^2\mathbf{x}. \quad (9.31)$$

For the case of a chaotic attractor we see that $\Gamma^s_{qL} = 1$ for $D = 2$ and thus $D_{qs} \equiv 2$. This corresponds to the fact that almost all points in the basin of attraction go to the attractor; i.e., the basin of attraction is two-dimensional.

As an example, we now apply the Lyapunov partition function formalism to the case of the chaotic attractor of the generalized baker's map. As shown in Chapter 3, application of the map n times to the units square results in 2^n vertical strips of varying widths $\lambda_a^m \lambda_b^{n-m}$ ($m = 0, 1, 2, \ldots, n$), and the number of strips of width $\lambda_a^m \lambda_b^{n-m}$ is the binomial coefficient $Z(n, m)$. A point $\mathbf{x}_i$ for which $\mathbf{M}^n(\mathbf{x}_i)$ is in a strip of width $\lambda_a^m \lambda_b^{n-m}$ has

$$\lambda_1(\mathbf{x}_i, n) = \left(\frac{1}{\tilde{\alpha}}\right)^m \left(\frac{1}{\tilde{\beta}}\right)^{n-m},$$

$$\lambda_2(\mathbf{x}_i, n) = \lambda_a^m \lambda_b^{n-m}.$$

Applying $\mathbf{M}^{-n}$ to the vertical strip of width $\lambda_a^m \lambda_b^{n-m}$, we find the region of initial conditions that yield this vertical strip. This region is a horizontal strip of width $\tilde{\alpha}^m \tilde{\beta}^{n-m}$; see Figure 9.6. If we choose $\mathbf{x}_i$ randomly in the unit square, the probability of it falling in the initial strip of width $\tilde{\alpha}^m \tilde{\beta}^{n-m}$ is just the area of that strip, which is also $\tilde{\alpha}^m \tilde{\beta}^{n-m}$. Thus, using (9.29b), we have

$$\langle [\lambda_2^{D-1}(\mathbf{x}, n) \lambda_1(\mathbf{x}, n)]^{1-q} \rangle = \sum_{m=0}^{n} Z(n, m) \tilde{\alpha}^m \tilde{\beta}^{n-m}$$

$$\left[(\lambda_a^m \lambda_b^{n-m})^{D-1} \left(\frac{1}{\tilde{\alpha}}\right)^m \left(\frac{1}{\tilde{\beta}}\right)^{n-m} \right]^{1-q}$$

$$= \left(\frac{\tilde{\alpha}^q}{\lambda_a^{\tilde{\tau}}} + \frac{\tilde{\beta}^q}{\lambda_b^{\tilde{\tau}}} \right)^n,$$

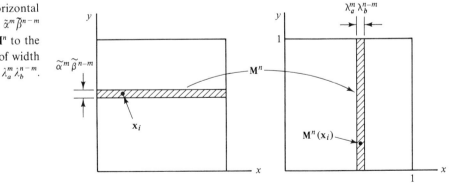

Figure 9.6 Horizontal strip of width $\tilde{\alpha}^m \tilde{\beta}^{n-m}$ maps under $\mathbf{M}^n$ to the vertical strip of width $\lambda_a^m \lambda_b^{n-m}$.

where we have used the basic property of the binomial coefficient $\Sigma_{m=0}^{n} Z(n,m) x^m y^{n-m} \equiv (x+y)^n$. Letting $n \to \infty$, we see that Γ_{qL} goes from zero to $+\infty$ at precisely the point where the previous result for the dimension, Eq. (9.25), holds.

In this section we have utilized the stretching properties of the dynamical system to obtain a partition function for the dimension spectrum on a chaotic attractor. It may be of interest to note that a very similar situation arises in the mixing of an impurity in an incompressible fluid undergoing large scale smooth chaotic flow. In that case the gradient of the impurity density tends to concentrate on a fractal, and the multifractal properties of the gradient measure can be obtained from a Lyapunov partition function formulation related to that presented here. (Note that, since the fluid is incompressible, there is no chaotic attractor for the impurity *particles*; it is the *gradient*, and not the particles themselves, which concentrates on a fractal.) See Ott and Antonsen (1988, 1989) and Városi *et al.* (1991). An experiment displaying this phenomenon has been carried out by Ramshankar and Gollub (1991).

9.4   Distribution of finite time Lyapunov exponents

Define the *finite time Lyapunov exponent* for an initial condition $\mathbf{x}$ as

$$h_i(\mathbf{x}, n) = \frac{1}{n} \ln \lambda_i(\mathbf{x}, n). \tag{9.32}$$

In this section we shall only consider chaotic ($h_1 > 0$) invariant sets of two-dimensional maps, and we will restrict attention to the largest exponent, $i = 1$.

Fujisaka (1983) introduces a 'q-order entropy spectrum'

$$H_q = \frac{1}{1-q} \lim_{n \to \infty} \frac{1}{n} \langle \exp[n(1-q)h_1(\mathbf{x}, n)] \rangle, \tag{9.33}$$

where $\langle \cdots \rangle$ denotes an average over the relevant invariant measure. For $q \to 1$, by expanding the exponential and the logarithm for small $1 - q$, we have $\langle \exp[n(1-q)h_1(\mathbf{x}, n)] \rangle \simeq 1 + n(1-q)\langle h_1(\mathbf{x}, n) \rangle$ and $\ln \langle \exp[n(1-q)h_1(\mathbf{x}, n)] \rangle \simeq n(1-q)\langle h_1(\mathbf{x}, n) \rangle$. Thus, $H_1 = \lim_{n \to \infty} \langle h_1(\mathbf{x}, n) \rangle$, and H_1 is just the (infinite time) Lyapunov exponent applying for almost every $\mathbf{x}$ with respect to the measure,[6]

$$H_1 = h_1.$$

For hyperbolic attractors of two-dimensional maps with $h_1 > 0 > h_2$, we have that h_1 is the metric entropy of the measure, $h_1 = h(\mu)$ (see Section 4.5). In addition, one can argue on the basis of the result by Newhouse (1986) that for $q = 0$, Eq. (9.31) gives the topological entropy h_T,

$$H_0 = h_T.$$

Thus, for general values of the index q, we may regard the quantity H_q as generalizing the topological and metric entropies to a continuous spectrum of entropies. Note that the scaling with n of the Lyapunov partition function for the constant Jacobian determinant case is directly implied by a knowledge of H_q. In particular, from (9.29') the relevant quantity is the average of a power of the largest Lyapunov number,

$$\langle [\lambda_1(\mathbf{x}, n)]^\sigma \rangle = \langle \exp[n\sigma h_1(\mathbf{x}, n)] \rangle \sim \exp[n\sigma H_{1-\sigma}]. \tag{9.34}$$

Among other reasons, it is useful for calculating averages, such as that in (9.33), to introduce a distribution function for the finite time Lyapunov exponents $h_1(\mathbf{x}, n)$. We denote this distribution function $P(h, n)$. For $\mathbf{x}$ randomly chosen with respect to the invariant measure under consideration, the quantity $P(h, n)\,dh$ is, by definition, the probability that the finite time exponent $h_1(\mathbf{x}, n)$ falls in the range h to $h + dh$. Thus, for example

$$\langle \exp[n\sigma h_1(\mathbf{x}, n)] \rangle = \int \exp(n\sigma h) P(h, n)\,dh. \tag{9.35}$$

For hyperbolic sets with a one-dimensional unstable manifold through each point on the set, we can argue that $\lambda_1(\mathbf{x}, n)$ is produced by the multiplication of n scalar numbers (as opposed to matrix multiplications (4.31)), the n numbers being the expansion factors separating infinitesimally nearby points on the unstable manifold on each of the n iterates of the map. Thus, by (9.32), $h_1(\mathbf{x}, n)$ may be regarded as an average over n numbers (the logarithms of the expansion factors mentioned in the previous sentence). In accordance with the chaotic nature of the orbits, we regard these numbers as, in some sense, random. In this case, it follows that, for large n, the distribution function $P(h, n)$ is asymptotically of the general form (e.g., Ellis (1985)),

$$P(h, n) \sim \frac{1}{[2\pi n G''(\bar{h})]^{1/2}} \exp[-nG(h)], \tag{9.36}$$

where the minimum value of the function G is zero and occurs at $h = \bar{h}$; $G(\bar{h}) = 0$, $G'(\bar{h}) = 0$, $G''(\bar{h}) > 0$. See Figure 9.7. Note that expansion of G around $\bar{h}$ yields a normal distribution (this is the 'central limit theorem'),

$$P(h, n) \simeq \frac{1}{[2\pi n G''(\bar{h})]^{1/2}} \exp\left[-\frac{1}{2}nG''(\bar{h})(h - \bar{h})^2\right], \tag{9.36'}$$

where the standard deviation of h is $\xi = [nG''(\bar{h})]^{-1/2}$. Thus, for large n, Eq. (9.36') is an excellent approximation to $P(h, n)$ for values of h that are within several standard deviations from $\bar{h}$. We emphasize, however, that in calculating integrals such as (9.35), we shall see that the dominant contribution comes from values of h that deviate from $\bar{h}$ by an amount independent of n. Thus, for these h-values $(h - \bar{h})/\xi \sim n^{1/2}$. Hence, the dominant contribution comes from values of h that deviate from $\bar{h}$ by a number of standard deviations which approach infinity like $n^{1/2}$ as n goes to infinity. Under these circumstances, the normal distribution (9.36′) is inadequate for use in (9.35), and we must use the more accurate so-called 'large deviation form', Eq. (9.36). Note that the function $G(h_1)$ depends on the particular map and measure under consideration.

Note that for $n \to +\infty$, the distribution $P(h, n)$ becomes more and more peaked about $h = \bar{h}$ and becomes a delta function, $\delta(h - \bar{h})$ at $n = \infty$. Thus, almost every initial point with respect to the measure yields the exponent $\bar{h}$, and we therefore have that

$$\bar{h} = h_1.$$

We emphasize that, while $P(h, n)$ approaches a delta function, the finite n deviations of $P(h, n)$ from the delta function (as well as from the normal approximation (9.36′)) are crucial for the multifractional properties of the measure.

Let us now use (9.36) to obtain the large n scaling of $\langle [\lambda_1(\mathbf{x}, n)]^\sigma \rangle$. Substituting into (9.35), we have

$$\langle [\lambda_1(\mathbf{x}, n)]^\sigma \rangle \simeq \int \exp\{-n[G(h) - \sigma h]\} \frac{dh}{[2\pi n G''(h)]^{1/2}}.$$

For large n, the dominant contribution to this integral comes from the vicinity of the minimum of the function $G(h) - \sigma h$ occurring at $h = h_\sigma$ given by (cf. Figure 9.7)

$$G'(h_\sigma) = \sigma. \tag{9.37}$$

Figure 9.7 Schematic of the function $G(h)$.

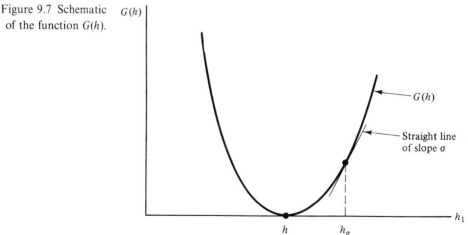

Thus, we have that

$$\langle [\lambda_1(\mathbf{x}, n)]^\sigma \rangle \sim \exp\{-n[G(h_\sigma) - \sigma h_\sigma]\}. \tag{9.38}$$

In terms of the quantity H_q, Eqs. (9.38) and (9.33) yield

$$H_q = [\sigma^{-1}G(h_\sigma) - h_\sigma]_{\sigma = 1 - q}.$$

We can also use (9.38) to obtain the dimension spectrum D_q from (9.29′) (Grassberger *et al.*, 1988). The Lyapunov partition function becomes

$$\Gamma_{qL}(D, n) \sim \exp\{n[(D - 1)(1 - q)\ln|J| - G(h_\sigma) + \sigma h_\sigma]\},$$

with $\sigma = (2 - D)(1 - q)$. Since Γ_{qL} goes to $+\infty$ or zero as $n \to \infty$ according to whether the term in square brackets in the exponential is positive or negative, we have that D_q is determined by requiring this term to be zero. Hence, we obtain the following transcendental equation for D_q,

$$(D_q - 1)(1 - q)\ln|J| = [G(h_\sigma) - \sigma h_\sigma]|_{\sigma = (2 - D_q)(1 - q)} \tag{9.39}$$

for the case where the determinant of the Jacobian matrix is a constant.

Equations (9.37)–(9.39) show that a knowledge of the function G is sufficient to obtain H_q and D_q. Thus, it is of interest to discuss how G can be determined numerically. In particular, we consider the case of a chaotic attractor and its natural measure. A possible procedure might be as follows. We randomly choose many initial conditions in the basin of attraction and iterate each one, say 100 times, so that the orbits are now on the attractor and are distributed in accord with the attractor's natural measure. We treat these as our new initial conditions and iterate each one, along with the corresponding tangent map, n times to obtain $h_1(\mathbf{x}_1, n)$. We then make a histogram approximation to $P(h, n)$. From (9.36) we have

$$G(h) \simeq -\frac{1}{h}\{\ln P(h, n) + \ln n + \tfrac{1}{2}\ln[2\pi G''(h)]\}. \tag{9.40}$$

As a first approximation we can neglect the last term for large n. Using the histogram approximation for P, Eq. (9.40) gives an approximation to G. Although we would like to make n very large, since (9.36) is only an asymptotic result for large n, in practice we are limited in how large we can make n. This is because, as n is made large the width of $P(h, n)$ narrows and more initial conditions are necessary to obtain good statistics (i.e., many $h_1(x_i, n)$-values in each histogram bin) away from $h = \bar{h}$. Thus, if n is too large, the computational cost becomes inordinate. Given that we obtain $G(h)$ at some 'reasonable' value of n (reasonable from the point of view of computational cost and of our feeling that n is sufficiently large), we can test how well the n-value used corresponds to the asymptotic regime by increasing n, say by 50%, and again calculating $G(h)$. In the asymptotic regime, we have that, according to (9.40), our result for the two numerical calculations of $G(h)$ should be essentially the same (independent of n). If this is so, then we have good indication that n is large enough. Numerical calculations along these lines have been done by Grassberger *et al.* (1988) and Városi *et al.* (1991). For example, in the case treated by the latter authors, reliable results for $G(h)$ were deemed to have been obtained for n-values between 30 and 50, and computer limitations prevented exploring values of n much above 50.

Finally, we wish to emphasize that our argument in obtaining (9.36) has invoked the assumption of hyperbolicity. It is of interest to note that for two-dimensional nonhyperbolic Hamiltonian maps, important modifications to (9.36) have been found (Sepúlveda *et al.*, 1989; Horita *et al.*, 1990) due to the KAM tori bounding the relevant ergodic chaotic region. Basically what happens is that chaotic orbits can wander close to the bounding KAM tori, and, when they do so, they can become stuck there for a long time. Since the Lyapunov exponents on the tori are $h_1 = h_2 = 0$ (hence nonhyperbolic), initial conditions $\mathbf{x}_i$ that lead to orbits which spend a very long time near the tori wind up having abnormally small values of $h_1(\mathbf{x}_i, n)$. This, in turn, results in an enhancement of $P(h, n)$ at low h and this enhancement has a temporal power law behavior qualitatively different from the time dependence in (9.36). We note, however, that this modification of (9.36) occurs only at low h and that the form (9.36) apparently remains valid for $h > \bar{h} = h_1$.

9.5 Unstable periodic orbits and the natural measure

In this section we show that the natural measure on a chaotic attractor of an invertible, hyperbolic, two-dimensional map can be expressed in terms of the infinite number of unstable periodic orbits embedded within the attractor. More specifically the principal result can be stated as follows. Let $\mathbf{x}_{jn}$ denote the fixed points of the n-times iterated map, $M^n(\mathbf{x}_{jn}) = \mathbf{x}_{jn}$, and let $\lambda_1(\mathbf{x}_{jn}, n)$ denote the magnitude of the expanding eigenvalue of the Jacobian matrix $\mathbf{DM}^n(\mathbf{x}_{jn})$. (Note that each $\mathbf{x}_{jn}$ is on a periodic orbit whose period is either n or a factor of n.) Then, the natural measure of an area S is given by

$$\mu(S) = \lim_{n \to \infty} \sum_{\mathbf{x}_{jn} \in S} \frac{1}{\lambda_1(\mathbf{x}_{jn}, n)} \qquad (9.41)$$

with the summation taking place over all fixed points of $\mathbf{M}^n$ in S. The interpretation of (9.41) is that, for large n, there is a small region about $\mathbf{x}_{jn}$ which covers a fraction $1/\lambda_1(\mathbf{x}_{jn}, n)$ of the natural measure, such that orbits originating from this small region closely follow the orbit originating from $\mathbf{x}_{jn}$ for n iterates. Before deriving (9.41) and demonstrating the above interpretation of it, we state some of the resulting consequences:

(i) If we let S cover the whole attractor ($\mu(S) \equiv 1$), then we obtain the result (Hannay and Ozorio de Almeida, 1984)

$$\lim_{n \to \infty} \sum_j \frac{1}{\lambda_1(\mathbf{x}_{jn}, n)} = 1, \tag{9.42}$$

where the sum is over all fixed points on the attractor. (This result is important for certain considerations of quantum chaos which we discuss in Chapter 10.)

(ii) The Lyapunov exponents are given by

$$h_{1,2} = \lim_{n \to \infty} \frac{1}{n} \sum_j \frac{1}{\lambda_1(\mathbf{x}_{jn}, n)} \ln \lambda_{1,2}(\mathbf{x}_{jn}, n). \tag{9.43}$$

(iii) The quantity H_q defined in Eq. (9.33) may be expressed as

$$H_q = \frac{1}{1-q} \lim_{n \to \infty} \frac{1}{n} \ln \left\{ \sum_j [\lambda_1(\mathbf{x}_{jn}, n)]^{-q} \right\}. \tag{9.44}$$

In particular, for $q = 0$ we obtain a result for the topological entropy (Katok, 1980)

$$h_{\mathrm{T}} = \lim_{n \to \infty} \frac{1}{n} \ln \bar{N}(n), \tag{9.45}$$

where $\bar{N}(n)$ denotes the number of fixed points of $\mathbf{M}^n$. Thus, the topological entropy gives the exponential increase of the number of fixed points of $\mathbf{M}^n$; i.e., $\bar{N}(n) \sim \exp(nh_{\mathrm{T}})$. (Referring to Eq. (9.42), we see that the number of terms in the sum in (9.42) increases exponentially with n, but that this exponential increase is exactly compensated for by the general exponential decrease of the terms $1/[\lambda_1(\mathbf{x}_j, n)]$ so that the resulting sum is 1.)

(iv) The Lyapunov partition function for a chaotic attractor, Eq. (9.29b), can be expressed in terms of the fixed points of $\mathbf{M}^n$ as

$$\Gamma_{q\mathrm{P}}(D, n) = \sum_j \frac{[\lambda_2(\mathbf{x}_{jn}, n)]^{(D-1)(1-q)}}{[\lambda_1(\mathbf{x}_{jn}, n)]^q}, \tag{9.46}$$

where we have replaced the subscript L (for Lyapunov) in (9.29) with the subscript P (for periodic orbits), and we call $\Gamma_{q\mathrm{P}}(D, n)$ the periodic orbits partition function. Again we claim that D_q is determined by the transition of the $n \to \infty$ limit of the periodic orbits partition function from 0 to ∞.

The above results show that a knowledge of the periodic orbits yields a great deal of information on the properties of the attractor. For more information on this topic see Katok (1980), Auerbach *et al.* (1987), Morita *et al.* (1987), Grebogi *et al.* (1987e, 1988b), and Cvitanović and Eckhardt (1991).

Equations (9.41)–(9.46) apply for chaotic attractors. Similar results can be obtained for nonattracting chaotic sets (Grebogi *et al.*, 1988b). For example, the result analogous to (9.42) is

$$\lim_{n \to \infty} \exp\left(\frac{n}{\langle \tau \rangle}\right) \sum_j \frac{1}{\lambda_1(\mathbf{x}_{jn}, n)} = 1,$$

which yields an expression for the average decay time in terms of the periodic orbits (Kadanoff and Tang, 1984),

$$\frac{1}{\langle \tau \rangle} = -\lim_{n \to \infty} \left\{ \frac{1}{n} \ln \left[\sum_j \frac{1}{\lambda_1(\mathbf{x}_{jn}, n)} \right] \right\}. \tag{9.47}$$

We now give a derivation of the result (Eq. (9.41)) for the measure of an attractor in terms of the periodic orbits (equivalently, the fixed points). Our treatment follows that of Grebogi *et al.* (1988b).

Imagine that we partition the space into cells C_i, where each cell has as

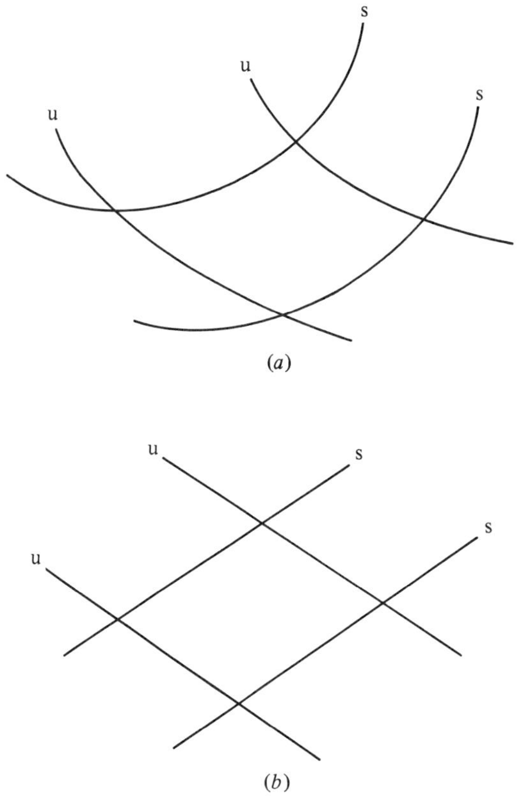

Figure 9.8 A cell in a partition of phase space. The letters s and u label stable and unstable manifold segments.

(a)

(b)

its boundaries stable and unstable manifold segments (Figure 9.8(a)). If the cells are taken to be small, the curvature of the boundaries will be slight and we can regard the cells as being parallelograms (Figure 9.8(b)). Consider a given cell C_k and sprinkle within it a large number of initial conditions distributed according to the natural measure of the attractor. Now imagine that we iterate these initial conditions n times. After n iterates, a small fraction of the initial conditions return to the small cell C_k. We assume the attractor to be ergodic and mixing. Thus, in the large n limit, the fraction of initial conditions that return is just the natural measure of the cell C_k, denoted $\mu(C_k)$.

Let $\mathbf{x}_0$ be an initial condition that returns and $\mathbf{x}_n$ its nth iterate. This is illustrated in Figure 9.9(a), where we take the stable direction as horizontal and the unstable direction as vertical. The line $ab(c'd')$ through $\mathbf{x}_0(\mathbf{x}_n)$ is a stable (unstable) manifold segment traversing the cell. Now take the nth forward iterate of ab and the nth backward iterate of $c'd'$. These map to $a'b'$ and cd as shown in Figure 9.9(b). Now consider a rectangle constructed by passing unstable manifold segments $e'f'$ through a' and $g'h'$ through b'. By the construction, the nth preimages of these segments are the unstable manifold segments ef and gh shown in Figure 9.9(c). Thus, we have constructed a rectangle $efgh$ in C_k such that all the points in $efgh$ return to C_k in n iterates. That is, $efgh$ maps to $e'f'g'h'$ in n iterates. The intersection of these two rectangles must contain a single saddle fixed point of the n times iterated map (cf. Figure 9.9(d)). Conversely, given a saddle fixed point, we can construct a rectangle of initial conditions $efgh$ which returns to C_k by closely following the periodic orbit which goes through the given fixed point j of $\mathbf{M}^n$ (the construction is the same as in Figures 9.9(a)–(c) except that $\mathbf{x}_0 = \mathbf{x}_n$). Thus, all initial conditions which return after n iterates lie in some long thin horizontal strip (like $efgh$) which contains a fixed point of the n times iterated map. We label this fixed point $\mathbf{x}_j$. Denoting the horizontal and vertical lengths of the sides of the cell C_k by ξ_k and η_k (cf. Figure 9.9(b)), we see that the initial strip $efgh$ has dimensions ξ_k by $[\eta_k/\lambda_1(\mathbf{x}_j, n)]$ and the final strip has dimensions $\xi_k \lambda_2(\mathbf{x}_j, n)$ by η_k (cf. Figure 9.9(d)). Since the dynamics is expanding in the vertical direction, the attractor measure varies smoothly in this direction. Since the cell is assumed small, we can treat the attractor as if it were essentially uniform along the vertical direction. Thus, the fraction of the measure of C_k occupied by the strip $efgh$ is $1/\lambda_{1j}$. Since, for $n \to \infty$, the fraction of initial conditions starting in C_k which return to it is $\mu(C_k)$, we have

$$\mu(C_k) = \lim_{n \to \infty} \sum_{\mathbf{x}_j \in C_k} \frac{1}{\lambda_1(\mathbf{x}_j, n)}. \tag{9.48}$$

Since we imagine that we can make the partition into cells as small as we wish, we can approximate any subset S of the phase space with reasonably smooth boundaries by a covering of such cells. Thus, we obtain the desired result,[7] Eq. (9.41).

9.6 Validity of the Lyapunov and periodic orbits partition functions for nonhyperbolic attractors

The arguments leading to our results for the Lyapunov partition function and for the periodic orbits partition function make use of the assumed hyperbolicity of the chaotic set. We note, however, that most attractors encountered in practice are not hyperbolic (e.g., the Hénon attractor; Figures 1.12 and 4.14), since they typically display tangencies between

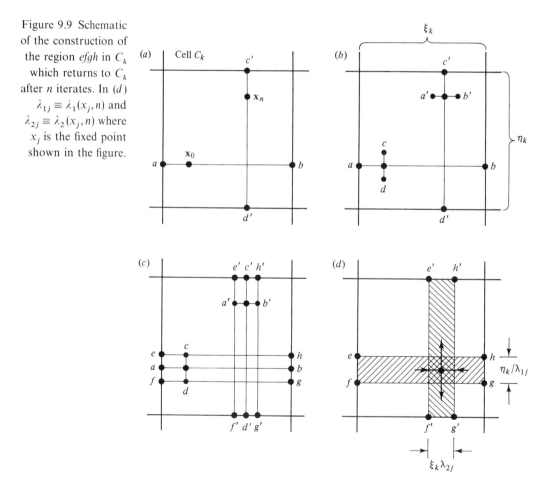

Figure 9.9 Schematic of the construction of the region $efgh$ in C_k which returns to C_k after n iterates. In (d) $\lambda_{1j} \equiv \lambda_1(x_j, n)$ and $\lambda_{2j} \equiv \lambda_2(x_j, n)$ where x_j is the fixed point shown in the figure.

their stable and unstable manifolds.[8] The question then arises as to what extent, if at all, the Lyapunov and periodic orbits partition functions can be used to obtain the D_q dimension spectra of typical nonhyperbolic two-dimensional maps such as the Hénon map. In this regard it has been conjectured (e.g., Grassberger *et al.* (1988), Ott *et al.* (1989a,b)) that, for typical nonhyperbolic attractors of two-dimensional maps, the Lyapunov and periodic orbits partition functions give the true D_q (i.e., that determined from the box-counting formula (3.14) or the partition function (9.14)) for values of q below the phase transition $q \leq q_T$. See Figure 9.2. (We have previously referred to $q \leq q_T$ as the 'hyperbolic range.')

An analytical example illustrating the above has been given by Ott *et al.* (1989b), who consider the invertible two-dimensional map

$$x_{n+1} = \begin{cases} \lambda x_n, & \text{for } y_n > \tfrac{1}{2}, \\ \tfrac{1}{2} + \lambda x_n, & \text{for } y \leq \tfrac{1}{2}, \end{cases} \tag{9.49a}$$

$$y_{n+1} = 4y_n(1 - y_n), \tag{9.49b}$$

where $\lambda < \tfrac{1}{2}$. For this map the true D_q is

$$D_q = \frac{2}{\ln(1/\lambda)} + \begin{cases} 1, & \text{for } q \leq q_T = 2, \\ \dfrac{q/2}{q-1}, & \text{for } q \leq q_T = 2. \end{cases} \tag{9.50}$$

Thus, D_q is the sum of the Cantor set dimension in x generated by (9.49a) (namely, $2/\ln(1/\lambda)$) and the logistic map D_q given by (9.12) and Figure 9.2(a). Ott *et al.* (1989a) also evaluate the dimension spectrum predicted by the periodic orbits partition function (we denote this prediction D_q') and the dimension spectrum predicted by the Lyapunov partition function (we denote this prediction D_q''). They obtain

$$D_q' = \frac{2}{\ln(1/\lambda)} + 1, \tag{9.51}$$

for all q, and

$$D_q'' = \begin{cases} \dfrac{\ln 2}{\ln(1/\lambda)} + 1, & \text{for } q \leq q_T, \\[2mm] \dfrac{1}{(q-1)}\left[\dfrac{\ln 2}{\ln(1/\lambda)} + 1 \right], & \text{for } q \geq q_T. \end{cases} \tag{9.52}$$

These results are illustrated in Figure 9.10. In accord with the conjecture, we see that (9.50)–(9.52) agree for $q \leq q_T$ but that they disagree outside this range. See, for example, Grassberger *et al.* (1988) and Ott *et al.* (1989a,b) for further discussion.

Problems

1. Derive Eq. (9.12).

2. Find D_q for the measure described in Problem 7 of Chapter 3 by using the partition function formalism. Sketch the corresponding $f(\alpha)$ labeling significant values on the vertical and horizontal coordinate axes.

3. Repeat Problem 2 above for the measure described in Problem 8 of Chapter 3.

4. Derive Eq. (9.21).

5. Taking the limit $q \to 1$, obtain Eq. (5.16) from Eq. (9.28). Similarly, use (9.31) to obtain (5.14).

6. Calculate H_q for the generalized baker's map.

7. Derive an expression for $P(h, n)$ for the generalized baker's map and show (using Stirling's approximation, Eq. (3.28)) that it is of the form given by Eq. (9.36). What is the function $G(h)$ for this example? Consider the case of constant determinant of the Jacobian matrix (namely $J = \lambda_a/\alpha = \lambda_b/\beta$), and verify that the equation for D_q in terms of G (Eq. (9.39)) agrees with the previously derived result, Eq. (3.23).

8. Using the symbolic dynamics representation to enumerate the periodic orbits of the generalized baker's map, obtain the periodic orbits partition function and verify that it yields Eq. (3.23).

Notes

1. Two review papers on this subject are those of Tél (1988) and Paladin and Vulpiani (1987).

2. To avoid confusion with the quantity α appearing as the argument of the singularity spectrum f, we shall henceforth replace the parameters α and β of the generalized baker's map (see Figure 3.4) by the symbols $\tilde{\alpha}$ and $\tilde{\beta}$.

Figure 9.10 Schematic plots of D_q, D_q' and D_q'' versus q for the map (9.49).

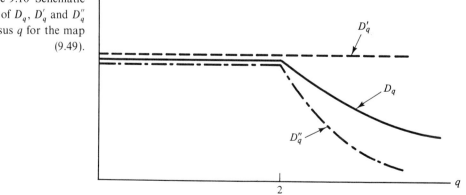

3. Phase transitions in the multifractal formalism have been discussed for example by Badii (1987), Cvitanović *et al.* (1988), Grassberger *et al.* (1988), Hata *et al.* (1987), Katzen and Procaccia (1987), Lopes (1990a,b), and Ott *et al.* (1989a).

4. Note that $\langle \tau \rangle$, the average decay time, is very different from the notationally similar quantity $\tau(q)$ appearing in (9.8).

5. It is suggested, at this point, that the reader refresh his memory by reviewing the material in Section 5.5.

6. Recall that we use the notation h_1 (i.e., without any arguments) to denote the value of $\lim_{n \to \infty} h_1(\mathbf{x}, n)$ assumed for almost every $\mathbf{x}$ with respect to the measure under consideration.

7. In the construction which we used in arriving at Eq. (9.48), we have made two implicit assumptions. Namely, we have assumed that the segment ab maps to a segment $a'b'$ which lies entirely within C_k, and we have assumed that the preimage of $c'd'$ is entirely within C_k. These situations might conceivably not hold if $\mathbf{x}_n$ is too close to a stable boundary or if $\mathbf{x}_0$ is too close to an unstable boundary. The point we wish to make here is that, for hyperbolic systems, the partition into cells can be chosen in such a way that $a'b'$ and $c'd'$ are always in C_k. See Grebogi *et al.* (1988b) for further discussion.

8. We note, however, that it is quite common for nonattracting chaotic sets encountered in practice to be hyperbolic. For example, this is apparently the case for the chaotic scattering invariant chaotic set shown in Figure 5.20(*a*). Note, from this figure, that the chaotic set appears to be the intersection of a Cantor set of stable manifold lines with a Cantor set of unstable manifold lines, and that these appear to intersect at angles that are well bounded away from zero, thus implying the absence of tangencies.

CHAPTER TEN

Quantum chaos

The description of physical systems via classical mechanics as embodied by Hamilton's equations (Chapter 7) may be viewed as an approximation to the exact description of quantum mechanics. Depending on the relevant time, length and energy scales appropriate to a given situation, one or the other of these descriptions may be the one that is most efficacious. In particular, if the typical wavelength in the quantum problem is very small compared to all length scales of the system, then one suspects that the classical description should be good. There is a region of crossover from the quantum regime to the classical regime where the wavelengths are 'small' but not extremely small. This crossover region is called the 'semiclassical' regime. In the semiclassical regime, we may expect quantum effects to be important, and we may also expect that the classical description is relevant as well. According to the 'correspondence principle,' quantum mechanics must go over into classical mechanics in the 'classical limit,' which is defined by letting the quantum wavelength approach zero. In a formal mathematical sense we can equivalently take the 'classical limit' by letting Planck's constant approach zero, $h \to 0$, with other parameters of the system held fixed. This limit is quite singular, and its properties are revealed by an investigation of the semiclassical regime. Particular interest attaches to the case where the classical description yields chaotic dynamics. In particular, one can ask, what implications does chaotic classical dynamics have for the quantum description of a system in the semiclassical regime? The field of study which addresses problems related to this question has been called quantum chaos. Quite apart from the fundamental problem of the correspondence principle, quantum chaos questions are of great practical importance because of the many physical systems that exist in the semiclassical regime.

The considerations we have been discussing above actually apply more generally than just to quantum wave equations. In Section 7.3.4 we have

seen how the short wavelength limit of a general wave equation yields a
system of ray equations which are Hamiltonian. The wave equation may
describe acoustic waves, elastic waves, electromagnetic waves, etc., or it
may be Schrödinger's equation for the quantum wavefunction. In the
latter case the rays are just the orbits of the corresponding classical
mechanics for the given Hamiltonian. Thus, the general question of the
short wavelength behavior of solutions of wave equations in relation to
solutions of the ray equations is essentially what is to be addressed in this
chapter.[1]

As an example, consider the case of a free point particle of mass m
bouncing in a closed hard-walled two-dimensional region with spatial
coordinates x and y (i.e., a 'billiard' system, as shown in Figure 7.24). In
this case Schrödinger's equation for the wavefunction $\psi_i(x, y)$, corresponding to a given energy level E_i, satisfies the Helmholtz equation,

$$\nabla^2 \psi_i + k_i^2 \psi_i = 0, \tag{10.1}$$

where $\nabla^2 = \partial^2/\partial x^2 + \partial^2/\partial y^2$ with $\psi_i(x, y) = 0$ on the walls. The solutions
of this problem give a discrete set of eigenfunctions ψ_i and corresponding
eigenvalues k_i in terms of which the energy levels are given by
$E_i = \hbar^2 k_i^2 / 2m$. Taking a typical wavelength to be $2\pi/k_i$, we see that the
semiclassical regime corresponds to large eigenvalues such that $k_i L \gg 1$,
where L is a typical scale size of the billiard (e.g., the radius of the end caps
for Figure 7.24(f)). Now say we consider the classical problem of an
electromagnetic wave in a two-dimensional cavity with perfectly conduc-
ting walls and shapes as shown in Figure 7.24. Taking the electric field to
be polarized in the z-direction, we again obtain Eq. (10.1) with ψ_i now
being the electric field and $k_i^2 = \omega_i^2/c^2$, where c is the speed of light and ω_i
is the resonant frequency of the ith cavity mode. Since the quantum
problem and the classical electromagnetic problem are mathematically
equivalent, knowledge of the short wavelength regime of one implies
knowledge of the short wavelength regime of the other. From now on our
discussion will be in the context of quantum mechanics.

We shall discuss three general classes of problems in this chapter:

Time-dependent Hamiltonians for systems with bounded orbits (i.e., the
orbits are confined to a finite region of phase space). An example of such a
problem is Eq. (10.1) applied to closed billiard domains.

Periodically driven systems in which the Hamiltonian is periodic in time.
An example of such a problem is the ionization of a hydrogen atom by a
microwave field.

Scattering problems in which the particle motions are unbounded (see
Section 5.4 for the classical case).

We conclude the introduction to this chapter by mentioning that more material on quantum chaos can be found in the books by Ozorio de Almeida (1988), Tabor (1989), Gutzwiller (1990), and Haake (1991), and in the review articles by Berry (1983), Eckhardt (1988b) and Jensen *et al.* (1991).

10.1 The energy level spectra of chaotic, bounded, time-independent systems

We consider systems that have a time-independent Hamiltonian so that the energy is a constant of the motion. We also assume that the classical orbits are bounded. That is, for any given energy of the system, all orbits remain within some sufficiently large sphere in the phase space (the radius of the sphere may increase with energy). To be specific, we consider the Hamiltonian

$$H(\mathbf{p},\mathbf{q}) = \frac{1}{2m}p^2 + V(\mathbf{q}), \tag{10.2}$$

where $\mathbf{p}$ and $\mathbf{q}$ are N-vectors and N is the number of degrees of freedom. The corresponding Schrödinger equation is

$$i\hbar\frac{\partial\psi}{\partial t} = -\frac{\hbar^2}{2m}\nabla^2\psi + V(\mathbf{q})\psi, \tag{10.3}$$

where ∇^2 now denotes the N-dimensional Laplacian,

$$\nabla^2 = \partial^2/\partial q_1^2 + \partial^2/\partial q_2^2 + \cdots + \partial^2/\partial q_N^2.$$

Assuming an energy eigenstate with energy level E_i, we have $\psi = \exp(-iE_i t/\hbar)\psi_i$, yielding the eigenvalue problem

$$\nabla^2\psi_i(\mathbf{q}) + \frac{2m}{\hbar^2}[E_i - V(\mathbf{q})]\psi_i(\mathbf{q}) = 0. \tag{10.4}$$

In accordance with our restriction to systems with bounded orbits of the classical Hamiltonian, we assume that (10.4) has a complete, *denumerable* set of eigenfunctions with corresponding energy levels. In this section we shall be interested in the behavior of the set of energy levels $\{E_i\}$ (i.e., the 'spectrum') in the semiclassical regime. In the next section (Section 10.2) we consider properties of the eigenfunctions. In both cases, particular interest will focus on the case where the classical dynamics is chaotic.

In our discussion of the spectra a key quantity will be the density of states $d(E)$ defined so that

$$\int_{E_a}^{E_b} d(E)\,\mathrm{d}E$$

is the number of states with energy levels between E_a and E_b. Thus,

$$d(E) = \sum_i \delta(E - E_i), \tag{10.5}$$

where we henceforth take the subscripts i such that

$$E_i \leq E_{i+1}.$$

That is, we label the levels in ascending order with respect to the numerical values of the E_i. Another quantity of interest will be the number of states with energies less than some value E,

$$N(E) = \int_{-\infty}^{E} d(E)\,\mathrm{d}E. \tag{10.6}$$

We call this quantity the cumulative density. Thus, $d(E)$ is a string of delta functions and $N(E)$ is a function which increases in steps of size one as E passes through each energy level, Figure 10.1. In the semiclassical limit one can also introduce a smoothed density of states,

$$\bar{d}(E) = \frac{1}{2\Delta} \int_{E-\Delta}^{E+\Delta} d(E)\,\mathrm{d}E, \tag{10.7}$$

and a corresponding smoothed cumulative density,

$$\bar{N}(E) = \int_{-\infty}^{E} \bar{d}(E)\,\mathrm{d}E. \tag{10.8}$$

The smoothing scale Δ will be taken to be much less than any typical energy of the *classical* system but much larger than $\hbar/T_{\min}$, where $T_{\min}$ is

Figure 10.1 The exact density function $d(E)$ and the exact cumulative density $N(E)$.

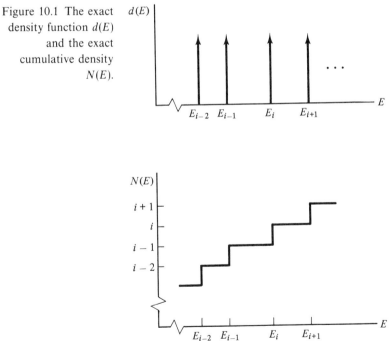

the shortest characteristic time for orbits of the classical problem. The reason for the restriction

$$\Delta \gtrsim h/T_{\min} \tag{10.9}$$

will become evident subsequently.

An expression for the average density of states $\bar{d}(E)$ is provided by Weyl's formula (see, for example, Gutzwiller (1990)) which is given as follows. The volume of classical phase space corresponding to system energies less than or equal to some value E is

$$v(E) = \int U(E - H(\mathbf{p}, \mathbf{q})) \, d^N \mathbf{p} \, d^N \mathbf{q}, \tag{10.10}$$

where $U(\ldots)$ denotes the unit step function:

$$U(x) \equiv 1 \text{ if } x > 0 \text{ and } U(x) \equiv 0 \text{ if } x \le 0.$$

Weyl's formula is equivalent to the statement that the average phase space volume occupied by a state is $(2\pi\hbar)^N$. Thus, the smoothed number of states with energies less than E is

$$\bar{N}(E) = v(E)/(2\pi\hbar)^N, \tag{10.11}$$

and, since the smoothed density of states is $\bar{d}(E) = d\bar{N}(E)/dE$, we have

$$\bar{d}(E) = \Omega(E)/(2\pi\hbar)^N, \tag{10.12a}$$

$$\Omega(E) = dv/dE = \int \delta(E - H(\mathbf{p}, \mathbf{q})) \, d^N \mathbf{p} \, d^N \mathbf{q}, \tag{10.12b}$$

where we have made use of the identity $\delta(x) = dU(x)/dx$. For example, for the special case of two-dimensional billiards, $H(\mathbf{p}, \mathbf{q}) = p^2/2m$ for $\mathbf{q}$ in the billiard region, and we thus obtain

$$v(E) = \pi p^2 A = 2\pi m E A, \tag{10.13a}$$

$$\Omega(E) = 2\pi m A, \tag{10.13b}$$

$$\bar{d}(E) = mA/2\pi\hbar^2, \tag{10.13c}$$

where A is the area of the billiard. More generally for a smooth potential we obtain from (10.12)

$$\Omega(E) = m\sigma_N \int_{E > V(\mathbf{q})} \{2m[E - V(\mathbf{q})]\}^{(N/2)-1} \, d^N \mathbf{q} \tag{10.13d}$$

for any number of degrees of freedom N. Here σ_N is the area of a sphere of radius 1 in N-dimensional space (e.g., $\sigma_3 = 4\pi$ for $N = 3$). (The Weyl result (10.13), is correct to leading order in $1/\hbar$, and, in the case of billiards, for example, higher order corrections that depend on the perimeter of the billiard and its shape have also been calculated. These semiclassically small corrections will be ignored.) We note from (10.12a) that the spacing between two adjacent states is typically $O(\hbar^N)$, and thus the restriction

(10.9) ensures that in the semiclassical regime for $N \geq 2$ there are many states in our smoothing interval Δ. Also, we see from (10.12) that $\Omega(E)$, and hence $\bar{d}(E)$, is nearly constant over the energy range Δ, since we have taken Δ to be classically small. We emphasize that the Weyl result applies if we examine $d(E)$ on the coarse scale Δ. If we were to examine $d(E)$ on a finer scale (i.e., smooth $d(E)$ over an energy range less than Δ), then the resulting smoothed $d(E)$ would fluctuate about $\bar{d}(E)$. (These fluctuations will be the object of interest in Section 10.1.2.)

10.1.1 The distribution of energy level spacings

Consider two adjacent energy levels E_i and E_{i+1}. Their difference, the 'energy level spacing,' $S_i = E_{i+1} - E_i$, averaged over i in a band Δ about a central energy E is just the inverse of the smoothed density of states, $1/\bar{d}(E)$. The spacings S_i fluctuate about this average. If we look at many S_i in such a band, we can compute a distribution function for the energy level spacings. Rather than do this directly, we 'normalize out' the density of states so that the resulting distribution function does not depend on $\bar{d}(E)$. Since $\bar{d}(E)$ is system-dependent, we hope that by such a normalization the resulting distribution function will be system-independent. That is, the distribution function will be universal for a broad class of systems. To this end we replace the spectrum $\{E_i\}$ by a new set of numbers $\{e_i\}$ defined by

$$e_i = \bar{N}(E_i). \tag{10.14}$$

Here $\bar{N}(E)$ is given by (10.11). The numbers e_i by definition have an average spacing of 1. Thus, we can think of the set $\{e_i\}$ as a set of normalized energies for which the smoothed density is 1. Letting $s_i = e_{i+1} - e_i$, we seek the universal distribution $P(s)$ such that for a randomly chosen i, the probability that $s \leq s_i \leq s + ds$ is $P(s)\,ds$.

For the case of a semiclassical regime quantum system whose Hamiltonian yields a classical integrable system, Berry and Tabor (1977) show that $P(s)$ is universally the same and is a Poisson distribution,

$$P(s) = \exp(-s). \tag{10.15}$$

The next natural question to ask is, is there a universal distribution in the case where the classical mechanics is chaotic? Actually, this question has to be refined somewhat. In particular, in the absence of special symmetries, there is a difference in the distribution $P(s)$ between time-reversible classical dynamics (such as for the Hamiltonian $H = p^2/2m + V(\mathbf{q})$ corresponding to motion in a potential) and non-time-reversible classical dynamics (such as the case where one considers charged particle motion in the presence of a static magnetic field). A further complication arises in the 'mixed case' where both chaotic and KAM orbits are present in the phase space (e.g., Figure 7.13). In fact the problem of characterizing the universal properties of $P(s)$ in the mixed case is currently rather unsettled (for examples treating this case see Seligman *et al.* (1984) and Bohigas *et al.* (1990)). Thus, we shall restrict our discussion to the case of completely chaotic systems (Section 7.5) such as the billiards shown in Figures 7.24(d)–(g).

First we consider the time-reversible case as in the Schrödinger equation, Eq. (10.4). Introducing a set of real orthonormal basis functions $\{u_j(\mathbf{q})\}$, we can write $\psi_i(\mathbf{q}) = \Sigma_j c_j^{(i)} u_j(\mathbf{q})$, in terms of which we obtain the infinite matrix eigenvalue problem

$$\mathbf{H}\mathbf{c}_i = \lambda_i \mathbf{c}_i, \tag{10.16}$$

where $\mathbf{c}_i = (c_1^{(i)}, c_2^{(i)}, \ldots)$, the matrix $\mathbf{H}$ has elements $H_{lm} = -\int u_l(\mathbf{q})[\nabla^2 - (2m/h^2)V(\mathbf{q})]u_m(\mathbf{q})\,\mathrm{d}^N\mathbf{q}$, and $\lambda_i = 2mE_i/\hbar^2$. The matrix H_{lm} is clearly real and symmetric, $H_{lm} = H_{ml}$. As an example of a non-time-reversible case, consider the Hamiltonian for a charged particle of charge Q in a potential $V(\mathbf{q})$ but with a static magnetic field added. The Hamiltonian is now $(\mathbf{p} - Q\mathbf{A})^2/2m + V(\mathbf{q})$, where $\mathbf{A}(\mathbf{q})$ is the magnetic vector potential. Correspondingly, the Schrödinger equation (10.4) is replaced by $\nabla^2\psi - (iQ/\hbar)[\mathbf{A}\cdot\nabla\psi + \nabla\cdot(\mathbf{A}\psi)] + (2m/\hbar^2)[E - \tilde{V}(\mathbf{q})]\psi = 0$, where $\tilde{V}(\mathbf{q}) \equiv V(\mathbf{q}) + (Q^2/2m)\mathbf{A}^2(\mathbf{q})$. The big difference is that without the magnetic field the wave equation is real, whereas with the magnetic field the wave equation becomes complex. Introducing a basis, we again obtain (10.16), but now the matrix $\mathbf{H}$ is Hermitian with complex off-diagonal elements, $H_{lm} = H_{ml}^*$ where H_{ml}^* denotes the conjugate of H_{ml}.

In nuclear physics one typically has to deal with very complicated interacting systems. In 1951 Wigner introduced the idea that the energy level spectra of such complicated nuclear systems could be treated statistically, and he introduced a conjecture as to how this might be done. In particular, he proposed that the spectra of complicated nuclear systems have similar statistical properties to those of the spectral of ensembles of random matrices. That is, we take the matrix $\mathbf{H}$ of Eq. (10.16) to be drawn at random from some collection (the 'ensemble'). In order to specify the ensemble, it has been proposed that the following two statistical conditions on the probability distribution of the ensemble of matrices should be satisfied.

(1) *Invariance.* Physical results should be independent of the set of basis functions $\{u_i(\mathbf{q})\}$ used to derive (10.16). Thus, the probability distribution for the elements of $\mathbf{H}$ should be invariant to orthogonal transformations of $\mathbf{H}$ for the case of a time-reversible system and should be invariant to unitary transformations for the non-time-reversible case. That is, the probability distribution of matrix

elements, $\tilde{P}(\mathbf{H})$, should be unchanged if $\mathbf{H}$ is replaced by $\mathbf{O}^\dagger \mathbf{H} \mathbf{O}$ with $\mathbf{O}$ an orthogonal matrix, or by $\mathbf{U}^{*\dagger} \mathbf{H} \mathbf{U}$ with $\mathbf{U}$ a unitary matrix, in the reversible and non-reversible cases, respectively. Here the symbol $\dagger$ stands for transpose. (An orthogonal matrix is defined by $\mathbf{O}^\dagger = \mathbf{O}^{-1}$ with $\mathbf{O}$ real, while a unitary matrix satisfies $\mathbf{U}^{*\dagger} = \mathbf{U}^{-1}$.)

(2) *Independence*. The matrix elements are independent random variables. The distribution function $\tilde{P}(\mathbf{H})$ for the matrix $\mathbf{H}$ is then the product of distributions for the individual elements H_{lm} for $l \le m$ (the elements for $l > m$ are implied by the symmetry of $\mathbf{H}$ about its diagonal).

These two hypotheses can be shown to imply Gaussian distributions for each of the individual H_{lm} in the time-reversible case and for $\mathrm{Re}\,(H_{lm})$ and $\mathrm{Im}\,(H_{lm})$ in the non-time-reversible case. The widths for these Gaussians are the same for all the diagonal elements, while all the widths for the Gaussian distribution of the off-diagonal ($l \ne m$) elements are half of the widths for the diagonal elements. The ensemble for the time-reversible case is called the Gaussian orthogonal ensemble (GOE), while the ensemble in the nontime-reversible case is called the Gaussian unitary ensemble (GUE). (There is a third relevant type of ensemble, the Gaussian symplectic ensemble, which we shall not discuss here.) A detailed theory for the GOE and GUE matrix ensembles exists (Mehta (1967); see also Bohr and Mottelson (1969)) and yields the following results for the level spacing distributions,

$$P(s) \simeq (\pi/2)s \exp(-\pi s^2/4), \text{ for the GOE case,} \qquad (10.17a)$$

$$P(s) \simeq (32/\pi) \exp[-(4/\pi)s^2], \text{ for the GUE case.} \qquad (10.17b)$$

The results on the right-hand sides of (10.17a) and (10.17b) are exact for two by two matrices, and are also very good approximations to the results for the situation of interest here, namely, M by M matrices in the $M \to \infty$ limit.

Another result of the GOE and GUE random matrix theory is expressions for the 'spectral rigidity', $\Delta_{sr}(L)$. The spectral rigidity is defined as the mean square deviation of the best local fit straight line to the staircase cumulative spectral density over a normalized energy scale L,

$$\Delta_{sr}(L) = \min_{A,B} \left\{ \frac{\bar{d}(E)}{L} \int_{-L/2\bar{d}(E)}^{+L/2\bar{d}(E)} [N(E + \varepsilon) - (A + B\varepsilon)]^2 \, d\varepsilon \right\}.$$

The random matrix theory results are

$$\Delta_{sr}(L) = \frac{1}{\pi^2} \ln L + K_1, \text{ for the GOE case,} \qquad (10.18a)$$

$$\Delta_{sr}(L) = \frac{1}{2\pi^2} \ln L + K_2, \text{ for the GUE case,} \qquad (10.18b)$$

where K_1 and K_2 are constants. In the case of a Poisson process (appropriate to integrable systems),

$$\Delta_{sr}(L) = L/15. \tag{10.18c}$$

As we have already said, random matrix theory (e.g., Eqs. (10.17) and (10.18)) was originally motivated by the study of nuclei. More recently it has also been proposed that random matrix theory applies to the semiclassical spectra of quantum problems that are classically completely chaotic. This is, in a sense, a departure from the nuclear situation, since now the system can be quite simple (e.g., billiards). While there are some suggestive theoretical results supporting the random matrix conjecture for quantum chaos (e.g., see the review by Yukawa and Ishikawa (1989) and the paper on spectral rigidity by Berry (1985)), the validity of this conjecture and its range of applicability, if valid, remain unsettled. The main support for it comes from numerical experiments where some striking agreement with the conjecture is obtained (McDonald and Kaufman, 1979; Bohigas *et al.*, 1984; Seligman and Verbaarschot, 1985). Figure 10.2 shows a histogram approximation to $P(s)$ obtained by Bohigas *et al.* (1984) by numerically solving the Helmholtz equation in the billiard shape shown in the upper right inset. The classical motion in this shape is chaotic, and the resulting histogram appears to agree well with the GOE result (10.17a) shown as the solid curve in the figure, but is very different from the Poisson distribution shown in the figure as the dashed curve. Bohigas *et al.* also obtain excellent agreement of their numerically calculated spectral rigidity with the random matrix prediction Eq. (10.18a).

In applying the GOE or the GUE statistics to a chaotic situation it is important that any discrete symmetries to the problem be taken into account. For example, the geometry of the billiard shown in Figure 7.24(*d*) has two diagonal symmetry lines, as well as a vertical and a horizontal symmetry line. Solutions of the Helmholtz equation in this billiard can be broken into classes according to whether they are even or odd about the symmetry lines. The problem shown in the inset of Figure 10.2 with the wavefunction set to zero on the boundaries corresponds to the symmetry class of solutions of the Helmholtz equation in the full billiard of Figure 7.24(*d*) that are odd about the symmetry lines. The random matrix ensemble is constructed assuming no symmetries. Thus, the GOE statistics conjecture is not taken to apply to the full spectrum for the billiard of Figure 7.24(*d*) but does apply to the spectrum restricted to any given symmetry class of the problem. For example, the odd–odd symmetry class of Figure 10.2 corresponds to the billiard shown in the inset, and this reduced billiard (which is $\frac{1}{8}$ of the original billiard area) has no symmetries.

Figure 10.2 Numeric-
ally obtained
histogram of $P(s)$ for
the Helmholtz
equation solved in the
region shown in the
upper right inset
($R = 0.2$) compared
with the GOE result
Eq. (10.17a) (solid
curve) and the Poisson
distribution, Eq.
(10.15) (dashed curve)
(Bohigas *et al.*, 1984).

As seen in Figure 10.2, the principal gross qualitative difference
between the level spacing distributions for integrable and chaotic systems
is that $P(s)$ goes to zero as $s \to 0$ for the chaotic cases, but has its maximum
at $s = 0$ for the integrable case. This is a manifestation of the phenomenon
of 'level repulsion' for nonintegrable systems. In particular, say we
consider the variation of energy levels as a function of a parameter of the
system. Then the situation is as shown in Figure 10.3. In the integrable
case (Figure 10.3(*a*)), levels cross creating degeneracies at the crossings. In
the nonintegrable case, degeneracies are typically avoided (Figure
10.3(*b*)); the levels 'repel.' Thus, there is a tendency against having small
s-values in the nonintegrable case, and $P(0) = 0$.

For a chaotic problem with discrete symmetries, each symmetry class
yields an *independent* problem of the form (10.16). Thus, energy levels
from one class do not 'know' about energy levels of a different class, and
two such levels will, therefore, not repel each other. Hence, even though

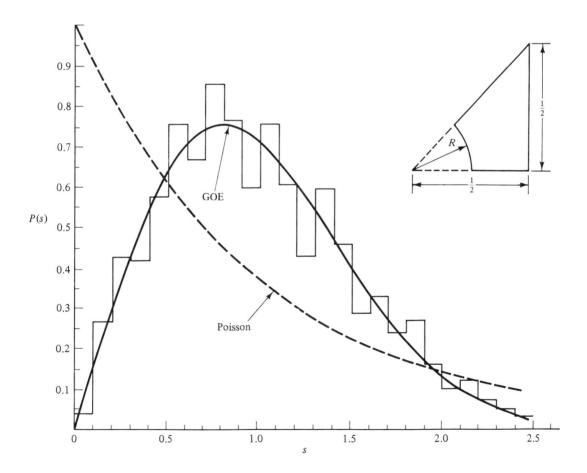

the problem is chaotic, crossings of levels as in Figure 10.3(a) typically occur, if the spectra of different symmetry classes are not separated.

10.1.2 Trace formula

An important and fundamental connection between the classical mechanics of a system and its semiclassical quantum wave properties is provided by trace formulae originally derived by Gutzwiller (1967, 1969, 1980) and by Balian and Bloch (1970, 1971, 1972).

We consider the Green function for the quantum wave equation corresponding to a classical Hamiltonian $H(\mathbf{p}, \mathbf{q})$,

$$H(-i\hbar\nabla, \mathbf{q})G(\mathbf{q}, \mathbf{q}'; E) - EG(\mathbf{q}, \mathbf{q}'; E) = -\delta(\mathbf{q} - \mathbf{q}'). \quad (10.19)$$

The eigenfunctions and eigenvalues (energy levels) satisfy

$$H(i\hbar\nabla, \mathbf{q})\psi_i(\mathbf{q}) = E_i\psi_i(\mathbf{q}), \quad (10.20)$$

and the orthonormality and completeness of the discrete set $\{\psi_i\}$ are respectively expressed by

$$\int \psi_i^*(\mathbf{q})\psi_j(\mathbf{q})\,\mathrm{d}^N\mathbf{q} = \delta_{ij}, \quad (10.21)$$

and

$$\sum_j \psi_j^*(\mathbf{q})\psi_j(\mathbf{q}) = \delta(\mathbf{q} - \mathbf{q}'). \quad (10.22)$$

Expressing G in terms of the complete basis $\{\psi_j\}$,

$$G(\mathbf{q}, \mathbf{q}'; E) = \sum_j c_j(\mathbf{q}', E)\psi_j(\mathbf{q}),$$

and using (10.21), Eq. (10.19) yields $(E_j - E)c_j(\mathbf{q}', E) = -\psi_j^*(\mathbf{q}')$ or

$$G(\mathbf{q}, \mathbf{q}'; E) = \sum_j \frac{\psi_j^*(\mathbf{q}')\psi_j(\mathbf{q})}{E - E_j}. \quad (10.23)$$

Figure 10.3 Behavior of two adjacent energy levels with variation of a system parameter (horizontal axis) for (a) an integrable case and (b) a nonintegrable case.

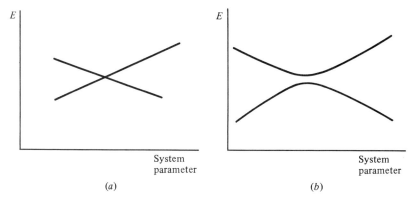

The above is singular as E passes through E_j for each j. To define this singularity we make use of causality. This leads to replacing E by $E + i\varepsilon$, where ε goes to zero through positive values (see Figure 10.4). That is,

$$\frac{1}{E - E_j} \to \lim_{\varepsilon \to 0^+} \frac{1}{(E + i\varepsilon) - E_j}$$

$$= P\frac{1}{E - E_j} - i\pi\delta(E - E_i), \qquad (10.24)$$

where $P(1/x)$ signifies that, when the function $1/x$ is integrated with respect to x, the integral is to be taken as a principal part integral at the singularity $x = 0$. (See standard texts on quantum mechanics for a discussion of Eq. (10.24).) Taking the imaginary part of (10.23) and using (10.24), we have

$$\text{Im } G(\mathbf{q}, \mathbf{q}'; E) = -\pi \sum_j \psi_j^*(\mathbf{q}')\psi_j(\mathbf{q})\delta(E - E_j),$$

which upon setting $\mathbf{q} = \mathbf{q}'$ and integrating yields

$$\text{Im } \int G(\mathbf{q}, \mathbf{q}; E) \, \mathrm{d}^N\mathbf{q} = -\pi \sum_j \delta(E - E_j).$$

Thus, from (10.5)

$$d(E) = -\frac{1}{\pi} \text{Im}[\text{Trace } (G)], \qquad (10.25)$$

where

$$\text{Trace } (G) \equiv \int G(\mathbf{q}, \mathbf{q}; E) \, \mathrm{d}^N\mathbf{q}.$$

Hence we obtain an exact formula for the density of states $d(E)$ in terms of the trace of the Green function.

We now wish to obtain a semiclassical approximation to G for use in (10.25). From now on we specialize to Hamiltonians of the form $H(\mathbf{p}, \mathbf{q}) = (p^2/2m) + V(\mathbf{q})$ for which the Green function satisfies the equation,

Figure 10.4 Illustration of Eq. (10.24) for integration over the real axis. (a) The term $1/[(E + i\varepsilon) - E_j]$ has a pole in E at $E_j - i\varepsilon$ labeled by the cross. (b) Letting $\varepsilon \to 0$ the integration path is deformed as shown. The term $P(E - E_j)^{-1}$ in (10.24) corresponds to the part of the path along the real axis, while the term $-i\pi\delta(E - E_i)$ results from integration around the infinitesmal semicircle skirting the pole.

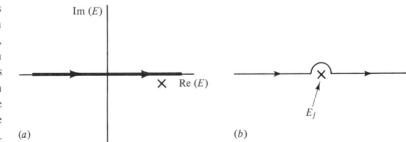

(a)   (b)

$$\left\{\nabla^2 + \frac{2m}{\hbar^2}[E - V(\mathbf{q})]\right\} G(\mathbf{q}, \mathbf{q}'; E) = \frac{2m}{\hbar^2} \delta(\mathbf{q} - \mathbf{q}'). \qquad (10.26)$$

Let $k' = \{(2m/\hbar^2)[E - V(\mathbf{q}')]\}^{1/2}$. We interpret $2\pi/k'$ as the wavelength of plane waves in the local region near the delta function. If $k'L \gg 1$, where L is a typical scale for the variation of $V(\mathbf{q})$ with $\mathbf{q}$, then one can choose a ball $\mathscr{R}$ in $\mathbf{q}$-space about the point $\mathbf{q} = \mathbf{q}'$ whose radius is small compared to L but is still many wavelengths across. In the region $\mathscr{R}$, the function $V(\mathbf{q})$ is nearly constant at the value $V(\mathbf{q}')$. Thus, to gain insight, consider the Green function G_0 for the case where $V(\mathbf{q})$ is constant everywhere at the value $V(\mathbf{q}')$:

$$[\nabla^2 + (k')^2]G_0(\mathbf{q}, \mathbf{q}'; E) = (2m/\hbar^2)\delta(\mathbf{q} - \mathbf{q}').$$

The solution of this problem is known for any number of degrees of freedom. In particular, for $N = 2$ and $N = 3$ we have

$$G_0(\mathbf{q}, \mathbf{q}'; E) = -\frac{\mathrm{i}}{4}\frac{2m}{\hbar^2} H_0^{(1)}(k'|\mathbf{q} - \mathbf{q}'|) \text{ for } N = 2, \qquad (10.27)$$

$$G_0(\mathbf{q}, \mathbf{q}'; E) = -\frac{2m}{\hbar^2}\frac{\exp(\mathrm{i}k'|\mathbf{q} - \mathbf{q}'|)}{4\pi|\mathbf{q} - \mathbf{q}'|} \text{ for } N = 3, \qquad (10.28)$$

where $H_0^{(1)}$ is the zero order Hankel function of the first kind. To interpret the $N = 2$ case we expand G_0 for $k'|\mathbf{q} - \mathbf{q}'| \gg 1$ (i.e., for observation points many wavelengths distant from the delta function source). The large argument approximation of $H_0^{(1)}$ is

$$H_0^{(1)}(k'|\mathbf{q} - \mathbf{q}'|) \sim \left(\frac{2}{\pi}\right)\exp\left(-\frac{\mathrm{i}\pi}{4}\right)\frac{\exp[\mathrm{i}k'|\mathbf{q} - \mathbf{q}'|]}{\sqrt{k'|\mathbf{q} - \mathbf{q}'|}}.$$

Thus, in both the cases $N = 2$ and $N = 3$, the Green function is of the form

$$G_0 \propto \frac{\exp[\mathrm{i}k'|\mathbf{q} - \mathbf{q}'|]}{|\mathbf{q} - \mathbf{q}'|^{(N-1)/2}}.$$

That is, G_0 is an outward propagating cylindrical ($N = 2$) or spherical ($N = 3$) wave originating from the point $\mathbf{q} = \mathbf{q}'$.

Thus, we have the following picture for the Green function. Since $k'|\mathbf{q} - \mathbf{q}'| \gg 1$ on the boundary of $\mathscr{R}$, waves leaving the region $\mathscr{R}$ may be thought of as local plane waves (the wavelength is much shorter than the radius of curvature of the wavefronts). Thus, the geometrical optics ray approximation (also called the eikonal approximation) is applicable for $\mathbf{q}$ outside $\mathscr{R}$. In $\mathscr{R}$ the Green function G consists of two parts. One is the local homogeneous potential contribution G_0 (which for points $\mathbf{q}$ too near $\mathbf{q}'$ cannot be approximated using geometrical optics). The other part consists of geometrical optics contributions from ray paths that have left $\mathscr{R}$ but then return to it after bouncing around in the potential. For points $\mathbf{q}$

outside $\mathscr{R}$, the Green function G consists of a sum of geometrical optics contributions from each of the ray paths connecting $\mathbf{q}'$ to $\mathbf{q}$,

$$G(\mathbf{q}, \mathbf{q}' ; E) \simeq \frac{1}{\hbar^{(N+1)/2}} \sum_{j=1}^{\infty} a_j(\mathbf{q}, \mathbf{q}' ; E) \exp\left[\frac{i}{\hbar} S_j(\mathbf{q}, \mathbf{q}' ; E) + i\phi_j\right], \quad (10.29)$$

where j labels a ray path (Figure 10.5). In (10.29), the quantity S_j is the action along path j,

$$S_j(\mathbf{q}, \mathbf{q}' ; E) = \int_{\mathbf{q}'}^{\mathbf{q}} \mathbf{p}(\mathbf{q}) \cdot d\mathbf{q}\big|_{\text{along path } j}, \quad (10.30)$$

ϕ_j is a phase factor (e.g., see Littlejohn (1986) and Maslov and Fedoriuk (1981)), which will not figure in an essential way in our subsequent considerations, and a_j is the wave amplitude whose determination takes into account the spreading or convergence of nearby rays. We emphasize that S_j, a_j and ϕ_j are independent of $\hbar$ and are determined purely by consideration of the classical orbits.

To calculate a_j in two spatial dimensions ($N = 2$), consider Figure 10.6. In this figure we show two infinitesimally separated rays originating from the source point $\mathbf{q}'$; l denotes distance along the ray; the radius r_0 is chosen so that the circle lies in $\mathscr{R}$ and satisfies $k'r_0 \gg 1$; $ds(l)$ denotes the differential arclength along the wavefront (perpendicular to the rays). By conservation of probability flux we have

$$|a(r_0)|^2 \dot{q}(r_0) \, ds(r_0) = |a(l)|^2 \dot{q}(l) \, ds(l),$$

where $\dot{q} = |\partial H/\partial \mathbf{p}|$ is the particle speed. Thus, we have

$$|a(l)| = |a(r_0)| \left[\frac{\dot{q}(r_0)}{\dot{q}(l)} \frac{ds(r_0)}{ds(l)}\right]^{1/2} \quad (10.31)$$

Since r_0 lies within $\mathscr{R}$ and satisfies $k'r_0 \gg 1$, we can find $a(r_0)$ by using the large argument approximation of $H_0^{(1)}(k'r_1)$, which gives the following result for $|\mathbf{q} - \mathbf{q}'| = r_1$,

$$G_0 \sim -\frac{m}{\hbar^2} \frac{\exp(i\pi/4)}{(2\pi)^{1/2}} \frac{\exp(ik'r_1)}{(k'r_1)^{1/2}}.$$

Figure 10.5 Illustration of some ray paths (labeled by j in the summation of Eq. (10.29)) in a billiard.

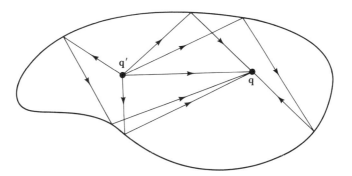

Comparing this with (10.29) we obtain

$$a(r_0) = \frac{m \exp{(i\pi/4)}}{2\pi r_0} \{2m[E - V(\mathbf{q}')]\}^{-1/4}. \tag{10.32}$$

Knowing $a(r_0)$ we can use (10.31) to calculate $a_j(\mathbf{q}, \mathbf{q}'; E)$ at any point $\mathbf{q}$ along a classical (ray) trajectory. Special consideration of wave effects, not included in the geometrical optics ray picture, is necessary at points where $ds(l) = 0$ (i.e., caustics and foci; e.g., see Ozorio de Almeida (1988)). Note that, for chaotic trajectories, nearby orbits separate exponentially, with the consequence that $ds(r_0)/ds(l)$, and hence also $a(l)$, on average decrease exponentially with the distance l along the orbit (ray).

The quantity of interest appearing in (10.25) should more properly be written as

$$\lim_{\mathbf{q} \to \mathbf{q}'} \mathrm{Im}\,[G(\mathbf{q}, \mathbf{q}'; E)].$$

For $\mathbf{q}$ very close to $\mathbf{q}'$, there is a short path directly from $\mathbf{q}$ to $\mathbf{q}'$, plus many long indirect paths from $\mathbf{q}$ to $\mathbf{q}'$ (cf. Figure 10.7). Each gives a contribution to $\lim_{\mathbf{q} \to \mathbf{q}'} \mathrm{Im}\,[G]$. For the short direct path, the geometrical optics approximation is not valid, but we may use G_0 to obtain this contribution. For the indirect paths the geometrical optics approximation is valid. Thus we write

$$d(E) = d_0(E) + \tilde{d}(E) \tag{10.33}$$

where $d_0(E)$ and $\bar{d}(E)$ represent the direct and indirect contributions,

$$d_0(E) = -\frac{1}{\pi} \int \lim_{\mathbf{q} \to \mathbf{q}'} \mathrm{Im}\,[G_0(\mathbf{q}, \mathbf{q}'; E]\, d^N\mathbf{q}, \tag{10.34}$$

$$\bar{d}(E) = -\frac{1}{\pi \hbar^{(N+1)/2}} \mathrm{Im}\left\{ \int \sum_j a_j(\mathbf{q}, \mathbf{q}; E) \exp[\mathrm{i}S_j(\mathbf{q}, \mathbf{q}; E)/\hbar + \mathrm{i}\phi_j]\, d^N\mathbf{q} \right\}. \tag{10.35}$$

Figure 10.6 Schematic illustrating how a_j is obtained.

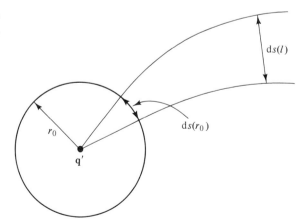

We claim that $d_0(E)$ given by (10.34) gives the Weyl result for $\bar{d}(E)$. If this is so, then, by (10.33), the quantity $\tilde{d}(E)$ given by (10.35) yields the fluctuations of $d(E)$ about its smoothed average $\bar{d}(E)$. We now explicitly verify that $d_0(E) = \bar{d}(E)$ for the case $N = 3$. Using (10.28) we have for $k'|\mathbf{q} - \mathbf{q}'| \ll 1$

$$G_0(\mathbf{q}, \mathbf{q}'; E) \approx -\frac{2m}{\hbar^2}\frac{1 + ik'|\mathbf{q} - \mathbf{q}'|}{4\pi|\mathbf{q} - \mathbf{q}'|},$$

so that

$$\lim_{\mathbf{q} \to \mathbf{q}'} \mathrm{Im}[G_0(\mathbf{q}, \mathbf{q}'; E)] = -\frac{mk'}{2\pi\hbar^2} = -\left(\frac{2m}{\hbar^2}\right)^{3/2}\frac{[E - V(\mathbf{q})]^{1/2}}{4\pi}.$$

From (10.34) this yields

$$d_0(E) = \left(\frac{2m}{\hbar^2}\right)^{2/3}\frac{1}{4\pi^2}\int_{E \geq V(\mathbf{q})} [E - V(\mathbf{q})]^{1/2}\,d^3\mathbf{q}.$$

Comparing this result with (10.12a) and (10.13d) with $\sigma_3 = 4\pi$, we indeed verify that $d_0(E)$ is the Weyl result for $\bar{d}(E)$.

We now focus our attention on obtaining the semiclassical expression for the fluctuation about $\bar{d}(E)$, namely $\tilde{d}(E)$. Since the semiclassical regime corresponds to very small $\hbar$, the factor $\exp(iS_j/\hbar)$ in the integrand of (10.35) varies very rapidly with $\mathbf{q}$. Thus one may use the stationary phase approximation to evaluate the integral. The stationary phase condition is $\nabla S_j(\mathbf{q}, \mathbf{q}; E) = 0$ or

$$[\nabla_{\mathbf{q}} S_j(\mathbf{q}, \mathbf{q}'; E) + \nabla_{\mathbf{q}'} S_j(\mathbf{q}, \mathbf{q}'; E)]_{\mathbf{q} = \mathbf{q}'} = 0.$$

From (10.30) this yields $\mathbf{p}(\mathbf{q}) - \mathbf{p}'(\mathbf{q}) = 0$ where $\mathbf{p}'(\mathbf{q}) \equiv \mathbf{p}(\mathbf{q}')|_{\mathbf{q}' = \mathbf{q}}$. Pictures of ray paths for which the stationary phase condition $\mathbf{p}(\mathbf{q}) = \mathbf{p}'(\mathbf{q})$ is (a) not satisfied and (b) satisfied are shown in Figure 10.8. We see that the stationary phase condition selects out classical periodic orbits. Thus, we have the important result that (10.35) reduces to a sum over all periodic orbits of the classical problem.

Figure 10.7 Illustration of the direct path and one of the infinite number of indirect paths for a billiard.

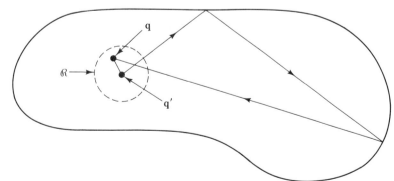

Carrying out the full integration in (10.35) is technically involved (see Gutzwiller (1990) for an exposition), and so we shall merely quote the result of the further analysis. There are three cases:

(1) Unstable periodic orbits.
(2) Isolated stable periodic orbits.
(3) Nonisolated stable periodic orbits.

Case (3) is the situation for integrable systems[2] (Berry and Tabor, 1976). For cases (1) and (2), the following result is obtained,

$$\tilde{d}(E) = \frac{1}{\pi h} \sum_k \frac{T_k}{[\det(\mathbf{M}_k - \mathbf{I})]^{1/2}} \exp\left\{ i\left[\frac{\tilde{S}_k(E)}{h} + \tilde{\phi}_k \right] \right\}. \quad (10.36)$$

We refer to a single traversal of a closed ray path as a 'primitive' periodic orbit. In (10.36) the index k labels the periodic orbits, and the summation includes both primitive and nonprimitive periodic orbits (i.e., multiple traversals are assigned a k label). The quantities appearing in (10.36) are as follows. $\tilde{S}_k(E) = \oint \mathbf{p} \cdot d\mathbf{q}$ where the integral is taken around periodic orbit k and represents the action for this orbit. T_k is the primitive period of orbit k and $\mathbf{M}_k$ is the linearized stability matrix for the Poincaré map of periodic orbit k. If orbit k is the rth round trip of some shorter periodic orbit k', then $T_k = T_{k'}$, $\tilde{S}_k = r\tilde{S}_{k'}$, $\mathbf{M}_k = \mathbf{M}_{k'}^r$, and $\tilde{\phi}_k = r\tilde{\phi}_{k'}$.

Perhaps the most interesting aspect of the semiclassical result for $\tilde{d}(E)$ is that it implies that the periodic orbits lead to oscillations of $d(E)$ with energy. Expanding a single term in the sum about some energy E_0, we have

$$\exp\left[\frac{i\tilde{S}_k(E)}{h} \right] \approx \exp\left[\frac{i\tilde{S}_k(E_0)}{h} + \frac{i(\partial \tilde{S}_k/\partial E)(E - E_0)}{h} \right]$$

$$= \exp\left[\frac{i\tilde{S}_k(E_0)}{h} + \frac{i\tilde{T}_k(E - E_0)}{h} \right],$$

Figure 10.8(a) The stationary phase condition is not satisfied. (b) The stationary phase condition is satisfied.

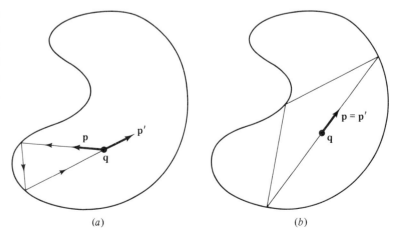

(a)   (b)

where we have made use of the classical expression for the period of a periodic orbit in terms of its action, $\partial \tilde{S}_k / \partial E = \tilde{T}_k = r T_k$. Thus, we see that the periodic orbit k contributes a term to $\tilde{d}(E)$ that oscillates with energy with energy period

$$\delta E_k \sim 2\pi \hbar / \tilde{T}_k. \tag{10.37}$$

Recall from (10.12a) that $\bar{d}(E) \sim \hbar^{-N}$. Hence, the number of levels per oscillation, $(\delta E_k)\bar{d}(E)$, is $O(1/\hbar^{N-1})$, which becomes very large for small $\hbar$ (provided $N > 1$). Thus we see that there is long-range clustering (i.e., long compared to the average level spacing) of the energy levels.

Also, as we include longer and longer period orbits in the sum, we see from (10.37) that the resolved scale of $\tilde{d}(E)$ becomes shorter. Hence, if we only desire a representation of $\tilde{d}(E)$ smoothed over some scale $\overline{\delta E}$ (where $\overline{\delta E}$ is shorter than the scale Δ of the smoothing used to obtain $\bar{d}(E)$), then we do not need to include all the periodic orbits in the summation (10.36); we only need include the *finite* number of periodic orbits whose period is not too large,

$$\tilde{T}_k < 2\pi \hbar / \overline{\delta E}.$$

(This justifies the restriction on Δ given by Eq. (10.9).)

In general, it is usually not possible to use the semiclassical trace formula for $\tilde{d}(E)$ to resolve a large number of individual levels of the spectrum (delta functions of $d(E)$). Indeed, even if all the amplitude, action, and phase quantities in the sum in (10.36) could be found for the infinite number of periodic orbits, it is still not clear that the exact delta function density (Figure 10.1(a)) would be recovered because the convergence of the sum in (10.36) is unlikely. Nevertheless, some work, which uses the semiclassical trace formula as a starting point, attempts to eliminate the problem of divergences (Berry and Keating, 1990; Tanner *et al.*, 1991; Sieber and Steiner, 1991). This apparently results in a systematic method for obtaining semiclassical approximations to individual energy levels of a classically chaotic system purely in terms of the properties of unstable classical periodic orbits. Such results for classically chaotic systems may be viewed as analogous to the well-known Bohr–Sommerfeld procedure for quantizing stable periodic orbits of classically integrable systems.

An important result which lends some theoretical support to the random matrix hypothesis (Section 10.1.1) was obtained by Berry[3] (1985). He used the trace formula (10.36) and the periodic orbits sum rule for chaotic systems given by Eq. (9.42) to show that the spectral rigidity of classically chaotic systems indeed satisfies the random matrix predictions (10.18a) and (10.18b) for L out to some maximum scale

$$L \ll L_{\max} \sim (\hbar/T_{\min})\bar{d}(E)$$

(cf., Eqs. (10.9) and (10.37)).

10.2   Wavefunctions for classically chaotic, bounded, time-independent systems

Say we consider a classical system which is ergodic on the energy surface, $E = H(\mathbf{p}, \mathbf{q})$, with Hamiltonian $(p^2/2m) + V(\mathbf{q})$. Let $f(\mathbf{p}, \mathbf{q})$ be the distribution function of the system such that the fraction of time that a typical orbit spends in some differential volume of phase space $d^N\mathbf{p}\, d^N\mathbf{q}$ located at the phase space point $(\mathbf{p}, \mathbf{q})$ is $f(\mathbf{p}, \mathbf{q}) d^N\mathbf{p}\, d^N\mathbf{q}$. Since the orbit is on the $E = H(\mathbf{p}, \mathbf{q})$ energy surface, $f(\mathbf{p}, \mathbf{q})$ must be of the form $C(\mathbf{p}, \mathbf{q})\delta(E - H(\mathbf{p}, \mathbf{q}))$. Since $f(\mathbf{p}, \mathbf{q})$ does not depend on time, it must be solely a function of isolating constants of the motion. Since we assume the orbit is ergodic on the $E = H(\mathbf{p}, \mathbf{q})$ energy surface, the only isolating constant is $H(\mathbf{p}, \mathbf{q})$ itself. Thus $C(\mathbf{p}, \mathbf{q})$ can be taken to be independent of $\mathbf{p}$ and $\mathbf{q}$ (i.e., it is just a constant), and, noting the normalization $\int f\, d^N\mathbf{p}\, d^N\mathbf{q} \equiv 1$, we have

$$f(\mathbf{p}, \mathbf{q}) = \frac{\delta[E - H(\mathbf{p}, \mathbf{q})]}{\int \delta[E - H(\mathbf{p}, \mathbf{q})]\, d^N\mathbf{p}\, d^N\mathbf{q}}. \tag{10.38}$$

This classical result leads to several natural conjectures (Berry, 1977) concerning the form of the eigenfunctions of the Schrödinger equation for such a system. In particular, integrating (10.38) over $\mathbf{p}$ we obtain $\int f\, d^N\mathbf{p} \sim [E - V(\mathbf{q})]^{(N/2)-1} U(E - V(\mathbf{q}))$. Thus, motivated by the correspondence principle, Berry conjectures that in the semiclassical limit the eigenfunctions satisfy

$$\overline{|\psi_i(\mathbf{q})|^2} \sim [E_i - V(\mathbf{q})]^{(N/2)-1} U(E_i - V(\mathbf{q})), \tag{10.39}$$

where the overbar signifies a coarse graining average in $\mathbf{q}$ over spatial scales small compared to the spatial size of the ergodic region, but large compared to a quantum wavelength. (Indeed we know that the exact $|\psi_i(\mathbf{q})|^2$ must oscillate on the wavelength scale.) For the case of a billiard, (10.38) reduces to

$$\overline{|\psi_i(\mathbf{q})|^2} \approx 1/A, \tag{10.40}$$

where A is the accessible area of the billiard.

Another conjecture comes from the observations that the Hamiltonian, $(p^2/2m) + V(\mathbf{q})$, and, from (10.38), also $f(\mathbf{p}, \mathbf{q})$ depend only on the magnitude of $\mathbf{p}$ and not on its direction. Thus, at any given point $\mathbf{q}$, it is equally likely that we observe an orbit through $\mathbf{q}$ in any direction. On the other hand, while the direction of $\mathbf{p}$ at $\mathbf{q}$ is uniformly distributed, its magnitude is fixed, $|\mathbf{p}| = \{2m[E - V(\mathbf{q})]\}^{1/2}$. In the semiclassical limit it is reasonable to view a wavefunction locally as a superposition of plane waves. Identifying the wavenumber $\mathbf{k}$ of these local plane waves with the classical particle momentum via, $\mathbf{p} = \hbar\mathbf{k}$, one arrives at the conjecture that, for a classically chaotic system, at any given point, ψ is a superposition of a large number of local plane waves, uniformly distributed in angle, with $|\mathbf{k}|$ fixed, and with random amplitudes and phases. This has the conjectured consequence (Berry, 1977) that, in the semiclassical regime, ψ is a Gaussian random function with local intensity fluctuations governed by the probability distribution,[4]

$$P(\psi) = [2\pi\overline{\psi^2}]^{-1/2} \exp(-\tfrac{1}{2}\psi^2/\overline{\psi^2}). \qquad (10.41)$$

Numerical experiments checking this result have been performed by McDonald and Kaufman (1988) using the stadium billiard of Figure 7.24(f).

Heller (1984) examined short wavelength numerical solutions of the Helmholtz equation (also in the stadium billiard) and found rather striking deviations from the random eigenfunction conjecture. In particular, he observed that wavefunctions often tend to have pronounced enhanced amplitudes along the paths of unstable periodic orbits. Different eigenfunctions exhibit enhancements along different periodic orbits, and can also have enhancements along more than one such orbit. Figure 10.9 from Heller's paper illustrates this phenomenon, which Heller calls 'scars.' More recently Bogomolny (1988) and Berry (1989) have

Figure 10.9 Contour plots of the intensity of semiclassical regime eigenfunctions for the stadium billiard. Dark regions have the highest intensity. Periodic orbit paths along which the high intensity regions apparently lie are also shown (Heller, 1984).

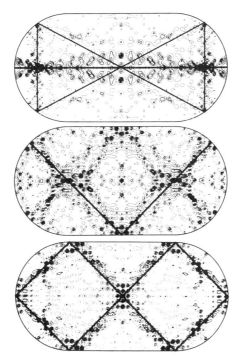

utilized the semiclassical Green function (10.29) to show theoretically that an energy band average of many eigenfunctions displays the effect of scarring, while Antonsen *et al.* (1991) consider the statistics of scarring associated with individual eigenfunctions and periodic orbits.

10.3　Temporally periodic systems

In our previous discussion we have assumed that the Hamiltonian has no explicit time dependence. In this section we consider the case where the Hamiltonian varies periodically with time. Physical examples of this type where the interplay of classical chaos and quantum mechanics is potentially important occur when sinusoidally varying electromagnetic fields act on atoms (Bayfield and Koch, 1974; Casati *et al.*, 1986; Jensen *et al.*, 1991; Meerson *et al.*, 1979) or molecules (Blümel *et al.*, 1986a,b) or on electrons on the surface of a conductor (Blümel and Smilansky, 1984; Jensen, 1984). Another potential experimental system involves a Josephson junction (Graham *et al.*, 1991). So far detailed experimental results on such temporally periodic forced systems have only been obtained for the case of a hydrogen atom initialized in a highly excited state and subjected to a microwave field (Bayfield and Koch (1974); for reviews see Bayfield (1987) and Jensen *et al.* (1991)). The important issue here is the possible ionization of the atom by the field. The experiments reveal a regime where the results are well described by the classical chaotic motion of the electron in the electric field of the wave plus the Coulomb electric field of the proton. In addition, another regime is also observed (Bayfield and Sokol, 1988; Galvez *et al.*, 1988; Blümel *et al.*, 1991; Arndt *et al.*, 1991) where the quantum effects appear to suppress the probability of ionization relative to the classical prediction.

This suppression of classical chaotic transport by quantum wave effects appears to be a fundamental consideration in time-dependent quantum chaos problems. The effect was first seen, and is most easily understood, within the context of the kicked rotor problem (Figure 7.3) which, in the classical case, leads to the standard map, Eq. (7.15), whose behavior we have discussed in Section 7.3.1. In the remainder of this section we shall restrict our discussion to this one instructive example.

The Hamiltonian for the kicked rotor is $H(p_\theta, \theta, t) = p_\theta^2/2\tilde{I} + K\cos\theta\Delta_\tau(t)$, where $\Delta_\tau(t) \equiv \Sigma_n \delta(t - n\tau)$ (cf. Eq. (7.14)). Replacing p_θ by the angular momentum operator $-i\hbar\partial/\partial\theta$, yields the time-dependent Schrödinger equation,

$$i\hbar\frac{\partial\psi(\theta,t)}{\partial t} = -\frac{\hbar^2}{2\tilde{I}}\frac{\partial^2\psi(\theta,t)}{\partial\theta^2} + K\cos\theta\Delta_\tau(t)\psi(\theta,t). \qquad (10.42)$$

Normalizing time to the period of the kicking τ via $\hat{t} \equiv t/\tau$, and normalizing h to $\tilde{I}/\tau$ via $\hat{h} \equiv h\tau/\tilde{I}$, we obtain

$$i\hat{h}\frac{\partial \psi}{\partial \hat{t}} = \frac{\hat{h}^2}{2}\frac{\partial^2 \psi}{\partial \theta^2} + \hat{K}\cos\theta\Delta(\hat{t})\psi, \qquad (10.43)$$

where $\hat{K}$ is the normalized kicking strength, $\hat{K} \equiv K\tau/\tilde{I}$, and $\Delta(\hat{t}) \equiv \Sigma_n \delta(\hat{t} - n)$. From (10.43) we see that the problem depends on two dimensionless parameters, $\hat{K}$ and $\hat{h}$. In contrast, the classical problem (7.15) involves only the single dimensionless parameter $\hat{K}$ which characterizes the kicking strength.[5] Hence, the dimensionless parameter $\hat{h}$ may be regarded as characterizing the strength of the quantum effects, and the semiclassical limit corresponds to $\hat{h} \ll 1$ (assuming $\hat{K} \sim O(1)$). (Alternatively, (10.43) follows from (10.42) by setting $\tilde{I} = 1$ and $\tau = 1$.) In what follows we shall drop the circumflexes over t, h, and K, and henceforth when we write t, h, and K we shall mean the normalized quantities formerly denoted $\hat{t}$, $\hat{h}$, and $\hat{K}$.

Equation (10.43) can be dealt with as follows. Let

$$\psi_{n\pm}(\theta) = \lim_{\varepsilon \to 0^+} \psi(\theta, n \pm \varepsilon),$$

where $\varepsilon \to 0^+$ signifies that the limit to zero is taken with ε positive. Thus, ψ_{n+} and ψ_{n-} denote the wavefunction just before and just after the application of the kick at $t = n$. Considering the small range of times near a kick, $n - 0^+ \le t \le n + 0^+$, we may neglect the term $\partial^2\psi/\partial\theta^2$, so that we have

$$ih\partial\psi/\partial t \equiv K\cos\theta\delta(t - n)\psi.$$

Integrating this from $t = n - 0^+$ to $t = n + 0^+$, we obtain

$$\psi_{n+}(\theta) = \psi_{n-}(\theta)\exp[-i(K/h)\cos\theta]. \qquad (10.44)$$

In the time interval $n + 0^+ \le t \le (n+1) - 0^+$ the delta function term is zero, $\Delta(t) \equiv 0$, and ψ obeys the equation,

$$i\frac{\partial\psi}{\partial t} = -\frac{h}{2}\frac{\partial^2\psi}{\partial\theta^2}. \qquad (10.45)$$

Since $\psi(\theta)$ is periodic in θ, $\psi(\theta) = \psi(\theta + 2\pi)$, we can solve (10.45) by introducing a Fourier series in θ

$$\psi(\theta, t) = \frac{1}{(2\pi)^{1/2}}\sum_{l=-\infty}^{+\infty}\phi(l, t)\exp(il\theta), \qquad (10.46a)$$

$$\phi(l, t) = \frac{1}{(2\pi)^{1/2}}\int_0^{2\pi}\exp(-il\theta)\psi(\theta, t)\,d\theta. \qquad (10.46b)$$

Thus, (10.45) becomes $i\partial\phi(l,t)/\partial t = (hl^2/2)\phi(l,t)$, and we obtain the result,

$$\phi_{(n+1)-}(l) = \phi_{n+}(l)\exp(-ihl^2/2). \qquad (10.47)$$

The representation (10.46a) is particularly nice in this context, because the Fourier basis functions $\exp(il\theta)$ are eigenfunctions of the angular momentum operator $-i\hbar\partial/\partial\theta$. Thus, the momenta are quantized at values $l\hbar$, and the expected value of p_θ^2 is

$$\overline{p_\theta^2} \equiv \int_0^{2\pi} \psi^*(-i\hbar\partial/\partial\theta)^2\psi \, d\theta = \hbar^2 \sum_{l=-\infty}^{+\infty} l^2|\phi(l,t)|^2. \qquad (10.48)$$

For large enough K, we saw in Chapter 7 that the classical kicked rotor yields diffusion in momentum (Eqs. (7.42)–(7.44)),

$$\langle p^2/2 \rangle \approx Dn \qquad (10.49)$$

where $D \approx K^2/4$. What happens quantum mechanically in the case of small $\hbar$? To try to answer this question one can integrate the Schrödinger equation numerically. A good way to do this is to utilize (10.44), (10.46) and (10.47). That is, advance ψ_{n-} to ψ_{n+} by multiplying by $\exp[-i(K/\hbar)\cos\theta]$; then take the Fourier transform, Eq. (10.46b) (using a fast Fourier transform algorithm) to obtain $\phi_{n+1}(l)$; then obtain $\phi_{(n+1)-}(\theta)$ by multiplying by $\exp(-i\hbar l^2/2)$; then inverse transform back to get $\psi_{(n+1)-}(\theta)$, Eq. (10.46a); and successively repeat this string of operations. The first numerical solution of this problem was done by Casati *et al.* (1977) (see also Hogg and Huberman (1982)). The result they obtained was rather surprising and is schematically illustrated in Figure 10.10. Using (10.48) they plotted $\overline{p_\theta^2}$ versus n starting from an initial zero momentum ($l = 0$) state, $\psi(\theta, 0^+) = (2\pi)^{-1/2}$. They found that for typical small values of $\hbar$ and typical large values of K, the quantum calculated momentum indeed diffused,[6] just as in the classical case (10.49), but only for a finite time, denoted n_* in the figure. When t becomes of order n_* quantum effects evidently arrest the diffusion and $\overline{p_\theta^2}$ remains bounded as $t \to \infty$. As $\hbar$ is made smaller, n_* becomes larger and so does the maximum attained value of $\overline{p_\theta^2}$. Thus the evolution looks classical for a longer time when $\hbar$ is decreased. Nevertheless, if we wait long enough, the quantum limitation of the classical chaotic diffusion eventually manifests itself. This 'localization' phenomenon has been claimed to be the explanation of the observed reduction in the microwave field ionization rate of hydrogen atoms mentioned at the beginning of this section.

A nice explanation of the quantum suppression of classical momentum diffusion in the rotor problem has been suggested in the paper of Fishman *et al.* (1982) and is discussed below.

Since (10.43) is periodic in time, then, according to Floquet's theorem, it can be represented as a superposition of solutions of the form

$$\exp(-i\omega t)w_\omega(\theta, t) = \exp(-i\omega t)\sum_l u_\omega(l, t)\exp(il\theta), \qquad (10.50)$$

where w_ω and u_ω are periodic in t with the period of the driving force. Since in our normalization the period is 1, we have $w_\omega(\theta, t) = w_\omega(\theta, t + 1)$ and $u_\omega(l, t) = u_\omega(l, t + 1)$. We can regard a solution of the form (10.50) as being exactly analogous to a Bloch wave for the time-independent Schrödinger equation in a spatially periodic potential. Here, however, the potential is time periodic rather than space periodic. Substitution of (10.50) into the Schrödinger equation for the quantum rotor (10.43) produces an eigenvalue problem. In particular, using Eqs. (10.44), (10.46) and (10.47), we can write the evolution equation for $\psi_{n+}(\theta)$ as

$$\psi_{(n+1)+}(\theta) = \mathscr{L}[\psi_{n+}(\theta)], \qquad (10.51)$$

where the operator $\mathscr{L}$ is unitary and is given by

$$\mathscr{L}[f(\theta)] = \frac{1}{2\pi} \sum_l \int_0^{2\pi} d\theta' \exp[-iR(\theta, \theta', l)] f(\theta'),$$

$$R(\theta, \theta', l) = (K/\hbar)\cos\theta + l(\theta - \theta') + (\hbar l^2/2).$$

($\mathscr{L}$ is unitary because the operator on the right-hand side of (10.42) is Hermitian.) Now since (10.50) is also a solution of our Schrödinger equation, we obtain from (10.51)

$$\exp(-i\omega)w_{\omega+}(\theta) = \mathscr{L}[\omega_{\omega+}(\theta)], \qquad (10.52)$$

where $w_{\omega+}(\theta) \equiv \lim_{t\to 0^+} w_\omega(\theta, t)$. That is, $w_{\omega+}(\theta)$ is an eigenfunction of the unitary operator $\mathscr{L}$, and $\exp(-i\omega)$ is the associated eigenvalue. Since the magnitudes of the eigenvalues of a unitary operator are 1, the

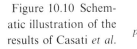

Figure 10.10 Schematic illustration of the results of Casati *et al.*

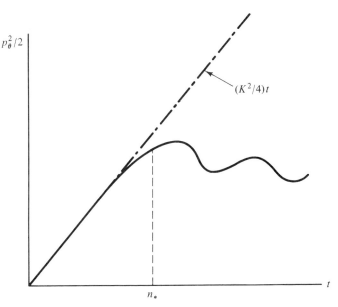

quantities ω are real. In general, the eigenvalue spectrum for a general problem such as that specified by (10.52) can be either discrete or continuous or a combination of the two types. Part of the result of Grempel *et al.* is that the spectrum in the case of the kicked rotor is essentially discrete. That is, we may label the eigenvalues and the associated eigenfunctions using an integer valued index j such that a solution of the Schrödinger equation may be written as a discrete sum,

$$\psi_n(\theta, t) = \sum_j A_j \exp(i\omega_j t) w_j(\theta, t), \qquad (10.53a)$$

or in the Fourier transform (momentum) representation

$$\phi_n(l, t) = \sum_t a_j \exp(i\omega_j t) u_j(l, t), \qquad (10.53b)$$

where in (10.53a) and (10.53b) $w_j = w_{\omega_j}$ and $u_j = u_{\omega_j}$, and we assume that the functions $w_j(\theta, t)$ and $u_j(l, t)$ form a complete orthonormal basis in θ and l (at any fixed time t).

Grempel *et al.* present a strong argument that the eigenfunctions are localized in momentum space (i.e., in l). In particular, they claim that, for a given j, the $u_j(l, t)$ will on average decay with l roughly exponentially away from a central region $l \sim \bar{l}_j$,

$$u_j(l, t) \sim \exp\left(-\frac{\hbar|l - \bar{l}_j|}{\Delta_{\mathrm{L}}}\right), \qquad (10.54)$$

where Δ_{L} is the 'localization length' of the eigenfunction and $\Delta_{\mathrm{L}}/\hbar$ is large when $\hbar \ll 1$. In order to see the consequences of (10.54), say we take an initial condition at a particular momentum $\hbar l_0$,

$$\psi(\theta, 0^+) = \frac{1}{(2\pi)^{1/2}} \exp(i l_0 \theta).$$

Then the coefficient a_j in (10.53b) is

$$a_j = u_j^*(l_0, 0^+). \qquad (10.55)$$

Thus, by (10.54) only eigenfunctions whose centers $\bar{l}_j$ are within a momentum range Δ_{L} of the initial level $\hbar l_0$ are appreciably excited. Hence, the expected value of $(p_\theta - \hbar l_0)^2$ cannot become much larger than Δ_{L}^2. We therefore interpret the quantum limitation to the growth of $\overline{p_\theta^2}$ in Figure 10.10 as occurring when $\overline{p_\theta^2} \sim \Delta_{\mathrm{L}}^2$.

In order to argue the momentum localization of the eigenfunctions, Grempel *et al.* recast the eigenvalue equation (10.52) in the following form:

$$T_l \bar{u}(l) + \sum_{r \neq 0} U_r \bar{u}(l + r) = \varepsilon \bar{u}(l), \qquad (10.56)$$

where $\bar{u}(l) \equiv \frac{1}{2}[u_\omega(l, 0^+) + u_\omega(l, 0^-)]$ (i.e., $\bar{u}(l)$ is the average of u_ω just before and just after the kick), $T_l = \tan(E_l/2)$, $E_l = \omega - (\hbar l^2/2)$, U_r is the *r*th Fourier coefficient in the Fourier expansion of $U(\theta) = -\tan(\frac{1}{2}K\cos\theta)$, and $\varepsilon = -U_0$. Next they note that (10.56) is of the same form as the equation for the quantum wavefunction on a time-independent, one-dimensional discrete (spatial) lattice. In this analogy l is the spatial location of a lattice site, ε is the energy level (eigenvalue), $\bar{u}(l)$ is the wavefunction value at lattice site l corresponding to energy level ε, U_r is the 'hopping element' to the *r*th neighbor, and T_l is the potential energy at site l. In the case where T_l is independent of l, we have a spatially homogeneous lattice. The solutions of the eigenvalue problem (10.56) give propagating waves which travel from $l = -\infty$ to $l = +\infty$. That is, the eigenstates are of the form $\bar{u}(l) \sim \exp[ik(\varepsilon)l]$ for ε in appropriate energy bands. Such solutions are called 'extended', as opposed to localized solutions which decay to zero both at $l \to +\infty$ and $l \to -\infty$. As a model of what happens when random impurities are introduced into the lattice, one can consider the case where T_l is an externally given random number. In this case, it has been shown that, essentially for any degree of randomness of T_l, the $\bar{u}(l)$ are exponentially localized as in Eq. (10.54). In the condensed matter context, this has the consequence that an electron in the lattice is quantum mechanically spatially localized, and the medium becomes an insulator. This phenomenon is called Anderson localization (Anderson, 1958). For large enough l, the quantity $T_l = \tan[\omega - (\hbar l^2/2)]$ varies rapidly with l in an erratic way. Grempel *et al.* contend that T_l may then be thought of as essentially random, even though it is a known deterministic function of l. Thus, by the analogy with Anderson localization, they conclude that the quantum kicked rotor should have localized momentum wavefunctions.

The localization length Δ_L of (10.54) can be roughly estimated as follows (Chirikov *et al.*, 1981). By (10.54) and (10.55) those modes most strongly excited are localized around momenta within Δ_L of $p_\theta = 0$ (assuming an initial condition $\psi(\theta, 0^+) = (2\pi)^{-1/2}$). Hence, the effective number of eigenfunctions excited by the initial condition is $\Delta_L/\hbar$. Each eigenfunction has an associated eigenvalue $\exp(-i\omega_j)$. Thus, the ω_j must be taken to lie in $[0, 2\pi]$. Since there are of order $\Delta_L/\hbar$ excited eigenfunctions, the typical frequency spacing between adjacent ω_j values is $\delta\omega \sim 2\pi/(\Delta_L/\hbar)$. For $n \leq /\delta\omega$, the system does not yet 'know' about the individual eigenfunctions, and the quantum nature of the problem is therefore not felt. Thus, for $n \leq 1/\delta\omega$ we expect that $\overline{p_\theta^2}$ increases linearly with time as in the classical problem. At $n \approx 1/\delta\omega$ the existence of discrete localized eigenfunctions becomes felt, and the quantum diffusion is arrested. Thus,

$$n_* \sim 1/\delta\omega \sim \Delta_L/\hbar,$$

where n_* is the turnover time shown in Figure 10.10. In addition, at the turnover, the characteristic spread will be the localization length Δ_L. Since classical diffusion with diffusion coefficient D applies for times up to n_*, we have

$$Dn_* \sim \Delta_L^2.$$

Using $n_* \sim \Delta_L/h$ and $D \approx K^2/4$, the above yields the desired estimate of Δ_L,

$$\Delta_L \sim K^2/h, \tag{10.57}$$

and $n_* \sim K^2/h^2$. Thus, as previously noted, Δ_L and n_* increase as h is reduced (the time interval of classical behavior lengthens).

The above argument giving the estimate (10.57) implies that very good coherence of the quantum waves must be maintained for the relatively long time n_*. Thus, one might expect that the delicate interference effects leading to localization could be strongly affected by noise. Ott *et al.* (1984a) consider the effect of noise by adding a small fluctuating component of root mean square size σ to the strength of the delta function kicks. They find that localization is destroyed when

$$\sigma \gtrsim h^2/K. \tag{10.58}$$

For small h, this level of noise produces only a tiny increase in the classical diffusion, but completely removes the quantum limitation at time n_*. That is, with noise of magnitude (10.58), the quantum evolution is such that $\overline{p_\theta^2} \sim 2nD$ for all time, just as in the classical case. Experimental observations on ionization of atoms by microwave fields also apparently show evidence of the destruction of localization by noise (Blümel *et al.*, 1991; Arndt *et al.*, 1991).

10.4 Quantum chaotic scattering

For scattering problems one is interested in unbounded orbits in phase space and a time-independent Hamiltonian. Due to the rapidly developing nature of the field of quantum chaotic scattering we shall be very brief. Thus, we shall limit the discussion to one particular result (Blümel and Smilansky, 1988) which seems particularly interesting. (Some further works and references on quantum chaotic scattering are Blümel (1991), Smilansky (1992), Jalabert *et al.* (1990), Cvitanović and Eckhardt (1989), Gaspard and Rice (1989b,c), and Gutzwiller (1983).)

Blümel and Smilansky consider the scattering matrix for an incoming state I scattering to an outgoing state I'. By using the semiclassical approximation for the scattering matrix element $S_{II'}(E)$, they find that classical chaos in the scattering problem results in enhanced fluctuations of $S_{II'}(E)$ with variation of E. Specifically they consider the correlation function,

$$C_{II'}(\varepsilon) = \langle S_{II'}^*(E)S_{II'}(E + \varepsilon)\rangle,$$

where the angle brackets denote an energy average over a range that is classically small, but quantum mechanically large in that it includes many wiggles of $S_{II'}(E)$. They find the important result that, in the semiclassical approximation, $C_{II'}(\varepsilon)$ is related to the classically obtained probability distribution $P(E, T)$, where $P(E, T)\,dT$ is the classical probability that a randomly chosen orbit experiences a delay time (defined on p. 170) between T and $T + dT$. Their result is

$$C_{II'}(\varepsilon) \sim \int dT P(E, T)\exp(i\varepsilon T/\hbar). \qquad (10.59)$$

This provides an interesting direct connection between a classical quantity $P(E, T)$ and an inherently quantum quality $C_{II'}(\varepsilon)$. In particular, (10.59) implies that, if $P(E, T)$ decays exponentially with T (cf. Eq. (5.1)), then $C_{II'}(\varepsilon)$ has a Lorentzian shape.

Problems

1. Obtain (10.13d) from (10.12b).
2. Consider a rectangular billiard of dimension a by b. Obtain the energy levels and plot some of them versus b with a held fixed, thus verifying that, for this integrable billiard, energy levels do not repel (Figure 10.3(a)).
3. Verify for $N = 2$ that $d_0(E)$ given by (10.34) yields the Weyl result. Hint: for small z the expansion of the Hankel function is $H_0^{(1)}(z) = 1 + (2i/\pi)[\ln(z/2) + \gamma] + O(z^2 \ln z)$, where γ is a constant.
4. Obtain the evolution of $\overline{p_\theta^2}$ for the quantum kicked rotor for the case where $\hbar = 4\pi$. In particular, show that $\overline{p_\theta^2}$ is proportional[6] to n^2 (for large n).
5. Derive Eq. (10.56) for the quantum kicked rotor.

Notes

1. One might call the subject of this chapter 'wave chaos' rather than 'quantum chaos' to emphasize this generality, but we shall adhere to the more traditional terminology.
2. Another, somewhat exceptional, case giving nonisolated periodic orbits is the chaotic 'stadium' billiard of Figure 7.24(f). In that case there is a continuous one parameter family of neutrally stable periodic orbits bouncing directly back and forth between the two straight parallel segments of the perimeter of the stadium.

3. Berry's considerations are built on work by Hannay and Ozorio de Almeida (1984).

4. That is, for randomly chosen $\mathbf{q}$, the probability that an eigen wavefunction value lies between ψ and $\psi + d\psi$ is $P(\psi)\,d\psi$. Equation (10.41) is for the class where ψ is real; i.e., the Hamiltonian is time-reversible. For the non-time-reversible case, an equation of the Gaussian form (10.41) holds separately for the real and imaginary parts of ψ. Although the distribution (10.41) is Gaussian, a random superposition of plane waves does lead to significant spatial correlations (O'Conner *et al.*, 1987; Blümel *et al.*, 1992).

5. In (7.15) we set $\tau/\tilde{I} = 1$ so that K and $\hat{K}$ are identical.

6. For special values of h, called quantum resonances, Izraelev and Shepelyansky (1979) show that $\overline{p_\theta^2}$ increases quadratically (rather than linearly) with time. Thus, the behavior is like the acceleration of a free particle under a constant force (rather than diffusion in momentum). The quantum resonances occur when h is a rational number times 4π (see Problem 4). The typical behavior, which occurs when h is not on a resonance, is diffusive (Figure 10.10).

References

H.D.I. Abarbanel, R. Brown and J.B. Kadtke, 'Prediction in Chaotic Nonlinear Systems: Methods for Time Series with Broadband Fourier Spectra,' *Phys. Rev. A* **41**, 1742 (1990).

R.C. Adler, A.C. Konheim and M.H. McAndrew, 'Topological Entropy,' *Trans. Am. Math. Soc.* **114**, 309 (1965).

V.S. Afraimovich, V.V. Bykov and L.P. Silnikov, 'On the Origin and Structure of the Lorenz Attractor,' *Sov. Phys. Dokl.* **22**, 253 (1977).

G. Ahlers and R.L. Behringer, 'Evolution of Turbulence from the Rayleigh–Bénard Instability,' *Phys. Rev. Lett.* **40**, 712 (1978).

P.W. Anderson, 'Absence of Diffusion in Certain Random Systems,' *Phys. Rev.* **103**, 1492 (1958).

D.V. Anosov, 'Geodesic Flows on a Compact Riemann Manifold of Negative Goedesic Curvature,' *Proc. Steklov Inst. Math.* **90**, 1 (1967).

T.M. Antonsen, E. Ott and Q. Chen, 'Statistics of Wavefunction Scars' Preprint (1991).

F.T. Arecchi, R. Meucci, G. Puccioni and J. Tredicce, 'Experimental Evidence of Subharmonic Birfurcations, Multistability and Turbulence in a Q-Switched Gas Laser,' *Phys. Rev. Lett.* **49**, 1217 (1982).

H. Aref, 'Stirring by Chaotic Advection,' *J. Fluid Mech.* **143**, 1 (1984).

H. Aref and S. Balachandar, 'Chaotic Advection in a Stokes Flow,' *Phys. Fluids* **29**, 3515 (1986).

M. Arndt, A. Buchleitner, R. N. Mantegna and H. Walther, 'Experimental Study of Quantum and Classical Limits in Ionization of Rubidium Rydberg Atoms,' *Phys. Rev. Lett.* **67**, 2435 (1991).

A. Arnéodo, P. Coullet and C. Tresser, 'Oscillators with Chaotic Behavior: An Illustration of a Theorem by Shilnikov,' *J. Stat. Phys.* **27**, 171 (1982).

V.I. Arnold, 'Small Denominators and Problems of Stability of Motion in Classical and Celestial Mechanics,' *Russ. Math. Surveys* **18.6**, 85 (1963).

V.I. Arnold, 'Instability of Dynamical Systems with Several Degrees of Freedom,' *Sov. Math. Dokl.* **5**, 581 (1964).

V.I. Arnold, 'Small Denominators, I: Mappings of the Circumference Into Itself,' *AMS Transl. Series 2*, **46**, 213 (1965).

V.I. Arnold, *Mathematical Methods of Classical Mechanics* (Springer, New York, 1978).

V.I. Arnold, *Geometrical Methods in the Theory of Ordinary Differential Equations* (Springer, New York, 1982).

V.I. Arnold and A. Avez, *Ergodic Problems of Classical Mechanics* (W.A. Benjamin, New York, 1968).

D. Auerbach, P. Cvitanović, J.-P. Eckmann, G. Gunaratne and I. Procaccia, 'Exploring Chaotic Motion Through Periodic Orbits,' *Phys. Rev. Lett.* **58**, 2387 (1987).

R. Badii, 'Conservation Laws and Thermodynamic Formalism for Dissipative Dynamical Systems,' Doctoral dissertation, University of Zurich (1987).

R. Badii and A. Politi, 'Renyi Dimensions from Local Expansion Rates,' *Phys. Rev. A* **35**, 1288 (1987).

J. Balatoni and A. Renyi, *Pub. Math. Inst. Hungarian Acad. Sci.* **1**, 9 (1956).

R. Balian and C. Bloch, 'Distribution of Eigenfrequencies for the Wave Equation in a Finite Domain: I, II and III,' *Ann. Phys.* **60**, 401 (1970); *ibid.* **64**, 271 (1971); *ibid.* **69**, 76 (1972).

P.M. Battelino, C. Grebogi, E. Ott and J.A. Yorke, 'Chaotic Attractors on a Three-Torus and Torus Break-Up,' *Physica D* **39**, 299 (1989).

J.E. Bayfield, 'Studies of the Sinusoidally Driven Weakly Bound Atomic Electron in the Threshold Region for Classically Stochastic Behavior,' in *Quantum Measurement and Chaos*, p. 1, edited by R.E. Pike and S. Sarkar (Plennum, New York, 1987).

J.E. Bayfield and P.M. Koch, 'Multiphoton Ionization of Highly Excited Atoms,' *Phys. Rev. Lett.* **33**, 258 (1974).

J.E. Bayfield and D.W. Sokol, 'Excited Atoms in Strong Microwaves: Classical Resonances and Localization in Experimental Final State Distributions,' *Phys. Rev. Lett.* **61**, 2007 (1988).

G. Benettin, L. Galgani, A. Giorgilli and J.-M. Strelcyn, 'Lyapunov Characteristic Exponents for Smooth Dynamical Systems and for Hamiltonian Systems: A Method for Computing All of Them. Part 2: Numerical Application,' *Meccanica* **15**, 21 (1980).

A. Ben-Mizrachi, I. Procaccia and P. Grassberger, 'Characterization of Experimental (Noisy) Strange Attractors,' *Phys. Rev. A* **29**, 975 (1984).

A. Ben-Mizrachi, I. Procaccia, N. Rosenberg, A. Schmidt and H.G. Schuster, 'Real and Apparent Divergences in Low Frequency Spectra of Nonlinear Dynamical Systems,' *Phys. Rev. A* **31**, 1830 (1985).

R. Benzi, G. Paladin, G. Parisi and A. Vulpiani, 'Characterization of Intermittency in Chaotic Systems,' *J. Phys. A* **17**, 3521 (1984).

P. Bergé, M. Dubois, M. Manneville and P. Pomeau, 'Intermittency in Rayleigh–Bénard Convection,' *J. Phys. (Paris) Lett.* **41**, L341 (1980).

P. Bergé, Y. Pomeau and C. Vidal, *Order Within Chaos* (Wiley, New York, 1986).

M.V. Berry, 'Regular and Irregular Semiclassical Wavefunctions,' *J. Phys. A* **12**, 2083 (1977).

M.V. Berry, 'Regular and Irregular Motion' in *Topics in Nonlinear Dynamics* edited by S. Jorna, (*Am. Inst. Phys. Conf. Proc.* **46**, 1978). (Reprinted in MacKay and Meiss (1987).)

M.V. Berry, 'Semiclassical Mechanics of Regular and Irregular Motion,' in *Chaotic Behavior of Deterministic Systems. Les Houches Summer School 1981*, edited by R.H.G. Helleman and G. Ioos (North-Holland, Amsterdam, 1983).

M.V. Berry, 'Semiclassical Theory of Spectral Rigidity,' *Proc. Roy. Soc.* (*London*) *A* **400**, 229 (1985).

M.V. Berry, 'Quantum Scars of Classical Closed Orbits in Phase Space,' *Proc. Roy. Soc.* (*London*) *A* **423**, 219 (1989).

M.V. Berry and J.P. Keating, 'A Rule for Quantizing Chaos?', *J. Phys. A* **23**, 4839 (1990).

M.V. Berry and M. Tabor, 'Closed Orbits and the Regular Bound Spectrum,' *Proc. Roy. Soc.* (*London*) *A* **349**, 101 (1976).

M.V. Berry and M. Tabor, 'Level Clustering in the Regular Spectrum,' *Proc. Roy. Soc.* (*London*) *A* **356**, 375 (1977).

O. Biham and W. Wenzel, 'Characterization of Unstable Periodic Orbits in Chaotic Attractors and Repellers,' *Phys. Rev. Lett.* **63**, 819 (1989).

G.D. Birkhoff, 'On the Periodic Motions of Dynamical Systems,' *Acta Mathematica* **50**, 359 (1927). (Reprinted in MacKay and Meiss (1987)).

S. Bleher, C. Grebogi, E. Ott and R. Brown, 'Fractal Boundaries for Exit in Hamiltonian Dynamics,' *Phys. Rev. A* **38**, 930 (1988).

S. Bleher, C. Grebogi and E. Ott, 'Bifurcation to Chaotic Scattering,' *Physica D* **46**, 87 (1990).

S. Bleher, E. Ott and C. Grebogi, 'Routes to Chaotic Scattering,' *Phys. Rev. Lett.* **63**, 919 (1989).

R. Blümel, 'Quantum Chaotic Scattering,' in *Directions in Chaos*, Vol. 3, edited by H. Bai-Lin, D.H. Feng, J.-M. Yuan (World Scientific, Singapore, 1991).

R. Blümel, A. Buchleitner, R. Graham, L. Sirko and U. Smilansky, 'Dynamical Localization in the Microwave Interaction of Rydberg Atoms: The Influences of Noise,' *Phys. Rev. A* **44**, 4521 (1991).

R. Blümel, I.H. Davidson, W.P. Reinhardt, H. Lin and M. Sharnoff, 'Quasilinear Ridge Structures in Water Waves,' *Phys. Rev. A* (1992) (to be published).

R. Blümel, S. Fishman and U. Smilansky, 'Excitation of Molecular Rotation by Periodic Microwave Pulses: A Testing Ground for Anderson Localization,' *J. Chem. Phys.* **84**, 2604 (1986a).

R. Blümel, S. Fishman, M. Griniasti and U. Smilansky, 'Localization in the Quantum Description of the Periodically Perturbed Rotor,' in *Quantum Chaos and Statistical Nuclear Physics*, edited by T.H. Seligman and H. Nishioka, Lecture Notes in Physics, **263** (Springer, Berlin, 1986b).

R. Blümel and U. Smilansky, 'Quantum Mechanical Suppression of Classical Stochasticity in the Dynamics of Periodically Disturbed Surface Electrons,' *Phys. Rev. Lett.* **52**, 137 (1984).

R. Blümel and U. Smilansky, 'Classical Irregular Scattering and Its Quantum-Mechanical Implications,' *Phys. Rev. Lett.* **60**, 477 (1988).

E.B. Bogomolny, 'Smoothed Wave Functions of Chaotic Quantum Systems,' *Physica D* **31**, 169 (1988).

O. Bohigas, M.J. Giannoni and C. Schmit, 'Characterization of Chaotic Quantum Spectra and Universality of Level Fluctuation Laws,' *Phys. Rev. Lett.* **52**, 1 (1984).

O. Bohigas, S. Tomsovic and D. Ullmo, 'Classical Transport Effects on Chaotic Levels,' *Phys. Rev. Lett.* **65**, 5 (1990).

A. Bohr and B.R. Mottelson, *Nuclear Structure*, Vol. 1 (Benjamin, New York, 1969) Appendix 2C, pp. 294–301.

T. Bohr and D. Rand, 'The Entropy Function for Characteristic Exponents,' *Physica D* **25**, 387 (1987).

T. Bohr and T. Tèl, 'The Thermodynamics of Fractals,' in *Directions in Chaos*, Vol. 2, edited by H. Bai-Lin (World Scientific, Singapore, 1988).

A. Bondeson, E. Ott and T.M. Antonsen, 'Quasiperiodically Forced Pendula and Schrödinger Equations with Quasiperiodic Potentials: Implications of their Equivalence,' *Phys. Rev. Lett.* **55**, 2103 (1985).

P.T. Bonoli and E. Ott, 'Toroidal and Scattering Effects on Lower Hybrid Wave Propagation in Tokamaks,' *Phys. Fluids* **25**, 359 (1982).

T. C. Bountis, 'Period Doubling, Bifurcations, and Universality in Conservative Systems,' *Physica D* **3**, 577 (1981).

R. Bowen, 'Markov Partitions for Axiom A Diffeomorphisms,' *Amer. J. Math.* **92**, 725 (1970).

R. Bowen and D. Ruelle, 'Ergodic Theory of Axiom A Flows,' *Invent. Math.* **79**, 181 (1975).

A. Brandstater, J. Swift, H.L. Swinney, A. Wolf, J.D. Farmer, E. Jen and J.P. Crutchfield, 'Low Dimensional Chaos in a Hydrodynamic System,' *Phys. Rev. Lett.* **51**, 1441 (1983).

A. Brandstater and H.L. Swinney, 'Strange Attractors in Weakly Turbulent Couette–Taylor Flow,' *Phys. Rev. A* **35**, 2207 (1987).

P. Bryant, R. Brown and H.D.I. Abarbanel, 'Lyapunov Exponents from Observed Time Series,' *Phys. Rev. Lett.* **65**, 1523 (1990).

P. Bryant and C. Jefferies, 'Bifurcations of a Forced Magnetic Oscillator Near Points of Resonance,' *Phys. Rev. Lett.* **53**, 250 (1984).

L.A. Bunimovich, 'On Ergodic Properties of Nowhere Dispersing Billiards,' *Comm. Math. Phys.* **65**, 295 (1979).

T.L. Carroll, L.M. Pecora and F.J. Rachford, 'Chaotic Transients and Multiple Attractors in Spin Wave Experiments,' *Phys. Rev. Lett.* **59**, 2891 (1987).

J.R. Cary and R.G. Littlejohn, 'Noncanonical Hamiltonian Mechanics and its Application to Magnet Field Line Flow,' *Ann. Phys.* **151**, 1 (1983).

J.R. Cary, J.D. Meiss and A. Bhattacharjee, 'Statistical Characterization of Periodic Measure Preserving Mappings,' *Phys. Rev. A* **23**, 2744 (1981).

G. Casati, B.V. Chirikov, I. Guarneri and D.L. Shelpelyansky, 'Dynamical Stability of Quantum "Chaotic" Motion in a Hydrogen Atom,' *Phys. Rev. Lett.* **56**, 2437 (1986).

G. Casati, B.V. Chirikov, F.M. Izraelev and J. Ford, 'Stochastic Behavior of a

Quantum Pendulum Under Periodic Perturbation,' in *Stochastic Behavior in Classical and Quantum Hamiltonian Systems*, (Lecture Notes in Physics **93** (Springer, 1977).

M. Casdagli, 'Nonlinear Prediction of Chaotic Time Series,' *Physica D* **35**, 335 (1989).

J. Chaiken, R. Chevray, M. Tabor and Q.M. Tan, 'Experimental Study of Lagrangian Turbulence in Stokes Flow,' *Proc. Roy. Soc. (Lond.) A* **408**, 165 (1986).

C. Chen, G. Györgyi and G. Schmidt, 'Universal Scaling in Dissipative Systems,' *Phys. Rev. A.* **35**, 2660 (1987).

Q. Chen, I. Dana, J.D. Meiss, N. Murray and I.C. Percival, 'Resonance and Transport in the Sawtooth Map,' *Physica D* **67**, 217 (1990a).

Q. Chen, M. Ding and E. Ott, 'Chaotic Scattering in Several Dimension,' *Phys. Lett. A* **115**, 93 (1990b).

Q. Chen, E. Ott and L. Hurd, 'Calculating Topological Entropies of Chaotic Dynamical Systems,' *Phys. Lett. A* **156**, 48 (1991).

B.V. Chirikov, 'A Universal Instability of Many-Dimensional Oscillator Systems,' *Phys. Rep.* **52**, 265 (1979).

B.V. Chirikov, F.M. Izrealev and D.L. Shepelyansky, 'Dynamical Stochasticity in Classical and Quantum Mechanics,' *Sov. Sci. Rev.* **2C**, 209 (1981).

B.V. Chirikov and D.L. Shepelyansky, 'Correlation Properties of Dynamical Chaos in Hamiltonian Systems,' *Physica D* **13**, 395 (1984).

P. Collet and J.P. Eckmann, *Iterated Maps of the Interval as Dynamical Systems* (Berkhauser, Boston, 1980).

P. Collet, J.L. Lebowitz and A. Parzio, 'The Dimension Spectrum of Some Dynamical Systems,' *J. Stat. Phys.* **47**, 609 (1987).

P. Constantin, C. Foias, B. Nicolaenko and R. Temam, 'Spectral Barriers and Inertial Manifolds for Dissipative Partial Differential Equations,' *J. Dyn. and Diff. Eq.* **1**, 45 (1989).

R. Courant and D. Hilbert, *Methods of Mathematical Physics*, Volume 2, third printing, (John Wiley, New York, 1966), p. 640.

J.P. Crutchfield, M. Nauenberg and J. Rudnick, 'Scaling for External Noise at the Onset of Chaos,' *Phys. Rev. Lett.* **46**, 933 (1981).

P. Cvitanović, *Universality in Chaos: A Reprint Selection* (Adam Hilger, Bristol, 1984).

P. Cvitanović and B. Eckhardt, 'Periodic Orbit Quantization of Chaotic Systems,' *Phys. Rev. Lett.* **63**, 823 (1989).

P. Cvitanović and B. Eckhardt, 'Periodic Orbit Expansions for Classical Smooth Flows,' *J. Phys. A* **24**, L237 (1991).

P. Cvitanović, G.H. Gunaratne and I. Procaccia, 'Topological and Metric Properties of Hénon-Type Strange Attractors,' *Phys. Rev. A* **38**, 1503 (1988).

P. Cvitanović, M.H. Jensen, L.P. Kadanoff and I. Procaccia, 'Renormalization, Unstable Manifolds and the Fractal Structure of Mode Locking,' *Phys. Rev. Lett.* **55**, 343 (1985).

P. Dahlqvist and G. Russberg, 'Existence of Stable Orbits in the x^2y^2 Potential,' *Phys. Rev. Lett.* **65**, 2837 (1990).

R.L. Devaney, *An Introduction to Chaotic Dynamical Systems* (Benjamin/Cummings, Menlo Park, 1986).

D. D'Humieres, M.R. Beasley, B. Huberman and A. Libchaber, 'Chaotic States and Routes to Chaos in the Forced Pendulum,' *Phys. Rev. A* **26**, 3483 (1982).

M. Ding, T. Bountis and E. Ott, 'Algebraic Escape in Higher Dimensional Hamiltonian Systems,' *Phys. Lett. A* **151**, 395 (1990a).

M. Ding, C. Grebogi, E. Ott and J.A. Yorke, 'Transition to Chaotic Scattering,' *Phys. Rev. A* **42**, 7025 (1990b).

W.L. Ditto, S.N. Rauseo and M.L. Spano, 'Experimental Control of Chaos,' *Phys. Rev. Lett.* **65**, 3211 (1990a).

W.L. Ditto, M.L. Spano, H.T. Savage, S.N. Rauseo, J. Heagy and E. Ott, 'Experimental Observation of a Strange Nonchaotic Attractor,' *Phys. Rev. Lett.* **65**, 533 (1990b).

M.F. Doherty and J.M. Ottino, 'Chaos in Deterministic Systems: Strange Attractors, Turbulence, and Applications in Chemical Engineering,' *Chem. Eng. Sci.* **43**, 139 (1988).

T. Dombre, U. Frisch, J.M. Greene, M. Hénon, A. Mehr and A.M. Soward, 'Chaotic Streamlines in the ABC Flows,' *J. Fluid Mech.* **167**, 353 (1986).

E. Doron, U. Smilansky and A. Frenkel, 'Experimental Demonstration of Chaotic Scattering of Microwaves,' *Phys. Rev. Lett.*, **65**, 3072 (1990).

U. Dressler and G. Nitsche, 'Controlling Chaos Using Time Delay Coordinates,' *Phys. Rev. Lett.* **68**, 1 (1992).

B. Eckhardt, 'Irregular Scattering,' *Physica D* **33**, 89 (1988a).

B. Eckhardt, 'Quantum Mechanics of Classically Non-integrable Systems,' *Phys. Reports* **163**, 205 (1988b).

B. Eckhardt and H. Aref, 'Integrable and Chaotic Motion of Four Vortices II. Collision Dynamics of Vortex Pairs,' *Phil. Trans. Roy. Soc. London* **326**, 655 (1988).

J.-P. Eckmann, 'Roads to Turbulence in Dissipative Dynamical Systems,' *Rev. Mod. Phys.* **53**, 643 (1981).

J.-P. Eckmann, S. Oliffson Kamphorst, D. Ruelle and S. Ciliberto, 'Lyapunov Exponents from Time Series,' *Phys. Rev. A* **34**, 4971 (1986).

J.-P. Eckmann and D. Ruelle, 'Ergodic Theory of Chaos and Strange Attractors,' *Rev. Mod. Phys.* **57**, 617 (1985).

R.S. Ellis, *Entropy, Large Deviations and Statistical Mechanics* (Springer-Verlag, New York, (1985)).

D.F. Escande, 'Large Scale Stochasticity in Hamiltonian Systems,' *Physica Scripta* **T2/1**, 126 (1982).

R. Eykholt and D.K. Umberger, 'Relating the Various Scaling Exponents to Characterize Fat Fractals in Nonlinear Dynamical Systems,' *Physica D* **30**, 43 (1988).

J.D. Farmer, 'Sensitive Dependence on Parameters in Nonlinear Dynamics,' *Phys. Rev. Lett.* **55**, 351 (1985).

J.D. Farmer, E. Ott and J.A. Yorke, 'The Dimension of Chaotic Attractors,' *Physica D* **7**, 153 (1983).

J.D. Farmer and J.J. Sidorowich, 'Predicting Chaotic Time Series,' *Phys. Rev. Lett.* **59**, 845 (1987).

M.J. Feigenbaum, 'Quantitative Universality for a Class of Nonlinear Transformations,' *J. Stat. Phys.* **19**, 25 (1978).

M.J. Feigenbaum, 'Universal Behavior in Nonlinear Systems,' *Los Alamos Science* **1**, 4 (1980a) (Reprinted in Cvitanović (1984)).

M.J. Feigenbaum, 'The Transition to Aperiodic Behavior in Turbulent Systems,' *Comm. Math. Phys.* **77**, 65 (1980b).

M.J. Feigenbaum and B. Hasslacher, 'Irrational Decimation and Path Integrals for External Noise,' *Phys. Rev. Lett.* **49**, 605 (1982).

M.J. Feigenbaum, L.P. Kadanoff and S.J. Shenker, 'Quasiperiodicity in Dissipative Systems: A Renormalization Group Analysis,' *Physica* **5D**, 370 (1982).

M. Feingold, L.P. Kadanoff and O. Piro, 'Passive Scalars, Three-Dimensional Volume Preserving Maps, and Chaos,' *J. Stat. Phys.* **50**, 529 (1988).

J.M. Finn, 'The Destruction of Magnetic Surfaces in Tokamaks by Current Perturbations,' *Nucl. Fusion* **15**, 845 (1975).

S. Fishman, D.R. Grempel and R. Prange, 'Quantum Chaos, Recurrence and Anderson Localization,' *Phys. Rev. Lett.* **49**, 509 (1982).

T.B. Fowler, 'Application of Stochastic Control Techniques to Chaotic Nonlinear Systems,' *IEEE Trans. Automatic Control* **24**, 201 (1989).

U. Frisch and G. Parisi, 'On the Singularity Structure of Fully Developed Turbulence,' in *Turbulence and Predictability in Geophysical Fluid Dynamics and Climate Dynamics*, edited by M. Ghil, R. Benzi and G. Parisi (North-Holland, New York, 1985), pp. 84–8.

H. Fujisaka, 'Statistical Dynamics Generated by Fluctuations of Local Lyapunov Exponents,' *Prog. Theor. Phys.* **70**, 1264 (1983).

H. Fujisaka and M. Inoue, 'Statistical-Thermodynamics Formalism of Self-Similarity,' *Prog. Theor. Phys.* **77**, 1334 (1987).

E.J. Galvez, B.E. Sauer, L. Moorman, P.M. Koch and D. Richards, 'Microwave Ionization of *H* Atoms: Breakdown of Classical Dynamics for High Frequences,' *Phys. Rev. Lett.* **61**, 2011 (1988).

P. Gaspard and S.A. Rice, 'Scattering from a Classical Chaotic Repeller,' *J. Chem. Phys.* **90**, 2225 (1989a).

P. Gaspard and S.A. Rice, 'Semiclassical Quantization of the Scattering from a Classical Chaotic Repeller,' *J. Chem. Phys.* **90**, 2242 (1989b).

P. Gaspard and S.A. Rice, 'Exact Quantization of the Scattering from a Classically Chaotic Repeller,' *J. Chem. Phys.* **90**, 2255 (1989c).

T. Geisel, J. Wagenhuber, P. Niebauer and G. Obermair, 'Chaotic Dynamics of Ballistic Electrons in Lateral Superlattices and Magnetic Fields,' *Phys. Rev. Lett.* **64**, 1581 (1990).

T. Geisel, A. Zacherl and G. Radons, 'Generic $1/f$ Noise in Chaotic Hamiltonian Dynamics,' *Phys. Rev. Lett.* **59**, 2503 (1987).

R.S. Gioggia and N.B. Abraham, 'Anomalous Mode Pulling, Instabilities and Chaos in a Single Mode, Standing Wave 3.39 $m\mu$ He–Ne Laser,' *Phys. Rev. A* **29**, 1304 (1984).

L. Glass and M.C. Mackey, *From Clocks to Chaos* (Princeton University Press, Princeton, 1988).

H. Goldstein, *Classical Mechanics*, 2nd ed. (Addison-Wesley, Reading, 1980).

J.P. Gollub and S.V. Benson, 'Many Routes to Turbulent Convection,' *J. Fluid Mech.* **100**, 449 (1980).

J.P. Gollub and H.L. Swinney, 'Onset of Turbulence in a Rotating Fluid,' *Phys. Rev. Lett.* **35**, 927 (1975).

R. Graham, M. Schlautmann and D.L. Shepelyansky, 'Dynamical Localization in Josephson Junctions,' *Phys. Rev. Lett.* **67**, 255 (1991).

P. Grassberger, 'On the Hausdorff Dimension of Fractal Attractors,' *J. Stat. Phys.* **26**, 173 (1981).

P. Grassberger, 'Generalized Dimensions of Strange Attractors,' *Phys. Lett. A* **97**, 227 (1983).

P. Grassberger, 'Generalizations of the Hausdorff Dimension of Fractal Measures,' *Phys. Lett. A* **107**, 101 (1985).

P. Grassberger, R. Badii and A. Politi, 'Scaling Laws for Invariant Measures on Hyperbolic and Nonhyperbolic Attractors,' *J. Stat. Phys.* **51**, 135 (1988).

P. Grassberger and I. Procaccia, 'Measuring the Strangeness of Strange Attractors,' *Physica D* **9**, 189 (1983).

C. Grebogi, E. Kostelich, E. Ott and J.A. Yorke, 'Multidimensioned Intertwined Basin Boundaries: Basin Structure of the Kicked Double Rotor,' *Physica D* **25**, 347 (1987a).

C. Grebogi, S.W. McDonald, E. Ott and J.A. Yorke, 'Final State Sensitivity: An Obstruction to Predictability,' *Phys. Lett. A* **99**, 415 (1983a).

C. Grebogi, S.W. McDonald, E. Ott and J.A. Yorke, 'Exterior Dimension of Fat Fractals,' *Phys. Lett. A* **110**, 1 (1985b).

C. Grebogi, H.E. Nusse, E. Ott and J.A. Yorke, 'Basic Sets: Sets that Determine the Dimension of Basin Boundaries,' in *Dynamical Systems* edited by J.C. Alexander, Lecture Notes in Math. **1342** (Springer-Verlag, Berlin, 1988a) pp. 220–50.

C. Grebogi, E. Ott, S. Pelikan and J.A. Yorke, 'Strange Attractors that Are Not Chaotic,' *Physica D* **13**, 261 (1984).

C. Grebogi, E. Ott, F. Romeiras and J.A. Yorke, 'Critical Exponents for Crisis-Induced Intermittency,' *Phys. Rev. A* **36**, 5365 (1987b).

C. Grebogi, E. Ott and J.A. Yorke, 'Chaotic Attractors in Crisis,' *Phys. Rev. Lett.* **48**, 1507 (1982).

C. Grebogi, E. Ott and J.A. Yorke, 'Fractal Basin Boundaries, Long-Lived Chaotic Transients, and Unstable–Unstable Pair Bifurcations,' *Phys. Rev. Lett.* **50**, 935 (1983b).

C. Grebogi, E. Ott and J.A. Yorke, 'Crises, Sudden Changes in Chaotic Attractors and Chaotic Transients,' *Physica D* **7**, 181 (1983c).

C. Grebogi, E. Ott and J.A. Yorke, 'Super-Persistent Chaotic Transients,' *Ergodic Theor. and Dyn. Sys.* **5**, 341 (1985a).

C. Grebogi, E. Ott and J.A. Yorke, 'Attractors on an N-Torus: Quasiperiodicity Versus Chaos,' *Physica D* **15**, 354 (1985c).

C. Grebogi. E. Ott and J.A. Yorke, 'Critical Exponents of Chaotic Transients in Nonlinear Dynamical Systems,' *Phys. Rev. Lett.* **57**, 1284 (1986a).

C. Grebogi, E. Ott and J.A. Yorke, 'Metamorphoses of Basin Boundaries in Nonlinear Dynamical Systems,' *Phys. Rev. Lett.* **56**, 1011 (1986b).

C. Grebogi, E. Ott and J.A. Yorke, 'Basin Boundary Metamorphoses: Changes in Accessible Boundary Orbits,' *Physica D* **24**, 243 (1987c).

C. Grebogi, E. Ott and J.A. Yorke, 'Chaos, Strange Attractors and Fractal Basin Boundaries in Nonlinear Dynamics,' *Science* **238**, 585 (1987d).

C. Grebogi, E. Ott and J.A. Yorke, 'Unstable Periodic Orbits and the Dimension of Chaotic Attractors,' *Phys. Rev. A* **36**, 3522 (1987e).

C. Grebogi, E. Ott and J.A. Yorke, 'Unstable Periodic Orbits and the Dimensions of Multifractal Chaotic Attractors,' *Phys. Rev. A* **37**, 1711 (1988b).

C. Grebogi, E. Ott and J.A. Yorke, 'Roundoff Induced Periodicity and the Correlation Dimension of Chaotic Attractors,' *Phys. Rev. A* **38**, 366 (1988c).

J.M. Greene, 'A Method for Determining a Stochastic Transition,' *J. Math. Phys.* **20**, 1183 (1979).

J.M. Greene and J.-S. Kim, 'The Calculation of Lyapunov Spectra,' *Physica D* **24**, 213 (1987).

J.M. Greene, R.S. MacKay, F. Vivaldi and M.J. Feigenbaum, 'Universal Behavior in Families of Area-Preserving Maps,' *Physica D* **3**, 468 (1981).

J. Guckenheimer and G. Buzyna, 'Dimension Measurements for Geostrophic Turbulence,' *Phys. Rev. Lett.* **51**, 1438 (1983).

J. Guckenheimer and P. Holmes, *Nonlinear Oscillations, Dynamical Systems, and Bifurcations of Vector Fields* (Springer-Verlag, New York, 1983).

I. Gumowski and C. Mira, *Recurrences and Discrete Dynamical Systems* Springer Lecture Notes in Mathematics, **809** (Springer-Verlag, Berlin, 1980).

G.H. Gunaratne, P.S. Linsay and M.J. Vinson, 'Chaos Beyond Onset: A Comparison of Theory and Experiment,' *Phys. Rev. Lett.* **63**, 1 (1989).

M.C. Gutzwiller, 'Phase Integral Approximation in the Momentum Space and the Bound States of an Atom: I and II,' *J. Math. Phys.* **8**, 1979 (1967), *ibid.* **10**, 1004 (1969).

M.C. Gutzwiller, 'Classical Quantization of a Hamiltonian with Ergodic Behavior,' *Phys. Rev. Lett.* **45**, 150 (1980).

M.C. Gutzwiller, 'Stochastic Behavior in Quantum Scattering,' *Physica D* **7**, 341 (1983).

M.C. Gutzwiller, *Chaos in Classical and Quantum Mechanics* (Springer, New York, 1990).

F. Haake, *Quantum Signatures of Chaos* (Springer, New York, 1991).

J. Hadamard, 'Les Surfaces à Courbures Opposées et Leurs Ligns Géodésiques,' *J. Math. Pure Appl.* **4**, 27 (1898).

T.C. Halsey, M.H. Jensen, L.P. Kadanoff, I. Procaccia and B.I. Shraiman, 'Fractal Measures and Their Singularities: The Characterization of Strange Sets,' *Phys. Rev. A* **33**, 1141 (1986).

S.M. Hammel, 'A Noise Reduction Method for Chaotic Systems,' *Phys. Lett. A* **148**, 421 (1990).

S. Hammel, C.K.R.T. Jones and J. Maloney, 'Global Dynamical Behavior of the Optical Field in a Ring Cavity,' *J. Opt. Soc. Am. B* **2**, 552 (1985).

S.M. Hammel, J.A. Yorke and C. Grebogi, 'Do Numerical Orbits of Chaotic Dynamical Processes Represent True Orbits?,' *J. Complexity* **3**, 136 (1987).

J.H. Hannay and A.M. Ozorio de Almeida, 'Periodic Orbits and a Correlation Function for the Semiclassical Density of States,' *J. Phys. A* **17**, 3429 (1984).

J.D. Hanson, 'Fat Fractal Scaling Exponent of Area Preserving Maps,' *Phys. Rev. A* **35**, 1470 (1987).

J.D. Hanson and J.R. Cary, 'Elimination of Stochasticity in Stellerators, *Phys. Fluids* **27**, 767 (1984).

J.D. Hanson, J.R. Cary and J.D. Meiss, 'Algebraic Decay in Self-Similar Markov Chains,' *J. Stat. Phys.* **39**, 327 (1985).

H. Hata, T. Morita, K. Tomita and H. Mori, 'Spectra of Singularities for the Lozi and Hénon Maps,' *Prog. Theor. Phys.* **78**, 721 (1987).

F. Hausdorff, 'Dimension und äusseres Mass,' *Math. Annalen.* **79**, 157 (1918).

R.H.G. Helleman, 'Self-Generated Chaotic Behavior in Nonlinear Dynamics,' in *Fundamental Problems in Statistical Mechanics*, 5, edited by E.G.D. Cohen (North Holland, Amsterdam 1980). (Reprinted in Cvitanović (1984).)

E.J. Heller, 'Bound-State Eigenfunctions of Classically Chaotic Hamiltonian Systems: Scars of Periodic Orbits,' *Phys. Rev. Lett.* **53**, 1515 (1984).

M. Hénon, 'A Two Dimensional Mapping with a Strange Attractor,' *Comm. Math. Phys.* **50**, 69 (1976).

H.G.E. Hentschel and I. Procaccia, 'The Infinite Number of Generalized Dimensions of Fractals and Strange Attractors,' *Physica D* **8**, 435 (1983).

J.E. Hirsch, M. Nauenberg and D.J. Scalapino, 'Intermittency in the Presence of Noise: A Renormalization Group Formulation,' *Phys. Lett.* **87A**, 391 (1982).

M.W. Hirsch and S. Smale, *Differential Equations, Dynamical Systems and Linear Algebra* (Academic Press, New York, 1974).

T. Hogg and B.A. Huberman, 'Recurrence Phenomena in Quantum Dynamics,' *Phys. Rev. A* **28**, 22 (1982).

T. Horita, H. Hata, R. Ishizaki and H. Mori, 'Long-Time Correlations and Expansion-Rate Spectra of Chaos in Hamiltonian Systems,' *Prog. Theor. Phys.* **83**, 1065 (1990).

G.-H. Hsu, E. Ott and C. Grebogi, 'Strange Saddles and the Dimensions of Their Invariant Manifolds,' *Phys. Lett. A* **127**, 199 (1988).

B. Hu and J. Rudnick, 'Exact Solution of the Feigenbaum Renormalization Group Equations for Intermittency,' *Phys. Rev. Lett.* **48**, 1645 (1982).

J.L. Hudson and J.C. Mankin, 'Chaos in the Belousov–Zhabotinskii Reaction,' *J. Chem. Phys.* **74**, 6171 (1981).

M. Iansiti, Q. Hu, R.M. Westervelt and M. Tinkham, 'Noise and Chaos in a Fractal Basin Boundary Regime of a Josephson Junction,' *Phys. Rev. Lett.* **55**, 746 (1985).

R. Ishizaki, T. Horita, T. Kobayashi and H. Mori, 'Anomalous Diffusion Due to Accelerator Modes in the Standard Map,' *Prog. Theor. Phys.* **85**, 1013 (1991).

F.M. Izraelev and D.L. Shepelyansky, 'Quantum Resonance for a Rotator in a

Nonlinear Periodic Field,' *Sov. Phys. Dokl.* **24**, 996 (1979).

M.V. Jacobson, 'Topological and Metric Properties of One Dimensional Endomorphisms,' *Sov. Math. Dokl.* **19**, 1452 (1978).

M.V. Jacobson, 'Absolutely Continuous Measures for One-Parameter Families of One-Dimensional Maps,' *Comm. Math. Phys.* **81**, 39 (1981).

R.A. Jalabert, H.V. Baranger and A.D. Stone, 'Conductance Fluctuations in the Ballistic Regime: A Probe of Quantum Chaos?', *Phys. Rev. Lett.* **65**, 2442 (1990).

M.H. Jensen, P. Bak and T. Bohr, 'Complete Devil's Staircase, Fractal Dimension and Universality of Mode-Locking Structures,' *Phys. Rev.* **50**, 1637 (1983).

M.H. Jensen, P. Bak and T. Bohr, 'Transition to Chaos by Interaction of Resonances in Dissipative Systems. I. Circle Maps,' *Phys. Rev. A* **30**, 1960 (1984).

M.H. Jensen, L.P. Kadanoff, A. Libchaber, I. Procaccia and J. Stavans, 'Global Universality at the Onset of Chaos: Results of a Forced Rayleigh–Bénard Experiment,' *Phys. Rev. Lett.* **55**, 2798 (1985).

R.V. Jensen, 'Stochastic Ionization of Surface-State Electrons: Classical Theory,' *Phys. Rev. A* **30**, 386 (1984).

R.V. Jensen, S.M. Susskind and M.M. Sanders, 'Chaotic Ionization of Highly Excited Hydrogen Atoms: Comparison of Classical and Quantum Theory with Experiment,' *Phys. Rpt.* **201**, 1 (1991).

P. Jung and P. Hänggi, 'Invariant Measure of a Driven Nonlinear Oscillator with External Noise,' *Phys. Rev. Lett.* **65**, 3365 (1990).

L.P. Kadanoff, 'Scaling for a Critical Kolmogorov–Arnol'd–Moser Trajectory,' *Phys. Rev. Lett.* **47**, 1641 (1981).

L.P. Kadanoff, 'Supercritical Behavior of an Ordered Trajectory,' *J. Stat. Phys.* **31**, 1 (1983).

L.P. Kadanoff and C. Tang, 'Escape from Strange Repellers,' *Natl. Acad. Sci. USA* **81**, 1276 (1984).

H. Kantz and P. Grassberger, 'Repellers, Semi-Attractors and Long-Lived Chaotic Transients,' *Physica D* **17**, 75 (1985).

J.L. Kaplan and J.A. Yorke, 'Preturbulence: A Regime Observed in a Fluid Flow Model of Lorenz,' *Comm. Math. Phys.* **67**, 93 (1979a).

J.L. Kaplan and J.A. Yorke, 'Chaotic Behavior of Multidimensional Difference Equations,' in *Functional Differential Equations and Approximations of Fixed Points*, edited by H.-O. Peitgen and H.-O. Walter, Lecture Notes in Mathematics, **730** (Springer, Berlin, 1979b), p. 204.

C.F.F. Karney, 'Long Time Correlations in the Stochastic Regime,' *Physica D* **8**, 360 (1983).

C.F.F. Karney, A.B. Rechester and R.B. White, 'Effect of Noise on the Standard Mapping,' *Physica D* **4**, 425 (1982).

A. Katok, 'Lyapunov Exponents, Entropy and Periodic Points for Diffeomorphisms,' *Pub. Math. IHES* **51**, 377 (1980).

D. Katzen and I. Procaccia, 'Phase Transitions in the Thermodynamic Formalism of Multifractals,' *Phys. Rev. Lett.* **58**, 1169 (1987).

A.N. Kolmogorov, 'A New Metric Invariant of Transitive Dynamical Systems and Automorphisms in Lebesgue Spaces,' *Dokl. Acad. Nauk SSSR* **119**, 861 (1958).

A.N. Kolmogorov, 'Preservation of Conditionally Periodic Movements with Small Change in the Hamiltonian Function,' *Dokl. Akad. Nauk SSSR* **98**, 527 (1954). (Reprinted in MacKay and Meiss (1987).)

E.J. Kostelich and J.A. Yorke, 'Noise Reduction in Dynamical Systems,' *Phys. Rev. A* **38**, 1649 (1988).

Z. Kovács and T. Tél, 'Thermodynamics of Irregular Scattering,' *Phys. Rev. Lett.* **64**, 1617 (1990).

F. Ladrappier, 'Some Relations Between Dimension and Lyapunov Exponent,' *Comm. Math. Phys.* **81**, 229 (1981).

D.P. Lathrop and E.J. Kostelich, 'Characterization of an Experimental Strange Attractor by Periodic Orbits,' *Phys. Rev. A* **40**, 4928 (1989).

Y.-T. Lau and J.M. Finn, 'Three Dimensional Kinematic Reconnection of Plasmoids,' *Astrophys. J* **366**, 577 (1991).

Y.-T. Lau and J.M. Finn, 'Dynamics of a Three-Dimensional Incompressible Flow with Stagnation Points,' *Physica D* (to be published) (1992).

Y.-T. Lau, J.M. Finn and E. Ott, 'Fractal Dimension in Nonhyperbolic Chaotic Scattering,' *Phys. Rev. Lett.* **66**, 978 (1991).

W. Lauterborn, 'Subharmonic Route to Chaos in Acoustics,' *Phys. Rev. Lett.* **47**, 1445 (1981).

M. Levi, 'Qualitative Analysis of Periodically Forced Relaxation Oscillations,' *Mem. Am. Math. Soc.* **214**, 1 (1981).

T.Y. Li and J.A. Yorke, 'Period Three Implies Chaos,' *Am. Math. Monthly* **82**, 985 (1975).

A. Libchaber and J. Maurer, 'Une Experience de Rayleigh–Bénard de Géométrie Réduite; Multiplication, Accrochage, et Démultiplication de Fréquencies,' *J. Phys. (Paris) Coll.* **41**, C3 (1980).

A.J. Lichtenberg and M.J. Lieberman, *Regular and Stochastic Motion* (Springer-Verlag, New York, 1983).

P.S. Linsay, 'Period Doubling and Chaotic Behavior in a Driven Anharmonic Oscillator,' *Phys. Rev. Lett.* **47**, 1349 (1981).

P.S. Linsay, 'An Efficient Method of Forecasting Chaotic Time Series Using Linear Interpolation,' *Phys. Lett. A* **153**, 353 (1991).

R.G. Littlejohn, 'The Semiclassical Evolution of Wavepackets,' *Phys. Rep.* **138**, 193 (1986).

A. Lopes, 'Dimension Spectra and a Mathematical Model for Phase Transition,' *Adv. Appl. Math.* **11**, 475 (1990a).

A. Lopes, 'Dynamics of Real Polynomials on the Plane and Triple Point Phase Transition,' *Math. Comp. Modelling* **13**, 117 (1990b).

E.N. Lorenz, 'Deterministic Nonperiodic Flow,' *J. Atmos. Sci.* **20**, 130 (1963).

R.S. MacKay, 'A Renormalization Approach to Invariant Circles in Area-Preserving Maps,' *Physica D* **7**, 283 (1983).

R.S. MacKay and J.D. Meiss, editors, *Hamiltonian Dynamical Systems: A Reprint Selection* (Adam Hilger, Bristol, 1987).

B.B. Mandelbrot, *The Fractal Geometry of Nature* (Freeman, San Francisco, 1982).

V.P. Maslov and M.V. Fedoriuk, *Semiclassical Approximation in Quantum Mechanics* (D. Reidel, Dordrecht, 1981).

P. Mattila, 'Hausdorff Dimension, Orthogonal Projections, and Intersections with Planes,' *Ann. Acad. Sci. Fenn. Ser. A*1, *Math.* **1**, 277 (1975).

R.M. May, 'Simple Mathematical Models with Very Complicated Dynamics,' *Nature* **261**, 459 (1976).

S.W. McDonald, C. Grebogi, E. Ott and J.A. Yorke, 'Fractal Basin Boundaries,' *Physica D* **17**, 125 (1985).

S.W. McDonald and A.N. Kaufman, 'Spectrum and Eigenfunctions for a Hamiltonian with Stochastic Trajectories,' *Phys. Rev. Lett.* **42**, 1189 (1979).

S.W. McDonald and A.N. Kaufman, 'Wave Chaos in the Stadium: Statistical Properties of Short-Wave Solutions of the Helmholtz Equation,' *Phys. Rev. A* **37**, 3067 (1988).

B.I. Meerson, E.A. Oks and P.V. Sasarov, 'Stochastic Instability of an Oscillator and the Ionization of Highly-Excited Atoms Under the Action of Electromagnetic Radiation,' *JETP Lett.* **29**, 72 (1979).

M.L. Mehta, *Random Matrices*, (Academic, New York, 1967).

J.D. Meiss and E. Ott, 'Markov–Tree Model of Intrinsic Transport in Hamiltonian Systems,' *Phys. Rev. Lett.* **55**, 2741 (1985).

M. Metropolis, M.L. Stein and P.R. Stein, 'On Finite Limit Sets for Transformations on the Unit Interval,' *J. Combinatorial Theory* **15**, 25 (1973).

J. Milnor and W. Thurston, 'On Iterated Maps of the Interval,' in *Dynamical Systems*, edited by J.C. Alexander, Lecture Notes in Mathematics, **1342** (Springer-Verlag, New York, 1987) pp. 465–563.

F.C. Moon, 'Fractal Boundary for Chaos in a Two State Mechanical Oscillator,' *Phys. Rev. Lett.* **53**, 962 (1984).

F.C. Moon, *Chaotic Vibrations* (John Wiley, New York 1987).

F.C. Moon and P.J. Holmes, 'A Magnetoelastic Strange Attractor,' *J. Sound Vib.* **65**, 285 (1979).

F.C. Moon and G.-X. Li, 'Fractal Basin Boundaries and Homoclinic Orbits for Periodic Motion in a Two-Well Potential,' *Phys. Rev. Lett.* **55**, 1439 (1985).

H. Mori, H. Hata, T. Horita and T. Kobayashi, 'Statistical Dynamics of Dynamical Systems,' *Prog. Theor. Phys.*, Supplement No. 99, 1 (1989).

T. Morita, H. Hata, H. Mori, T. Horita and K. Tomita, 'On Partial Dimensions and Spectra of Singularities of Strange Attractors,' *Prog. Theor. Phys.* **78**, 511 (1987).

J. Mork, J. Mark and B. Tromberg, 'Route to Chaos and Competition Between Relaxation Oscillations for a Semiconductor Laser with Optical Feedback,' *Phys. Rev. Lett.* **65**, 1999 (1990).

J. Moser, *Stable and Random Motions in Dynamical Systems* (Princeton Univ. Press, 1973).

M. Napiórkowski, 'Scaling of the Uncertainty Exponent,' *Phys. Rev. A* **33**, 4423 (1986).

M. Nauenberg and I. Rudnick, 'Universality and the Power Spectrum at the Onset of Chaos,' *Phys. Rev. B* **24**, 239 (1981).

S.E. Newhouse, 'Entropy and Volume as Measures of Orbit Complexity,' in *The Physics of Phase Space*, edited by Y.S. Kim and W.W. Zachary (Springer-Verlag, Berlin, 1986).

S. Newhouse, D. Ruelle and F. Takens, 'Occurrence of Strange Axiom A Attractors Near Quasiperiodic Flows on T^m ($m = 3$ or more), *Commun. Math. Phys.* **64**, 35 (1978).

D.W. Noid, S.K. Gray and S.A. Rice, 'Fractal Behavior in Classical Collisional Energy Transfer,' *J. Chem. Phys.* **85**, 2649 (1986).

H.E. Nusse and J.A. Yorke, 'A Procedure for Finding Numerical Trajectories on Chaotic Saddles,' *Physica D* **36**, 137 (1989).

P.W. O'Conner, J. Gehlen and E.J. Heller, 'Properties of Random Superpositions of Plane Waves,' *Phys. Rev. Lett.* **58**, 1296 (1987).

K. Ogata, *Modern Control Engineering*, 2nd ed. (Prentice Hall, Englewood Cliffs, 1990), pp. 778–84.

A.R. Osborne and A. Provenzale, 'Finite Correlation Dimension for Stochastic Systems with Power Law Spectra,' *Physica D* **35**, 357 (1989).

V.I. Oseledec, 'A Multiplicative Ergodic Theorem: Lyapunov Characteristic Numbers for Dynamical Systems,' *Trudy Mosk. Obsch.* **19**, 179 (1968). (*Trans. Mosc. Math. Soc.* **19**, 197 (1968).)

S. Ostlund, D. Rand, J. Sethna and E. Siggia, 'Universal Properties of the Transition from Quasiperiodicity to Chaos in Dissipative Systems,' *Physica D* **8**, 303 (1983).

E. Ott, 'Strange Attractors and Chaotic Motions of Dynamical Systems,' *Rev. Mod Phys.* **53**, 655 (1981).

E. Ott and T.M. Antonsen, 'Chaotic Fluid Convection and the Fractal Nature of Passive Scalar Gradients,' *Phys. Rev. Lett.* **61**, 2839 (1988).

E. Ott and T.M. Antonsen, 'Fractal Measures of Passively Convected Vector Fields and Scalar Gradients in Chaotic Fluid Flows,' *Phys. Rev. A* **38**, 3660 (1989).

E. Ott, T.M. Antonsen and J.D. Hanson, 'Effect of Noise on Time-Dependent Quantum Chaos,' *Phys. Rev. Lett.* **53**, 2187 (1984a).

E. Ott, C. Grebogi and J.A. Yorke, 'Theory of First Order Phase Transitions for Chaotic Attractors of Nonlinear Dynamical Systems,' *Phys. Lett. A* **135**, 343 (1989a).

E. Ott, C. Grebogi and J.A. Yorke, 'Controlling Chaos,' *Phys. Rev. Lett.* **64**, 1196 (1990a).

E. Ott, C. Grebogi and J.A. Yorke, 'Controlling Chaotic Dynamical Systems,' in *Chaos*, edited by D.K. Campbell (American Institute of Physics, New York, 1990b).

E. Ott and J.D. Hanson, 'The Effect of Noise on the Structure of Strange Attractors,' *Phys. Lett. A* **85**, 20 (1981).

E. Ott, T. Sauer and J.A. Yorke, 'Lyapunov Partition Functions for the Dimensions of Chaotic Sets,' *Phys. Rev. A* **39**, 4212 (1989b).

E. Ott, W.D. Withers and J.A. Yorke, 'Is the Dimension of Chaotic Attractors Invariant Under Coordinate Changes?', *J. Stat. Phys.* **36**, 687 (1984b).

E. Ott, E.D. Yorke and J.A. Yorke, 'A Scaling Law: How an Attractor's Volume Depends on Noise Level,' *Physica D* **16**, 62 (1985).

J.M. Ottino, *The Kinematics of Mixing: Stretching, Chaos, and Transport* (Cambridge University Press, New York, 1989).

A.M. Ozorio de Almeida, *Hamiltonian Systems: Chaos and Quantization* (Cambridge University Press, New York, 1988).

G. Paladin and A. Vulpiani, 'Anomalous Scaling Laws in Multifractal Objects,' *Phys. Rep.* **156**, 147 (1987).

B.-S. Park, C. Grebogi, E. Ott and J.A. Yorke, 'Scaling of Fractal Basin Boundaries Near Intermittency Transitions to Chaos,' *Phys. Rev. A* **40**, 1576 (1989).

S. Pelikan, 'A Dynamical Meaning of Fractal Dimension,' *Trans. Am. Math. Soc.* **292**, 695 (1985).

Ja. B. Pesin, 'Lyapunov Characteristic Exponents and Ergodic Properties of Smooth Dynamical Systems with an Invariant Measure,' *Sov. Math. Doklady* **17**, 196 (1976). (Reprinted in MacKay and Meiss (1987).)

J.M. Petit and M. Hénon, 'Satellite Encounter,' *Icarus* **60**, 536 (1986).

T. Poggio and F. Girosi, 'Regularization Algorithms for Learning that Are Equivalent to Multilayer Networks,' *Science* **247**, 978 (1990).

Y. Pomeau and P. Manneville, 'Intermittent Transition to Turbulence in Dissipative Dynamical Systems,' *Comm. Math. Phys.* **74**, 189 (1980).

R.R. Prasad, C. Meneveau and K.R. Sreenivasan, 'Multifractal Nature of the Dissipation Field of Passive Scalars in Fully Turbulent Flows,' *Phys. Rev. Lett.* **61**, 74 (1988).

R. Ramshankar and J.P. Gollub, 'Transport by Capillary Waves, Part II: Scalar Dispersion and Structure of the Concentration Field,' *Phys. Fluids A* **3**, 1344 (1991).

A. Rapisarda and M. Baldo, 'Coexistence of Regular and Chaotic Scattering in Heavy-Ion Collisions,' *Phys. Rev. Lett.* **66**, 2581 (1991).

A.B. Rechester and M.N. Rosenbluth, 'Electron Heat Transport in a Tokamak with Destroyed Magnetic Surfaces,' *Phys. Rev. Lett.* **40**, 38 (1978).

A.B. Rechester, M.N. Rosenbluth and R.B. White, 'Fourier Space Paths Applied to the Calculation of Diffusion for the Chirikov–Taylor Model,' *Phys. Rev. A* **23**, 2664 (1981).

A.B. Rechester and R.B. White, 'Calculation of Turbulent Diffusion for the Chirikov–Taylor Map,' *Phys. Rev. Lett.* **44**, 1586 (1980).

A. Renyi, *Probability Theory* (North Holland, Amsterdam, 1970).

R.W. Rollins and E.R. Hunt, 'Intermittent Transient Chaos at Interior Crises in the Diode Resonator,' *Phys. Rev. A* **29**, 3327 (1984).

F.J. Romeiras, A. Bondeson, E. Ott, T.M. Antonsen, Jr. and C. Grebogi, 'Quasiperiodic Forcing and the Observability of Strange Nonchaotic Attractors,' *Physica Scripta* **40**, 442 (1989).

F.J. Romeiras and E. Ott, 'Strange Nonchaotic Attractors of the Damped Pendulum with Quasiperiodic Forcing,' *Phys. Rev. A* **35**, 4404 (1987).

F.J. Romeiras, E. Ott, C. Grebogi and W.P. Dayawansa, 'Controlling Chaotic Dynamical Systems,' *Proc. Am. Control Conf.*, p. 1113 (1991).

V. Rom-Kedar, A. Leonard and S. Wiggins, 'An Analytical Study of Transport, Mixing and Chaos in an Unsteady Vortical Flow,' *J. Fluid Mech.* **214**, 347 (1990).

M.N. Rosenbluth, R.Z. Sagdeev, J.B. Taylor and G.M. Zaslavski, 'Destruction of Magnetic Surfaces by Magnetic Field Irregularities,' *Nuc. Fusion* **6**, 297 (1966).

O.E. Rössler, 'Chaotic Behavior in Simple Reaction Systems,' *Z. Naturforsch. A* **31**, 1168 (1976).

M.L. Roukes and O.L. Alerhand, 'Mesoscopic Junctions, Random Scattering and Strange Repellors,' *Phys. Rev. Lett.* **65**, 1651 (1990).

J. Roux, A. Rossi, S. Bachelert and C. Vidal, 'Representation of a Strange Attractor from an Experimental Study of Chemical Turbulence,' *Phys. Lett. A* **77**, 391 (1980).

D. Ruelle, *Thermodynamic Formalism* (Addison-Wesley, Reading, 1978).

D. Ruelle, *Chaotic Evolution and Strange Attractors* (Cambridge University Press, New York, 1989).

D. Ruelle and F. Takens, 'On the Nature of Turbulence,' *Commun. Math. Phys.* **20**, 167 (1971).

D.A. Russell, J.D. Hanson and E. Ott, 'Dimension of Strange Attractors,' *Phys. Rev. Lett.* **44**, 453 (1980).

R.Z. Sagdeev, D.A. Usikov and G.M. Zaslavsky, *Nonlinear Physics* (Harwood Academic Publishers, New York, 1990).

B. Saltzman, 'Finite Amplitude Convection as an Initial Value Problem. I,' *J. Atmos. Sci.* **19**, 329 (1962).

M. Sano and Y. Sawada, 'Measurement of the Lyapunov Spectrum from Chaotic Time Series,' *Phys. Rev. Lett.* **55**, 1082 (1985).

A.N. Sarkovskii, 'Coexistence of Cycles of a Continuous Map of a Line Into Itself,' *Ukr. Mat. Z.* **16**, 61 (1964).

G. Schmidt and J. Bialek, 'Fractal Diagrams for Hamiltonian Stochasticity,' *Physica D* **5**, 397 (1982). (Reprinted in MacKay and Meiss (1987).)

G. Schmidt and Q. Chen, 'Transition to Ergodicity in the Autonomous Fermi System' (preprint, 1991).

H.-G. Schuster, *Deterministic Chaos*, 2nd ed. (VCH, New York, 1988).

T.H. Seligman and J.J.M. Verbaarschot, 'Quantum Spectra of Classically Chaotic Systems with Time Reversal Invariance,' *Phys. Lett.* **108A**, 183 (1985).

T.H. Seligman, J.J.M. Verbaarschot and M.R. Zirnbaum, 'Quantum Spectra and Transition from Regular to Chaotic Classical Motion,' *Phys. Rev. Lett.* **53**, 215 (1984).

M. Sepúlveda, R. Badii and E. Pollak, 'Spectral Analysis of Conservative Dynamical Systems,' *Phys. Rev. Lett.* **63**, 1226 (1989).

R. Shaw, 'Strange Attractors, Chaotic Behavior, and Information Flow,' *Z. Naturforsch. A* **36**, 80 (1981).

R. Shaw, *The Dripping Faucet as a Model Chaotic System* (Aerial Press, Santa Cruz, 1984).

S.J. Shenker, 'Scaling Behavior in a Map of a Circle Onto Itself,' *Physica D* **5**, 405 (1982).

L.P. Shilnikov, 'A Contribution to the Problem of the Structure of an Extended Neighborhood of a Rough Equilibrium State of Saddle-Focus Type,' *Math. USSR Sb.* **10**, 91 (1970).

T. Shinbrot, E. Ott, C. Grebogi and J.A. Yorke, 'Using Chaos to Direct Trajectories to Targets,' *Phys. Rev. Lett.* **65**, 3250 (1990).

B. Shraiman, C.E. Wayne and P.C. Martin, 'Scaling Theory for Noisy Period-Doubling Transitions to Chaos,' *Phys. Rev. Lett.* **46**, 935 (1981).

M. Sieber and F. Steiner, 'Quantization of Chaos,' *Phys. Rev. Lett.* **67**, 1941 (1991).

C. Simó, 'On the Hénon–Pomeau Attractor,' *J. Stat. Phys.* **21**, 465 (1979).

R.H. Simoyi, A. Wolf and H.L. Swinney, 'One Dimensional Dynamics in a Multicomponent Chemical Reaction,' *Phys. Rev. Lett.* **49**, 245 (1982).

Ya. G. Sinai, 'On the Concept of Entropy of a Dynamical System,' *Dokl. Akad. Nauk SSSR* **124**, 768 (1959).

Ya. G. Sinai, 'Dynamical Systems with Elastic Reflections: Ergodic Properties of Dispersing Billiards,' *Russ. Math. Surveys* **25**, (2), 137 (1970).

Ya. G. Sinai, 'Gibbs Measures in Ergodic Theory,' *Russ. Math. Surveys* **27**, No. 4, 21 (1972).

Ya. G. Sinai, *Introduction to Ergodic Theory* (Princeton University Press, Princeton, 1976).

R.M. Sinclair, J.C. Hosea and G.V. Sheffield, 'A Method for Mapping a Toroidal Magnetic Field by Storage of Phase Stabilized Electrons,' *Rev. Sci. Instrum.* **41**, 1552 (1970).

J. Singer, Y.Z. Wang and H.H. Bau, 'Controlling a Chaotic System,' *Phys. Rev. Lett.* **66**, 1123 (1991).

S. Smale, 'Differentiable Dynamical Systems,' *Bull. Amer. Math. Soc.* **73**, 747 (1967).

U. Smilansky, 'The Classical and Quantum Theory of Chaotic Scattering,' in *Chaos and Quantum Physics*, edited by M.-J. Giannoni, A. Voros and J. Zinn-Justin (Elsevier, Amsterdam, 1992).

J.C. Sommerer, W.L. Ditto, C. Grebogi, E. Ott and M.L. Spano, 'Experimental Confirmation of the Theory of Critical Exponents of Crises,' *Phys. Lett: A* **153**, 105 (1991a).

J.C. Sommerer, W.L. Ditto, C. Grebogi, E. Ott and M.L. Spano, 'Experimental Confirmation of the Scaling Theory for Noise-Induced Crises,' *Phys. Rev. Lett.* **66**, 1947 (1991b).

J.C. Sommerer, E. Ott and C. Grebogi, 'Scaling Law for Characteristic Times of Noise-Induced Crisis,' *Phys. Rev. A* **43**, 1754 (1991c).

C. Sparrow, *The Lorenz Equations: Bifurcations, Chaos and Strange Attractors* (Springer-Verlag, New York, 1982).

K.R. Sreenivasan, 'Chaos in Open Flow Systems,' in *Dimensions and Entropies in Chaotic Systems*, edited by G. Mayer-Kress (Springer-Verlag, New York, 1986).

T.H. Stix, 'Radiation and Absorption via Mode Conversion in an Inhomogeneous Collision-Free Plasma,' *Phys. Rev. Lett.* **15**, 878 (1965).

H.L. Swinney and J.P. Gollub, 'Transition to Turbulence,' *Physics Today* **31**, (8), 41 (August, 1978).

M. Tabor, *Chaos and Integrability in Nonlinear Dynamics* (John Wiley and Sons, New York, 1989).

F. Takens, 'Detecting Strange Attractors in Turbulence,' in *Dynamical Systems and Turbulence*, edited by D.A. Rand and L.-S. Young, Springer Lecture Notes in Mathematics, **898** (Springer-Verlag, New York, 1980), pp. 366–81.

G. Tanner, P. Scherer, E.B. Bogomolny, B. Eckhardt and D. Wintgen, 'Quantum Eigenvalues from Classical Periodic Orbits,' *Phys. Rev. Lett.* **67**, 2410 (1991).

T. Tél, 'Fractals and Multifractals,' *Zeit. Naturforsch, A* **43**, 1154 (1988).

T. Tél, 'Transient Chaos,' in *Directions in Chaos*, Vol. 3, edited by H. Bai-Lin, D.H. Feng and J. M. Yuan (World Scientific, Singapore, 1991).

J. Testa, J. Pérez and C. Jefferies, 'Evidence for Universal Behavior of a Driven Nonlinear Oscillator,' *Phys. Rev. Lett.* **48**, 714 (1982).

Y. Ueda, 'Explosion of Strange Attractors Exhibited by Duffing's Equations,' in *Nonlinear Dynamics* edited by R.H.G. Helleman (New York Academy of Sciences, 1980).

D.K. Umberger and J.D. Farmer, 'Fat Fractals on the Energy Surface,' *Phys. Rev. Lett.* **55**, 661 (1985).

D.K. Umberger, J.D. Farmer and I.I. Satija, 'A Universal Strange Attractor Underlying the Quasiperiodic Transition to Chaos,' *Phys. Lett. A* **114**, 41 (1986).

F. Városi, T.M. Antonsen and E. Ott, 'The Spectrum of Fractal Dimensions of Passively Convected Scalar Gradients in Chaotic Fluid Flows,' *Phys. Fluids A* **3**, 1017 (1991).

S. Ya. Vyshkind and M.T. Rabinovich, 'The Phase Stochastization Mechanism and the Structure of Wave Turbulence in Dissipative Media,' *Sov. Phys. JETP* **44**, 292 (1976).

J.M. Wersinger, J.M. Finn and E. Ott, 'Bifurcations and Strange Behavior in Instability Saturation by Mode Coupling,' *Phys. Rev. Lett.* **44**, 453 (1980).

J.M. Wersinger, E. Ott and J.M. Finn, 'Ergodic Behavior of Lower Hybrid Decay Wave Ray Trajectories in Toroidal Geometry,' *Phys. Fluids* **21**, 2263 (1978).

K.G. Wilson, 'The Renormalization Group and Critical Phenomena I and II,' *Phys. Rev. B* **4**, 3174 (1971) and *ibid.*, p. 3184.

J. Wisdom, 'Chaotic Dynamics in the Solar System,' *Icarus* **72**, 241 (1987).

A. Wolf, J.B. Swift, H.L. Swinney and J.A. Vastano, 'Determining Lyapunov Exponents from a Time Series,' *Physica D* **16**, 285 (1985).

J.A. Yorke, C. Grebogi, E. Ott and L. Tedeschini-Lalli, 'Scaling Behavior of Windows in Dissipative Dynamical Systems,' *Phys. Rev. Lett.* **54**, 1095 (1985).

J.A. Yorke and E.D. Yorke, 'Metastable Chaos: The Transition to Sustained Chaotic Behavior in the Lorenz Model,' *J. Stat. Phys.* **21**, 263 (1979).

Z. You, 'Analytic and Numerical Studies of Unstable Manifolds,' Ph.D. dissertation, Department of Mathematics, University of Maryland (1991).

L.-S. Young, 'Dimension, Entropy, and Lyapunov Exponents,' *Ergodic Theory and Dyn. Systems*, **2**, 109 (1982).

T. Yukawa and T. Ishikawa, 'Statistical Theory of Level Fluctuations,' *Prog. Theor. Phys.*, *Supplement* **98**, 157 (1989).

G.M. Zaslavski, M. Yu. Zakharov, A.I. Neishtadt, R.Z. Sagdeev, D.A. Usikov and A.A. Chernikov, 'Multidimensional Hamiltonian Chaos,' *Sov. Phys. JETP* **69**, 885 (1989).

Index